CW01674924

'In this highly original contribution, leading anthropological scholars from the University of Cambridge provide a new and compelling approach to the history of anthropological ideas.... Insightful, succinct but also consistently challenging, I expect that these essays will inspire students of anthropology for years to come.'

Adam Reed, *University of St Andrews, UK*

'A useful antidote to the presentism of much current anthropological theorizing, this rich and variegated collection – which takes account of some of the deepest roots and freshest sprigs – especially reflects the influential view of the discipline from the venerable Cambridge tradition, which displays in these pages an impressively global and historically comprehensive reach.'

Michael Herzfeld, *Harvard University, USA*

Schools and Styles of Anthropological Theory

This book presents an overview of important currents of thought in social and cultural anthropology, from the 19th century to the present. It introduces readers to the origins, context and continuing relevance of a fascinating and exciting kaleidoscope of ideas that have transformed the humanities and social sciences, and the way we understand ourselves and the societies we live in today.

Each chapter provides a thorough yet engaging introduction to a particular theoretical school, style or conceptual issue. Together they build up to a detailed and comprehensive critical introduction to the most salient areas of the field. The introduction reflects on the substantive themes which tie the chapters together and on what the very notions of 'theory' and 'theoretical school' bring to our understanding of anthropology as a discipline.

The book tracks a core lecture series given at Cambridge University and is essential reading for all undergraduate students undertaking a course on anthropological theory or the history of anthropological thought. It will also be useful more broadly for students of social and cultural anthropology, sociology, human geography and cognate disciplines in the social sciences and humanities.

Matei Candea is a Lecturer in the Department of Social Anthropology at the University of Cambridge, UK.

Schools and Styles of Anthropological Theory

Edited by Matei Candea

Routledge
Taylor & Francis Group
LONDON AND NEW YORK

First published 2018
by Routledge
2 Park Square, Milton Park, Abingdon, Oxon OX14 4RN

and by Routledge
711 Third Avenue, New York, NY 10017

Routledge is an imprint of the Taylor & Francis Group, an informa business

British Library Cataloguing-in-Publication Data
A catalogue record for this book is available from the British Library

Library of Congress Cataloging-in-Publication Data
A catalog record for this book has been requested

ISBN: 978-1-138-22971-6 (hbk)
ISBN: 978-1-138-22972-3 (pbk)
ISBN: 978-1-315-38826-7 (ebk)

Typeset in Bembo
by Sunrise Setting Ltd, Brixham, UK

Contents

Figures

Introduction

Echoes of a conversation

Matei Candea

What is this book?

This book provides an overview of important currents of thought in social and cultural anthropology from the 19th century to the present. It offers a broad introduction to key theoretical schools and styles of this extended period. It gives some sense of their historical context and their interconnections and points of overlap. The primary focus is on developments in British, and to a lesser extent, American and French anthropological traditions, although the chapters also demonstrate the progressive interweaving of these traditions over the course of the 20th and 21st centuries. It will introduce readers to a fascinating and exciting kaleidoscope of ideas that have transformed the humanities and social sciences, and the way we all understand ourselves and the societies we live in today. The theories examined in these pages engage with some of the most fundamental questions anthropologists continue to ask today: What, if any, sort of freedom do human beings have? How can we explain and understand the regularities and the patterned nature of our collective lives? What is culture and what is society? What can our bodies, our minds and our technologies do, and what happens in their interaction? What are the sources, meanings and effects of the differences – in terms of identity, perspective or power – that run between and within human collectives? Is there a place for the study of non-humans in anthropology?

The chapters in this book track a longstanding core lecture series given at Cambridge University for social anthropology students, entitled 'Schools and Styles of Anthropological Theory'. While the lecture series is primarily aimed at undergraduates, it is attended by Masters students, and is often also audited by doctoral students. The aim of the lecture series is to provide a broad, accessible yet relatively sophisticated introduction to anthropological theory, and this is also the main aim of this book.

This book engages with the classic anthropological 'isms' (evolutionism, diffusionism, functionalism, structural-functionalism, structuralism, transactionalism, neo-Marxism, interpretivism, feminism, postcolonialism), frequently identified theories and theoretical schools (the Frankfurt School, the Manchester School, practice theory, actor-network theory), classic and more recent moments of theoretical rupture (the 'writing culture' moment, the ontological turn), and more diffuse reflections around particular conceptual problems such as the problem of historical thinking in anthropology (see chapter 5), the question of the extension and boundaries of fieldsites (see chapter 6) and the distinctive dynamics of the shaping and reshaping of anthropological concepts (see chapter 15). All of the above are treated here, albeit not all at chapter length or in the form of self-contained sketches. A number of chapters weave together accounts of shifts, tensions and transformations between two or more of the above, and some chapters return to the same school or style from different perspectives; most notably, for instance, postcolonial critiques in anthropology are evoked in chapters 1, 5, 6 and 12, rather than being subsumed in a single chapter. I will return to the organisation of the book and the chapters below.

While 'schools and styles' are its primary organising device, however, this book is not simply a list of theories. It is also a collective reflection on what anthropological theory is and how it changes. The authors in this book propose different explicit and implicit answers to that question. In this and in other ways, this book is best thought of as a conversation – at times an argument – rather than a single narrative.

The section 'Views from Cambridge?' gives some more background on the origin of this book and reflects on the particular kind of perspective on theory that is implied by a book based on a lecture course in one particular department. The four sections after this delve into more fundamental questions concerning what theory is and how to think about it. Along the way they elucidate some of the organisation of this book. Before we begin, however, one very general question needs to be answered especially – but not only – with undergraduate readers of this book in mind: Why bother engaging with the history of anthropological theory at all?

On learning to see theory

In some students, the very thought of a 'theory course' or a book based on such a course, will induce despondency or terror. Partly under the influence of increasing modularity in undergraduate teaching and with an attentive eye to students as customers whose tastes must be catered for, anthropological courses and introductions to anthropology have tended to veer towards catchy topics and titles. Sex and death, mystery and inequality, the strangest practices made familiar and your unexamined everyday life made strange: anthropology provides all of this in droves, and this is often where students are invited to begin. Theory, by contrast, seems tedious, lifeless and irrelevant; old theories even more so. Theory also seems, by contrast to those catchy topics and cases, essentially *difficult*.

Theory, this book will demonstrate, is none of these things. And this is true, crucially, because 'theory' is not a single, free-standing thing at all. Theories come in many shapes and sizes, and in anthropology at least, they are always intimately interwoven with practice and with particulars. Theory is already there, at the heart of the more immediately attractive or relevant-seeming arguments and cases with which students first encounter anthropology. Like Molière's character Monsieur Jourdain, who was surprised to be told that he had all his life been speaking in prose, readers of this book will swiftly realise that they have been doing theory all along.

Of course, this observation could lead to rendering theory meaningless in a different way. As I describe below, some recent schools of thought in anthropology and beyond would seek to do away with 'theory' altogether as a distinct topic. This book, by contrast, proposes that theory, as a distinct focus of study, still has an important role to play. Focusing on theory allows one to make explicit the conceptual issues that structure and underlie anthropological discussions and debates, and to see how these have shifted and changed through time. This overview of debates empowers newcomers to anthropology – and indeed seasoned anthropologists engaging upon a new topic or field of research – by allowing them to situate the works they encounter within a broader historical and discursive landscape. Learning to recognise the distinctive clues that suggest an author is writing in a particular school or style of thinking means learning to see their accounts, descriptions and cases as *arguments*, rather than simply statements of fact or articles of faith. It will be invaluable in helping students to critically detect assumptions, blind spots and shortcuts in the texts they read. But this work is not entirely negative. The critical exercise of detecting theoretical assumptions is just one of the skills that comes with a thorough knowledge of the history of theory. Another is that of imagining how, and to what effect, two radically different theoretical perspectives might apply to the same body of material. This, in turn, is the first step in learning to build one's own distinctive theoretical arguments.

There is a broader point here concerning the use of theory, and the use of this book, for the attention not only of newcomers to anthropology, but also for graduate students or indeed professional anthropologists embarking upon an original research project. A handful of the theories engaged in this book are lively contemporary positions that anthropologists writing today might explicitly espouse. The majority, however, are usually understood as belonging to history rather than to present debate. The most common reason such 'old theories' are usually invoked in anthropology is as a catalogue of errors, a list of conceptual shortcuts that we wish to avoid repeating. This is not in itself a bad reason to recall them. In particular, it is often possible with hindsight to build a historical context around theories that the actors themselves may not have seen, or seen too well – either way – that they would not themselves have considered as 'context' (see chapter 1). This in turn can provide powerful lessons for the present, in the form of errors and shortcuts to avoid. Important as it is, however, this cannot be the only reason for retrospection, or the only mode in which it occurs. Old theories can also be mined for new insights, particularly if we recognise those aspects of their problematics that still resonate.

Views from Cambridge?

This book is a collective endeavour of a somewhat unusual kind. Most edited books are the result of conferences or workshops. They represent a conversation that took place at one point in time over a few days. Encyclopaedias, including thematic ones, by contrast, are assembled by commissioning articles from scholars in a range of institutions, who have often not been in conversation at all. They seek to provide a comprehensive coverage of a discipline or subject area.

This book, by contrast, is the result of a much longer conversation. As its origin as a lecture series determines much about the form, content and 'voice' of the book, it is worth saying a few words about how a lecture series, and this one in particular, is organised in Cambridge. The 'Schools and Styles' lecture series has been running under this title for many decades along the same basic principle: lecturers and associates of the department of social anthropology are each called upon to give one or two lectures on a theoretical school, style or problem in which they have a particular interest or expertise. The selection is made by a paper coordinator, whose role is to ensure the balanced and comprehensive nature of the set as a whole. This oversight is, however, collective as well as individual: the core teaching staff of the department come together to discuss the content of every lecture series once a year. Coordinators present their proposed papers for the following years to the scrutiny of their colleagues, who will often comment on particular inclusions and exclusions. The department comes together in the same way to set examination papers, at which point again the balance of topics and the way in which they are treated is examined collectively. As a result of this process, the lecture series, and therefore this book, is a thoroughly collective endeavour. It represents an ongoing conversation between a group of colleagues with diverse interests about the history and state of anthropological theory. This – crucially – includes colleagues who are not represented as authors here, but who have been involved formally and informally in these conversations over the years.[1]

This conversation is longstanding but it is also perpetually changing. The chapters in this book reflect a moment: they are based on the lectures as given in the 2016 to 2017 academic year. As the personnel of the department changes and their interests shift, so does the content of the lectures, the theoretical schools they choose to lecture on, the overall outline of the course and, more broadly, the way in which 'theory' itself is portrayed and understood – more on this later. Individual and collective perspectives about what such a course should contain shift through time, tracking transformations in anthropological theory, and transformations in the Cambridge department. Some topics are enduring: I was lectured on structural-functionalism as an undergraduate nearly 20 years ago; I now give that lecture, which forms the basis of chapter 1 in this book. Needless to say, it is no longer the same lecture as the one that I once attended. As the same topic is taken up by different people, each rewrites the lectures more or less from scratch, sometimes drawing on the reading lists of their predecessors. Other topics represented here are new: chapter 9 on the Frankfurt School is based on a lecture given for the first time in the 2016 to 2017 academic year.

In sum, then, this book does not claim to be either exhaustive or representative of anthropological theory as a whole. As we shall see below, any such claim would be inherently meaningless. Like any other account of theory, this is an account from a particular time and place, and I have tried in the above to give a sense of where and when that is. This book is the result of the complex process through which a collective of scholars in an academic department put together a partly shared perspective on anthropological theory.

However, the sense in which this book gives a 'Cambridge perspective' on theory should be understood under the caveat that any such perspective is internally multiple and historically changing. Seen from outside, university departments are often caricatured as holding a particular line or representing a particular style, in an endless process of self-reproduction. Yet the briefest consideration of a university department's actual structure as a community of practice should demonstrate how unlikely this is to be the case. Some of the contributors in this book were trained in Cambridge and others were not. Some have been lecturing there for many years. Others joined the department much more recently. A number will be employed elsewhere by the time of publication. Thus, the reader should not be surprised to find radical differences in tone, style and approach between the chapters in this book. This book is the echo of a conversation that took place in Cambridge. It is not 'the Cambridge view', as there is no such thing.

What is theory?

As I noted above, contributors' views are diverse not only in their approach to particular theories, but in the more fundamental question of what 'theory' is. This book as a whole is best treated as a collective and multi-vocal answer to this question. It cannot be summed up in a few pithy lines. The rest of this introduction will, however, outline three longstanding threads to the general discussion about the nature of theory that runs through this book. The first concerns the 'external' problem: how, and to what effect, does one mark out theory from other things (method, data, practice, etc.) often subsumed in anthropology through a distinction between theory and ethnography? The second concerns the 'internal' question of how theory is subdivided (into schools, styles, paradigms, concepts, etc.). The third question asks what, if anything, is distinctively anthropological about anthropological theory.

These questions point to three demarcations that organise, in part, the subject matter and approach of this book: the theory/ethnography distinction, the device of grouping theory into 'paradigms' and indeed the device of treating anthropological theory as distinctive. None of these is self-evident, and this book, while relying on them to some extent, does not take them for granted. However, I will argue that all three of these conceptual devices can be and have been extremely productive tools for thinking, even though they are not philosophically tenable in some broader senses.

What, if anything, separates 'theory' from anything else? In particular, lines are often drawn between theory and method, on the one hand, and between theory and material (content, data, description, examples) on the other. For a substantial period in the history of anthropology (and in some quarters still today), theory was understood to stand apart from, and above, method and material. Fieldwork pointed to both of the latter terms: a technical procedure for gathering 'data' that would then be analysed and theorised. This speaks to the enduring division in anthropology between 'ethnography' (both in the sense of a fieldwork method, and in the sense of a written product) and 'theory'. This distinction draws on, and echoes, within our discipline, epistemological distinctions widespread throughout social science and indeed science more broadly: distinctions between description and explanation; and between the particular and the general. For evolutionists, some structural-functionalists such as Radcliffe-Brown (see chapter 1) and some structuralists such as Lévi-Strauss (see chapter 2), this conceptual division was also a division of labour: fieldworkers on the one hand, theorists on the other, had different roles and skillsets that would be found in the same person only coincidentally. To the fieldworker fell the task of accurately describing the way of life and customs of a people. To the theorist the – implicitly nobler – task of comparing, abstracting and generalising from this data in view of a theory. Even though the professional culture of anthropologists since the beginning of the 20th century mostly enjoined them to take on both of these roles, the sense in which these roles are distinct along the lines described above endures in backroom talk about one's own particular strengths and weaknesses in comparison with other anthropologists ('He's a fantastic ethnographer, but not much of a theorist', etc.).

Needless to say, these distinctions in anthropology between theory and method, and also between theory and data, are inherently political in more than one sense. They map the internal politics of the discipline, with its various implicit and explicit scales of value and accreditation. But they emerge also from anthropology's historical place in a global order of knowledge production in which metropolitan scholars theorised about data extracted from the colonies and the peripheries (see chapters 1 and 5). This reflects the broader point that, for much of the history of anthropology, as pithily summarised by Clifford Geertz: 'its subjects and its audience were not only separable but morally disconnected, that the first were to be described but not addressed, the second informed but not implicated' (1988: 132). 'Theory' played the role of a filter, through which anthropologists performed that miracle of one-way translation. As Chua and Mathur (forthcoming) recently noted, anthropologists' still-frequent reference to an 'us' position remains an enduring instance of this unequal global order of knowledge production. Anthropologists have frequently deconstructed the idea that there might be a culturally homogeneous (Western or EuroAmerican) 'us' to whom anthropology might be addressed. But an equally problematic implication is that of the univocality of anthropology as a discipline. That anthropological 'us' masks an unequal global academic landscape in which (new, exciting, cutting-edge) 'theory' is still looked for in the same old metropolitan centres. In this order of things, 'theory from the South' (Comaroff and Comaroff 2012) or 'from the East' (Howe and Boyer 2015) has been seen as requiring specific and explicit acknowledgement. A number of chapters in this book explore the way in which Marxist, feminist, post-colonial and other critical scholars have seen themselves as challenging the politics of anthropological knowledge production (see chapters 1, 5, 6, 8 and 12). In so doing, they have not only produced theory themselves, but also explored the politics of doing so. Of course, their own way with theory, and their own politics of knowledge production, have in turn become the subject of further critiques.

And yet, from another point of view, these critiques were merely reformulating a very old point embedded in the disciplinary structure of anthropology itself. Indeed, the birth of anthropology as a discipline coincided precisely with a challenge to the classic way of dividing theory from method on the one hand, theory from material on the other. With Malinowski's focus on long-term fieldwork came the recognition that questions of field method were inherently theoretical. Malinowskian functionalism was as much a methodological development as a theoretical one: new kinds of 'data' and new understandings of what 'data' might be, made old theoretical questions meaningless (see chapter 1). The point that ethnographic method is an inherently theoretical question has been a recurrent theme in anthropological discourse ever since; as Harri Englund shows in his discussion of the Manchester School's 'extended-case method', and the more recent development of 'multi-sited fieldwork' (see chapter 6). In a different vein, Bourdieu's moves towards a 'theory of practice', however highly theoretical they themselves ended up being, were premised on an explicit challenge to the usual way in which theory had been abstracted from descriptions of the flow of life (see chapter 4).

There was also another way in which anthropological scholarship, from the beginning, involved a challenge to the usual way of imagining the theory/method distinction. This was the profound sense in which the conceptual work of anthropology, from the early 20th century, if not indeed before, was intended as inherently disruptive and critical, challenging Western assumptions and established philosophical paradigms by showing the rationality of unfamiliar modes of thinking or the effectiveness and beauty of unfamiliar social arrangements. This was closely linked to Malinowski's challenge to the 'division of labour' model of anthropological research: fieldworkers, in the Malinowskian view, *had* to be theorists, and theorists *had* to be fieldworkers, because the engine of anthropological knowledge production was the experience of otherness in the course of field study (Kuper 1973: 32–33). Making 'defamiliarisation' into the core anthropological move meant, in turn, that successful fieldwork had to be transformative, if not indeed destructive of established theory, and creative of new theoretical perspectives.

That Malinowskian view has not gone unchallenged, and the nature of fieldwork has changed profoundly also over the past century (see chapter 6). But this view of fieldwork as transformative persists in contemporary anthropological attitudes to the ethnography–theory relation. In other, more self-consciously scientific disciplines, excitement and success tends to be attached to research that confirms clearly set out hypotheses, whereas – an often deplored fact – 'negative results' are rarely even published (see Granqvist 2015). In anthropology, by contrast, fieldwork has usually been seen as successful precisely at the moment when it proved unexpected, and exceeded theory. The role of fieldwork was in effect to produce that moment when the theoretical frames with which one had initially approached the problem revealed themselves to be inadequate. This model of anthropology as perpetual conceptual revolution has remained deeply anchored in anthropologists' ways of evaluating each other and themselves, even though this is not, of course, all that anthropologists do.[2]

One of the effects of this model is a particularly frequent fragmentation of theoretical perspectives, with each subsequent fieldworker feeling the need to break with a previous theoretical status quo. Hence the multiplicity of schools, styles, labels and 'turns' with which this introduction began. As much as a new 'framework', what is often at stake in these changes and shifts is a different set of cases and experiences. That is why so many of the chapters in this book are, in effect, as much a history of paradigmatic ethnographies as a history of theories.

Another effect of this model of permanent conceptual revolution is that anthropology, from the start, posed the question of the encounter with others' theories, long before 'theory from the South' was formulated as a problem in those terms. Certainly, there was always a positivistically inclined strand of anthropology that gave non-Western theories relatively short shrift. They featured mainly as elements of a factual reality to be explained by our own, definitionally superior, theories. But another, interpretive, vein that also ran through anthropology from the start of the discipline (see chapters 1 and 8) asked how another point of view on the world might transform, inform or challenge our own. This question, ever reinvented, took a more radical form with the 'writing-culture' critique of the 1980s, when anthropologists' own knowledge practices, modes of explanation and techniques of authorship came under more direct scrutiny (see chapter 8). Anthropologists' claim to be able to explain, organise and translate a diversity of cultural points of view was critically examined. An authoritative anthropological interpretation of others' perspectives was more clearly distinguished from a commitment to actually letting those others speak in their own voice. This being said, detractors noted that the 1980s critique itself was animated as much by resolutely 'metropolitan' high theory imported from literary studies and philosophy as by the actual transformative encounter with the voices of 'the other', and indeed often in practice led to a focus on the writing, rather than the doing, of ethnography (Handelman 1994).

A further (ontological) turn of that (epistemological) screw followed the observation that anthropologists' concern with the study of 'cultures' or points of view on the world carried an implicit imbalance that undermined its own relativist message (see chapter 14). With cultural relativism, everyone was entitled to their viewpoints, of course. But 'the world', or 'nature', remained out of the anthropological frame; a matter for biologists, physicists and the like. In other words, everyone had their culture, but the West, as it happened, also had the key to nature. The 'ontological turn' that emerged as a critique of that position is only the most recent (albeit perhaps the most radical) instance of the idea that anthropology's role is to provide conceptual disturbance to Western theories by taking non-Western ones seriously.

This ontological move chimed in with other developments at the turn of the 21st century, such as actor-network theory (see chapter 13), in attacking the very figure of theory itself.

Both of these schools (for that is what they have now become) come with the strong, explicit claim that they are not 'theories', in the sense of overarching systematic accounts of the world to be infirmed or confirmed through empirical study, but something else: conceptual techniques, recipes for innovation; in one word: heuristics (see chapter 14; Abbott 2004; Candea 2016). This appeal to heuristics has three main characteristics: it embraces fragmentariness and multiplicity (positions, concepts, arguments are not supposed to build up into an overarching scheme); it eschews repetition (there is no 'framework' to apply from one study to the next, only methodological principles for producing novelty each time); it presents itself as performative, a mode of doing rather than saying (each account is a practical or political intervention into the world, rather than a mere description of the world). While this turn to heuristics resolves many key issues relating to the definition of 'theory' (not least the question, examined in this section, of the distinction between theory and method, or theory and material), it brings with it its own inherent difficulties, as chapter 13 and especially chapter 14 point out.

In other words, the key difficulty in writing about theory today, in the way this book seeks to do, lies primarily in the fact that the schools and styles examined here are not simply different views on the world, they also imply different – sometimes radically different – visions of what theory is. To write about the history of theory is thus to write about a constantly changing object. In other words, this book is not just a conversation about the nature of theory: it is an account of, but also a contribution to, a longstanding anthropological conversation on that subject.

This way of seeing things gives us a better sense of what the object of this book is: not 'theory' *per se* as a clearly defined thing out there, but rather the problem of theory as a focus of debate and concern in the history of anthropology. Seen like this, there are powerful reasons to resist the dissolution of theory as a category. Dissolving the distinction between theory and ethnography is in many ways an attractive move, as is the recurrent and always popular move of sweeping away 'old dualisms'. And yet that move does tend to miss the point that 'dualisms', while not always philosophically tenable, have their pragmatic and conceptual uses. Throughout all of the debates briefly outlined above, the constantly reimagined, always unstable, and in many ways clearly fictional distinction between 'ethnography' and 'theory' has been put to constant work. Of course, this caveat is not incompatible with a heuristic view; quite the opposite: it points precisely to what the theory/ethnography distinction ought to be seen as; namely, a heuristic. For all its faults and limitations, the distinction between theory and ethnography has provided anthropology with an ever-renewed heuristic engine (Strathern 2011; Heywood forthcoming).

Strengths and weaknesses of the '-isms'

My invocation above on the vision of radically opposed schools and styles leads us straight to the second line of questioning that runs through this book, which concerns the internal subdivision of theory: What is it made of? The heuristic turn I have just described was in part a reaction against the classic narrative device of subdividing 'theory' into lumps – schools, theories or 'paradigms' – the '-isms' for short. That vision is so familiar as to have become common sense. Each '-ism' comes packaged with its leading figures, its internal dissenters, its key works, its historical moment, its classic questions, its characteristic type of data and forms of argument, its standard failures and critiques. It comes, in other words, with – as in that unavoidable undergraduate essay title – its 'strengths and weaknesses'.

As I noted at the outset, and as its title suggests, this book – and the lecture series it is based on – espouses paradigms as part of its mode of exposition. It does not do so exclusively, however, nor does it do so uncritically. Mostly, when '-isms' are invoked in these pages, they are invoked with the limits of their definition kept firmly in view. This book is as much about the shifts between schools and styles, the overlaps and connections across them and the tensions within them, as it is an account of schools and styles themselves. Nevertheless, as with the theory/ethnography distinction, paradigmatic views of theory, however – necessarily – partial they may be, have an important job to do.

The narrative device of subdividing theory into '-isms' has been until recently so pervasive that one might miss the fact that it has a history. As late as the 1950s, an author such as Nadel could still imagine that a single theoretical perspective could provide *The Foundations of Social Anthropology* (1951). Nadel's rigorous and sophisticated book rises above (or perhaps delves below) any vision of competing 'schools' in order to lay out the basic principles for the anthropological observation and interpretation of human behaviour and institutions. In this and in other ways, Nadel was an outlier, as the narrative of changing theories and competing schools was born with modern anthropology itself. Two scholars classically invoked as anthropological founding figures – Boas in the USA and Malinowski in the UK – teased out their own theoretical position through repeated attacks against 'evolutionism' (typically the first '-ism' in histories of anthropological theory: see chapter 1). By the time Nadel wrote *Foundations*, 'functionalism', 'diffusionism' and 'structuralism' had joined evolutionism as commonplace labels. And yet, as Nadel's book testifies, it was still possible in 1951 to imagine a fundamental common ground underlying this diversity of approaches. The point was forcefully put by Radcliffe-Brown, as an explicit critique of Malinowski's statement that he had created a functionalist 'school':

> There is no place in natural science for 'schools' in this sense, and I regard social anthropology as a branch of natural science. Each scientist starts from the work of his predecessors, finds problems which he believes to be significant, and by observation and reasoning endeavours to make some contribution to a growing body of theory. Co-operation amongst scientists results from the fact that they are working on the same or related problems. Such co-operation does not result in the formation of schools, in the sense in which there are schools of philosophy or of painting. There is no place for orthodoxies and heterodoxies in science. Nothing is more pernicious in science than attempts to establish adherence to doctrines. All that a teacher can do is to assist the student in learning to understand and use the scientific method. It is not his business to make disciples.
>
> (Radcliffe-Brown 1940: 1)

The vision of a discipline divided into radically different 'paradigms' that swept away Nadel and Radcliffe-Brown's hope of shared principles, follows loosely the understanding of scientific knowledge production introduced by Thomas Kuhn (1970). Kuhn first applied the notion of paradigm to the study of science. It contrasts directly and explicitly with the commonplace vision of science articulated by Radcliffe-Brown above. Kuhn argued that scientists work not simply with a shared method, but rather necessarily 'within' paradigms: relatively coherent bodies of theories with their particular assumptions, techniques and questions. These are organised around a few key 'paradigmatic' works that set the scene of the thinkable for subsequent scholars. From time to time, a new work or experiment ruptures the frame and a paradigm shift occurs. A new paradigm then emerges and from its new vantage point, as Ardener wrote of anthropological theory: 'for practical purposes textbooks that looked useful, no longer are; monographs that used to appear exhaustive now seem selective; interpretations which once looked full of insight now seem mechanical and lifeless' (Ardener 1971: 449, cited in Stocking 1984: 180).

With or without reference to Kuhn, this paradigmatic vision of theories as powerfully organising perspectives, 'within which' anthropologists of a particular time and place think and work, became for a while the key narrative device for recalling anthropological theory. The institutionalisation of this view within anthropology began with book-length reviews of subsequent theoretical perspectives on one particular topic, such as totemism (Lévi-Strauss 1963) and religion (Evans-Pritchard 1965) that could be seen as precursors for later accounts of '-isms' on the scale of anthropology as a whole (Kuper 1973; Layton 1997; Barnard 2000). Many of those classic overviews of anthropological theory are still staple general reading for theory courses today. They all present anthropological theory in the form of a sequence of paradigms, relatively small in number, that could be manageably treated by one author in a single-volume work.

This model is beginning to show signs of strain. For one thing, the sheer multiplication of '-isms', schools and turns over the past 30 or so years seems to have deterred scholars from following that example. Where recent developments could once be subsumed under 'postmodernism' (Barnard 2000: 158ff), the world of theory has moved on in such a frenzied way that postmodernism now seems a very long time ago. There now seem to be too many '-isms' to encompass in a single-authored book. In some respects, the encyclopaedia is taking over as a form from these classic overviews (Barnard and Spencer 1996; 2009). There, a multitude of schools and styles can be described through polished self-contained yet cross-referenced miniatures. Alternatively, authors have focused on accounts of particular bodies of theory such as feminism (Moore 1988) or actor-network theory (Latour 2005), or have experimented with telling the story of anthropological theory through a focus on one particular problematic such as the gift (Sykes 2005); approached theory through thematic readers (Moore and Sanders 2006); or, indeed, written collectively about the contemporary state of theory itself (Moore 2000; Boyer, Faubion and Marcus 2015). This multiplicity and fragmentation of approaches to theory marks the tension surrounding the '-isms' model as a narrative device for approaching anthropological theory.

Some may view the possible demise of the paradigmatic vision of anthropology with equanimity or even with relief. For, indeed, the limits of this narrative form are familiar (for a systematic critique of the idea that anthropology is made up of 'paradigms' in the Kuhnian sense, see Strathern 1987). The first set of concerns turns on what a paradigmatic view leaves out. This vision of anthropology as a sequence of '-isms' is inherently synchronic: it depicts multiple coherent theoretical viewpoints on the world, not unlike the 'cultures' described by interpretivist anthropologists (see chapter 8).[3] The most obvious problem with paradigms, and indeed with cultures, is that as soon as one looks closer, the neat list of self-contained perspectives dissolves into a cacophony of idiosyncratic viewpoints inhabited by particular people in particular contexts. The more clearly and coherently a paradigm is outlined, the harder it is to find it exemplified anywhere, except maybe in the programmatic writings of one or two key proponents. Imagining the history of theory as a sequence of paradigms tends to minimise both this internal diversity and the continuities and overlap between different paradigms; overlaps both conceptually and in terms of personnel. This is a point that many of the chapters stress, even as they invoke the common '-isms' of anthropological theory.

Second, a paradigmatic vision also tends to comfort a narrative of permanent revolution, whose attractions for anthropologists we have mentioned above. With each paradigm shift, a grand rupture or *caesura* is imagined to sweep away all that came before it. This 'caesurist' view (Pina-Cabral 2010) eliminates any sense that anthropology might be a shared or indeed a cumulative endeavour. It also imagines paradigms as essentially cast-offs in a great march forward. The future is open, the past a series of errors. Indeed, those neatly self-contained paradigms are almost always identified and labelled in retrospect, their limitations and blind-spots standing out against our putatively superior vantage points. Hence that recurrent question by students on theory courses: 'If *they* were functionalists, structuralists, Marxists and so forth – what are *we* now?' There is no answer to that question because its premise is mistaken. Even when 'they' did invoke these terms, 'they' were not simply functionalist or structuralists, at least not in their own perspective, but rather, as 'we' are now: open-minded scholars seeking to understand and/or explain the world in the best way they could. It is usually in retrospect – and often somewhat unfairly – that we can see them as theory-bound and context-bound, just as others will no doubt see us. A paradigmatic vision forecloses the sense of open-endedness, excitement and conceptual experimentation that existed in earlier periods of anthropology as it does now. The effect has been evocatively described as one of 'ruination' of previous theoretical approaches (Navaro-Yashin 2009); for an insightful analysis of the dynamics of paradigm shifts and the elicitation of perpetual novelty in the social sciences, see Abbott (2001).

A third key issue with the paradigmatic view is that it implies that theories come as a finite list of similar entities. The paradigmatic view seems to suggest that one might agree on an exhaustive master list of 'what all the theories are'. But this is a mistake, for reasons already explored above. What counts as a theory, an '-ism', depends on one's perspective. At what point a group of scholars can usefully be lumped together as saying essentially the same thing depends on the scale on which one considers them, and indeed on the measure of distance one takes from their position.[4] The resulting list of paradigms is always perspectival, lumping together things that some would hold apart, or distributing across different paradigms positions and perspectives that others might wish to hold together as one. There is no exhaustive set of theories, any more than there could be a complete list of the world's cultures. Totality in that respect is a mirage. It is thus no use arguing, in the abstract, that this or that '-ism' is missing in this or any other account of anthropological theory. The real question is what if anything this absence *does*, and how it relates to what the work is setting out to do. On this point as a general principle for critiquing the 'gaps' in theories, see chapter 1.

These three objections are important to bear in mind. However, they are primarily objections to the way paradigms are invoked in the daily cut and thrust of anthropological argument. In laying out the novelty of their own approach, authors frequently cede to the temptation of sorting 'the literature' into self-contained boxes, carefully stacked to produce a clear and obvious 'blind spot' that their particular case, argument or approach will, of course, illuminate. There, the invocation of paradigms can serve as a convenient rhetorical device for ignoring complexity. As with the question of 'gaps', the problem is not 'neatness' per se, but rather the uses to which it is put. It is a great virtue of a paradigmatic view, to be able to background complexity in order to make certain key relationships and differences stand out. The worry is that, in the day-to-day arguments of anthropologists, such backgrounding can become erasure.

This problem is less common, however, in works such as this one, that focus retrospectively on anthropological theory *per se* and are not required to set out their own stand quite so boldly. In such works, paradigmatic accounts of anthropological theory rarely stand alone. In practice, they are usually interwoven with a more historical view that foregrounds precisely the complexities, changes, shifts, internal dissensions, and external continuities that are backgrounded in a strictly paradigmatic view. In other words, the paradigmatic and the historical views together form a heuristic pair throughout the classic overviews of anthropological theory evoked above, as indeed they do in many chapters in this book.

In sum, the core point here parallels the one raised for the theory/ethnography distinction. 'Schools and Styles' are not invoked in this book as a simple list of objects out there in the world. Rather, they serve as a heuristic – a way of thinking through changes and stabilities in anthropological theory and focusing in and out of broader cross-cutting theoretical problems. It is also important to remember that, however useful the paradigm/history pair may be as a device for making sense of history, very different devices and visions of theory are possible. These can productively bring other things into view. In chapter 15, for instance, Strathern proposes a way of tracking the shifting patterns of fields of concepts (in this case, concepts surrounding 'personhood') that does away with the need to rely on a paradigmatic account of the history of anthropology.

A roadmap

Some of the chapters in this book focus primarily on a single paradigm or thinker, albeit with considerations about the broader critical conversation within which they were embedded. Others focus on a contrast or transition between two or more schools of thought. Some eschew paradigms altogether, to focus on a particular theoretical problem. Consequently, many of the chapters track overlapping periods of time, or cut across time in different ways; there is no single way to order these into a chronological sequence. It is an important part of the spirit of this book, and of the lecture series it is based on, that beyond some very broad elements of chronology,[5] there is no settled year-on-year sequence in which overlapping schools are introduced. There is, therefore, no single narrative way through it. Various narratives about precursors and followers, about principal and secondary branchings in the history of theory, are possible, so the contributors to this book do not all tell that story in the same way. Students are expected to build their own picture of the history of theory through independently assembling the cross-referenced picture given to them in the lectures, and through their own independent reading and research. We hope readers of this book will use it in this spirit also.

A few very broad elements of sequencing, however, may serve to give a roadmap to the contents of the book. The book opens with an account of the birth of anthropological theory in the midst of a multi-stranded international debate around notions of evolution, diffusion and function. Theoretical schools that are in important ways distinct (19th-century evolutionism, the diffusionist and relativist anthropology of Franz Boas in the USA, the functionalism of Malinowski in the UK and Durkheim in France and the structural-functionalism spearheaded by Radcliffe-Brown) are here evoked together, to highlight the way in which their differences were forged through interaction. This uncharacteristically long chapter serves as a background to the rest of the book; many of the schools and styles examined in later chapters can be read as critical reactions to, transformations or reappropriations of elements drawn from this initial theoretical conversation.

The first of these departures is the focus of chapter 2: the structuralist thinking of Claude Lévi-Strauss. This broke with a central tenet shared by the various schools examined in chapter 1, namely that of the organic analogy – the notion that societies or cultures might be thought of as similar to biological organisms. Lévi-Straussian structuralism explores a radically different way of thinking about culture: as a system of signs. In another sense, chapters 1 and 2 form a pair. For Lévi-Strauss's focus on structure on the one hand, and the very different notion of structure deployed by the structural-functionalists (see chapter 1), taken together map some of the key tensions and conceptual possibilities of mid-20th-century British and French anthropology.

Chapters 3 to 8 present a series of critical reactions to, and departures from, the mid-20th-century anthropological focus on structure in its various forms. Chapter 3 focuses on the way Marx's writing was rediscovered as a source of critical insight by a generation of anthropologists in the 1970s. Chapter 4 traces two theoretical paradigms that, at different moments and with different conceptual tools and implications, sought to recapture the importance of individual agency: transactionalism and practice theory. Chapter 5 examines diverse anthropological approaches to the problem of history (famously backgrounded in anthropological accounts that took 'structure' as their main reference point), and spans a very extended period, from the immediate critics of structural functionalism in the 1950s, through to re-theorisations of the notion of the 'event' in the late 20th century. Whilst chapter 5 examines time, chapter 6, by contrast, focuses on place and on two key attempts to rethink the problem of ethnographic location in anthropology beyond the classic single-sited method. Chapter 6 also stretches across the second half of the 20th century, but instead of tracing a long development, it focuses on two points at the beginning and end of that period: the Manchester School's method of 'extended-case' study, and the reinvention of 'multi-sited fieldwork' in the 1990s. Chapter 7 explores a relatively recent turn to the study of cognition in anthropology, which self-consciously tries to reverse Durkheim's original decision to split sociological from psychological and biological questions, and seeks to challenge the enduring distaste of sociocultural anthropologists for psychological and experimental explanations of human behaviour.

Chapter 8 takes up a theme often seen as neglected in structural approaches: namely, the question of meaning. This was explored in American anthropologist Clifford Geertz's interpretive anthropology, through an eclectic recombination of Boasian cultural anthropology with Weberian hermeneutics. Chapter 8 opens a new cycle, focusing on meaning, knowledge and power, in which the general theme of 'departures from structure' takes a more specifically 'post-structuralist' turn. Indeed, the second half of chapter 8 focuses on the 'Writing-Culture' critique, that drew Geertzian hermeneutics back onto the question of anthropological writing itself.

Chapter 9 traces some of the ways in which postmodernist concerns with fragmentariness in anthropology were pre-empted and exceeded by the 'Frankfurt School' writings of Adorno and Benjamin, in which Marxist analytics were transmogrified in disruptive and thought-provoking ways. Chapter 10 gives an overview of the many-stranded work of Michel Foucault at the intersection of questions of knowledge, power and ethics, and evokes its multiple anthropological legacies. Chapter 11 focuses on the shift from studies of the body to a phenomenologically inflected concern with 'embodiment', taking the reader through both the innovatory and problematic aspects of this. Chapter 12 traces the complex epistemic and political relationship between anthropology and feminism, from early structuralist-inspired work on the anthropology of women, through to vibrant contemporary debates around gender, power and difference.

Chapters 13 and 14 focus on two recent conceptual developments (actor-network theory in chapter 13 and the ontological turn in chapter 14) that share – as we saw above – a certain commitment to not being classified as theories in the usual sense. The volume closes with a retrospective account (chapter 15) of theoretical discussions surrounding the notion of personhood. These discussions spanned the entire historical period of the volume. This final chapter thus doubles up as a broader consideration of the changing ways in which anthropological concepts have operated and continue to operate. While chapters 13 to 15 certainly continue the themes of power, knowledge and meaning that run through the previous cycle, they also form a cycle of their own around the question of what might be gained, and what lost, by imagining that one has arrived at a point that is 'after theory'.

What's anthropology got to do with it?

In closing this introduction, let me turn to the third and final problem I evoked earlier: What is anthropological about anthropological theory? After all, much of what is written under the previous two headings applies to theory in the social sciences and humanities more generally. What, then, does it mean to focus specifically on *anthropological* theory? What, if anything, is that object? This question is particularly pertinent as many of the key themes and contrasts recur throughout the pages of this book – the question of the agentive freedom of the individual in relation to broader forces and structures – the problem of how to think about change and stability in the same framework; the question of what constitutes adequate explanation, translation, description or interpretation – are general problems for scholars in the social sciences and the humanities. In engaging these problems, anthropologists also engage in a much broader interdisciplinary conversation.

Thus, a number of the theorists encountered in this book (such as Marx, Weber, Durkheim, Adorno, Foucault, Bourdieu, Butler or Latour) are primarily, or in some cases, exclusively, remembered as sociologists or philosophers, rather than as anthropologists. Conversely, some anthropological theories began within the discipline and then radiated out into the humanities and social sciences – Lévi-Strauss's structuralism (see chapter 2) and Geertz's interpretivism (see chapter 8) are two commonly invoked examples. But Lévi-Strauss built his thinking in close conversation with the structural linguistics of Roman Jakobson, and Geertz arguably drew from literary theorists as much as he gave in turn to historians and others. Finally, a number of the theoretical schools and fields examined in this book – such as gender studies or actor-network theory – are explicitly and self-consciously invested in breaking down disciplinary boundaries.

'Anthropological theory', therefore, is no more self-evident as an object, than is 'theory' in general, or particular 'theories'. Cutting it out from the broader interdisciplinary conversation in which it is embedded is a particular decision. This decision can be justified, much as the invocation of paradigms was above, as a heuristic one: to focus on anthropological theory is not to ignore those broader interdisciplinary conversations, but precisely to provide a frame against which questions about those broader contexts can be posed clearly.

But one might suggest a more ambitious justification: namely, that there is, after all, something distinctive about theory as viewed from anthropology, something we have already encountered above. What is distinctive is less the content, or even the labels of the theories invoked, and more the way in which theory in general fits into the economy of the discipline; namely, in perpetual complementary tension with fieldwork.

Concretely, this means two things. First, that discussions of theory in anthropology are usually intimately interwoven with close, detailed accounts of particular cases. Case studies and examples recur throughout these chapters, but as Englund (chapter 6) points out, following Max Gluckman, ethnography can be more than mere 'illustration'. Working and thinking through cases transforms the very nature and substance of anthropological theory, keeps it grounded, thick and contextual in a way that is quite characteristic of the discipline. Second, the ethnography–theory pairing in anthropology leads to the particular dynamic I described above, in which unfamiliar realities and experiences are constantly drawn upon to reorient and distort theoretical frameworks and assumptions. Anthropology has often had an irreverent and refreshing way with theory.

Of course, what is at stake in these claims is the question of the broader distinctiveness of anthropology as a discipline – a question that I do not have the space to explore in more detail here. However, one of the ways in which this book can be read is precisely as a collective and historically minded answer to this question. This book is in part an account of how the distinctiveness of anthropology was built, not only at a particular historical moment, as anthropology emerged from a broader interdisciplinary field, but also throughout the history of the discipline, as theoretical borrowings from, and loans to, other disciplines nevertheless produced a distinctively anthropological conversation about theory. It is this conversation to which the present book provides an introduction.

Notes

1 They include, most proximally, Yael Navaro, who lectured on the series but was unfortunately unable to contribute to the book, and the other members of the department involved in discussions of teaching and lecture planning in 2016: Uradyn Bulag, Hildegard Diemberger, Paola Filippuci, Sian Lazar, Perveez Mody and Joel Robbins. More distantly, this includes all of the past lecturers on this course over the decades – too many to name – whose thinking and teaching fed into the collective and enduring conversation described here.

2 Particularly in regional or more thematically oriented scholarship, anthropological work does often follow a more cumulative pattern. But the fact that higher status often attaches to those 'generalist' revolutionary claims than to more cumulative regional ones, underscores my point.

3 This is hardly surprising. Kuhn was himself inspired by a longstanding conversation between anthropology, sociology and history on cultural relativism and the social construction of knowledge. The 'paradigms' of historians of science and the 'cultures' of anthropologists are sister concepts.

4 This relates to an insightful observation by sociologist Andrew Abbott: theory in the social sciences tends to have a fractal structure; distinctions on higher scales replicate on lower ones (Abbott 2001). The distinction between positivist and interpretivist approaches in anthropology (see chapters 1 and 8 in this book) is sometimes described as a distinction between whole schools (for instance, symbolic anthropology being marked as interpretivist, functionalist anthropology marked as positivist). But, on another scale, the distinction replicates within functionalism itself: some functionalist anthropologists (such as Malinowski) were more interested in interpretive questions surrounding 'the native's point of view' while others (structural-functionalists) focused on the positivist question of the maintenance of social structure. And again, within 'structural-functionalism' itself, there are those works in which questions of perspective and interpretation make a showing (such as in the early chapters of Evans-Pritchard's *The Nuer*) and those in which positivism ruled supreme.

5 And indeed, the odd extreme play with chronology can itself lead to interesting conceptual reflections and unexpected connections. For instance, for a few years, I taught structural-functionalism late in the series and just before actor-network theory, as a mini-series on 'social and anti-social anthropologies'. Readers might note a few resulting correspondences between chapters 1 and 13.

Bibliography

Abbott, Andrew 2001. *Chaos of Disciplines*. Chicago, IL: University of Chicago Press.

——— 2004. *Methods of Discovery: Heuristics for the Social Sciences*. New York, NY: W.W. Norton & Company.

Ardener, Edwin 1971. The new anthropology and its critics. *Man* 6 (3): 449 .

Barnard, Alan 2000. *History and Theory in Anthropology*. Cambridge: Cambridge University Press.

Barnard, Alan, and Jonathan Spencer (eds.). 1996. *Encyclopedia of Social and Cultural Anthropology*. London: Routledge.

——— (eds). 2009. *Routledge Encyclopedia of Social and Cultural Anthropology*. London; New York, NY: Routledge.

Boyer, Dominic, James D. Faubion and George E. Marcus 2015. *Theory Can be More Than It Used to Be: Learning Anthropology's Method in a Time of Transition*. Ithaca, NY: Cornell University Press.

Candea, Matei 2016. De deux modalités de la comparaison en anthropologie sociale. *L'Homme* 218: 183–218.

Chua, Liana, and Nayanika Mathur Forthcoming. *Who Are 'We'? Reimagining Alterity and Affinity in Anthropology*. New York, NY: Berghahn.

Comaroff, Jean, and John L. Comaroff 2012. Theory from the South: Or, how Euro-America is evolving toward Africa. *Anthropological Forum* 22 (2): 113–131.

Evans-Pritchard, E.E. 1965. *Theories of Primitive Religion*. Oxford: Clarendon Press.

Geertz, Clifford 1988. *Works and Lives: The Anthropologist as Author*. Palo Alto, CA: Stanford University Press.

Granqvist, Emma 2015. Why science needs to publish negative results. *Authors' Update*. March 2. https://www.elsevier.com/authors-update/story/innovation-in-publishing/why-science-needs-to-publish-negative-results.

Handelman, D. 1994. Critiques of anthropology: Literary turns, slippery bends. *Poetics Today* 15 (3): 341–381.

Heywood, P. Forthcoming. *After Difference: Queer Activism in Italy and Anthropological Theory*. Oxford: Berghahn.

Howe, Cymene, and Dominic Boyer 2015. Portable analytics and lateral theory. In Dominic Boyer, James D. Faubion and George E. Marcus (eds.), *Theory Can be More Than it Used to Be: Learning Anthropology's Method in a Time of Transition*, pp. 15–38. Ithaca, NY: Cornell University Press.

Kuhn, Thomas S. 1970. *The Structure of Scientific Revolutions*. Chicago, IL: University of Chicago Press.

Kuper, Adam 1973. *Anthropologists and Anthropology: The British School 1922–1972*. London: Allen Lane.

Latour, Bruno 2005. *Reassembling the Social: An Introduction to Actor-Network-Theory*. Oxford: Oxford University Press.

Layton, Robert 1997. *An Introduction to Theory in Anthropology*. Cambridge: Cambridge University Press.

Lévi-Strauss, Claude 1963. *Totemism*. Boston, MA: Beacon Press.

Moore, Henrietta 1988. *Feminism and Anthropology*. Minneapolis, MN: University of Minnesota Press.

Moore, Henrietta L. 2000. *Anthropological Theory Today*. London: Wiley.

Moore, Henrietta L, and Todd Sanders 2006. *Anthropology in Theory: Issues in Epistemology*. Malden, MA: Blackwell Publishing.

Nadel, S.F. 1951. *The Foundations of Social Anthropology*. London: Cohen & West.

Navaro-Yashin, Y. 2009. Affective spaces, melancholic objects: Ruination and the production of anthropological knowledge. *Journal of the Royal Anthropological Institute* 15 (1): 1–18.

Pina-Cabral, João de 2010. The door in the middle: six conditions for anthropology. In Deborah James, Evelyn Mary Plaice and Christina Toren (eds.), *Culture Wars: Context, Models and Anthropologists' Accounts*, pp. 152–169. EASA Series 12. New York, NY: Berghahn Books.

Radcliffe-Brown, A.R. 1940. On social structure. *The Journal of the Royal Anthropological Institute of Great Britain and Ireland* 70 (1): 1–12.

Strathern, Marilyn 1987. An awkward relationship: The case of feminism and anthropology. *Signs: Journal of Women, Culture and Society* 12 (2): 276–292.

——— 2011. Binary license. *Common Knowledge* 17 (1): 87–103.

Stocking, George W. 1984. Radcliffe-Brown and British social anthropology. In George W. Stocking Jr. (ed.), *Functionalism Historicized: Essays on British Social Anthropology*, pp. 131–191. Madison, WI: University of Wisconsin Press.

Sykes, Karen Margaret 2005. *Arguing with Anthropology: An Introduction to Critical Theories of the Gift*. London: Routledge.

Chapter 1

Severed roots

Evolutionism, diffusionism and (structural-)functionalism

Matei Candea

Introduction: Varieties of the organic analogy

This chapter provides an overview of the birth of anthropology as a discipline. Where ought such a story to begin? The beginning of 'modern anthropology' is often associated with a revolutionary break from earlier evolutionist assumptions. Conveniently, the turn of a century roughly marks the cut: '19th-century evolutionism' feels like a neatly self-contained precursor to 'modern anthropology' in its various forms. At that point, narratives often assume a national form and multiply the disciplines of anthropology: in the UK, the discipline is called *social anthropology* and begins with the two warring functionalisms of Bronislaw Malinowski and Alfred Reginald Radcliffe-Brown. In the USA, *cultural anthropology* begins with Franz Boas's diffusionist and relativist critiques of evolutionism. In France, Emile Durkheim and his nephew Marcel Mauss are a common point of reference for the theoretical origins of *ethnologie* as well as sociology.

This chapter aims to tell those stories as one story, giving an overview of these various beginnings in their interaction with each other, and with the evolutionism that preceded them. Much is gained by thinking of these theories together. In particular, the vision of a radical break from evolutionism – often featured in popular understandings of the discipline as a briefly sketched mess of nonsense and prejudice – leaves some questions unanswered: Why, having rejected evolutionism, did 20th-century anthropologists still talk of non-Western people as 'primitive'? Why, when functionalism took a turn towards structural-functionalism, did grand comparative schemes of decontextualised institutions seem to reappear on the scene? How come so many of the key concepts, questions and problematics that agitated 20th-century anthropology – animism, totemism, joking and avoidance relationships – were first identified by evolutionist authors? What is needed is a story that gives a sense of the distinctiveness of these schools, without losing sight of the continuities and interconnections between them.

The scale and scope of that story explains the uncharacteristic length of this chapter compared to the others in this book, to which it serves as an extended background. For, indeed, many if not all of the theoretical developments traced in the rest of this book are in some way or other taking off from, critiquing or rediscovering elements of one or more of the four classic schools of theory examined in this chapter.

As a frame for this complex story, it will be useful to begin with a – necessarily simplistic– thumbnail sketch of the way these schools are usually remembered. Evolutionists sought to reconstruct what they imagined to be the singular, progressive history of mankind, from the comparative study of the customs of contemporary and past peoples worldwide.

Boas proposed a programme that sought to retrace the movement of cultural traits in a more empirical and controlled way between adjoining cultures and raised the question of how these traits were integrated into distinct cultural wholes. Functionalists in the UK, following Bronislaw Malinowski, produced monographic studies based on intensive ethnographic work in single societies, showing the interrelation of different aspects of culture and social organisation. Structural-functionalists sharpened and narrowed this question down to an investigation of the structural equilibrium maintained by the interrelation of social institutions, and widened this enquiry back out as a comparative one – both moves centrally indebted to the writings of French sociologist Emile Durkheim.

The story of these theories has been told before – many times – and this chapter draws extensively on some classic accounts of the schools above (including Kuper 1973; 2005; Stocking 1983a; 1984b; 1991; Trautmann 1992b) as well as on the original texts themselves. Readers are encouraged to go back to that primary and secondary literature for a much more historically nuanced and detailed account of many of the shifts and transformations below.

One virtue of the condensed treatment these different schools get here, however, is to pull out two key features that can get lost in a more historically detailed and extensive account. The first is that, while the different national traditions are indeed in many ways distinct, and while there certainly was a break from the evolutionism that preceded them, all these schools nevertheless shared a particular vision of societies and cultures as units, entities on the ground, assemblages of parts. This vision echoes biology's vision of organisms. In some cases, the comparative reference to biology is explicit, in others it is implicit or even partly disavowed. The biology that anthropologists might have imagined as a parallel or counterpoint to their own discipline itself shifted over this period; notably, as we shall see, with the advent of Darwinian evolutionism, whose relationship to these various schools is more complex than one might initially suspect.

The organic analogy, as this vision is often called, is very old indeed. We find it, amongst many other places, in Plato's *Republic* (Lear 2006), and it is a powerful metaphor in that it leaves many options open. For biologists, the study of organisms could take the form of a structural comparison of different species, of the physiological description of the interaction of organs, or it could involve investigating the transformation of one organic form into another. This, in turn, can be imagined as a search for grand evolutionary laws, or for modest historical contingencies. In the same way, anthropologists of the mid-19th to the mid-20th century differed in whether they focused on the inter-relation and integration of parts of society and culture, on the comparison of these parts in different societies/cultures or on both; and they differed also in the questions they asked and the lessons they drew from these approaches. The positions were often violently opposed, but they shared a number of elements of the same language. More often than not, as we shall see, shifts and distinctions that are now seen as profoundly theoretical were made initially in terms of method (see introduction for a discussion of this distinction). This commonality becomes even more evident by contrast to the conceptual break introduced by mid-century French structuralism (see chapter 2) which mounted a full-scale attack against the organic analogy and proposed a radically different way of imagining the objects of anthropological study (Salmon 2013).

Starting from something these theories shared – the organic analogy – locates the beginning of anthropology not in one particular place, or in one well-defined time, but rather in the middle of an international conversation. British social anthropology, American cultural anthropology and French ethnology developed radically different assumptions about history, the nature of society and culture, the methods and purpose of anthropology – all of these in explicit contrast to earlier British and American evolutionists. But the point is precisely that all these differences were worked out in explicit contrast and conversation. There were differences between these schools, but differences between the schools were, in other words, relational. It is out of these relational differences that anthropology as we know it emerged.

This brings me to a final theme running through this chapter, namely that of the recurrences of contrasts and dichotomies, that get mapped and remapped in different ways in the debates between these schools, but ultimately stay stable as key tropes for discussion. These are contrasts between theory and method, between structure and process, between humanistic interpretation and scientific explanation. These contrasts were forged in the conversation explored in this chapter, and it is in these contrasts that – as the chapters below make clear – much of anthropology has since continued to dwell (cf. Abbott 2001).

The evolution (or development) of society

The thinkers remembered under the label of 19th-century evolutionism were a diverse and international collection of scholars, including Herbert Spencer, Henry Maine and E.B. Tylor in the UK, Lewis Henry Morgan in the USA, Johannes Bachofen in Switzerland, Karl Marx and Friedrich Engels in Germany, Auguste Comte in France. Their intellectual backgrounds, styles of writing, methods and positions differed in many ways, but they are grouped together because they shared a broad concern with the question of the transformation of human society, from a postulated 'primitive' form, through a set of (more or less) identifiable stages, to its 'modern' instantiation in 19th-century Europe and America. Most of them also shared the basic methodological principle that one might reconstruct and theorise about this historical process through the comparative study of historical societies and contemporary non-Western ones, imagined to exemplify earlier, pre-modern, stages in the story of human social progress.

The key to understanding both this anthropological evolutionism and its later critics is to note that despite its name, what is usually remembered today as 19th-century anthropological 'evolutionism' was not in any straightforward sense an offshoot of the work of Charles Darwin. If anything, the shoe is on the other foot. 'Evolution' as a general term to describe a grand historical transformation was popularised by philosopher, biologist and anthropologist Herbert Spencer, and was only retrospectively adopted by Darwin to describe the theory he had laid out in *The Origin of the Species* (1859).[1] The term evolution – etymologically rooted in the Latin for 'the unrolling of a scroll' – had been associated since the 18th-century with the process of ontogenesis: the development of an embryo or germ into its mature form ('Evolution, N.' 2017; see also Kuper 2005: 61). Evolution in this sense implied a vision of the development of an entity, through a fixed and recognisable set of stages, from its undifferentiated origin, to its mature form: a differentiated and complexly interrelated organism. This old ontogenetic meaning of 'evolution' implies, of course, that one already knows what this final form (the mature organism) will look like. This development is thus progressive, both in the sense of being gradual, and in the sense of going from a more basic to a more accomplished, indeed perfected, form.

Spencer used the notion of evolution (often interchangeably with 'development') in a number of different ways – a point to which I will return below – but the broadest and most famous way was in support of an extremely general philosophical vision, that imagined an analogous process of differentiation and integration applied universally to organisms, societies, and indeed, entities in general, taking them progressively (in both senses above) from lower to higher forms:

> Alike during the evolution of the Solar System, of a planet, of an organism, of a nation, there is progressive aggregation of the entire mass. . . . From the lowest living forms upwards, the degree of development is marked by the degree in which the several parts constitute a co-operative assemblage. The advance from those creatures which live on in each part when cut to pieces, up to those creatures which cannot lose any considerable part without death, . . . is an advance to creatures which, while more integrated in respect to their solidification, are also more integrated as consisting of organs that live for and by each other. The like contrast between undeveloped and developed societies, need not be shown in detail: the ever-increasing co-ordination of parts, is conspicuous to all.
>
> (Spencer 1867: 327–328)

The Darwinian 'evolution' of natural species as we think of it today is a very different process indeed to this grand unrolling of the cosmic scroll. It involves an account of the constant transformation and subdivision of species, that has no predetermined end-point, and therefore no recognisable set of stages that apply universally, nor any clear way to imagine ranking 'lower' and 'higher' forms. There is no progress here; simply transformations in form and function in relation to changing selective and environmental pressures. This may explain why Darwin only belatedly adopted Spencer's term evolution to describe the processes he had in mind. Indeed, for a number of later commentators, 'what is radical about Darwin's book is not its evolutionism but its materialism: species evolve according to processes that are entirely natural, chance-generated, and blind' (Ansell-Pearson, Miquel and Vaughan 2010; after Menand 2011). In Darwinian biology, it is this blind and random process that came to be called evolution, clearly distinct from the ontogenetic process ('development').

By contrast, 19th-century anthropological 'evolutionism', as usually remembered and criticised, presents an analogue of the ontogenetic view, applied to the development of human society through history. Many of these 'evolutionists' believed, as Radcliffe-Brown sarcastically put it of Morgan, 'not in evolution, but in progress . . . conceived as the steady and material improvement of mankind from crude stone implements and sexual promiscuity to the steam engines and monogamous marriage of Rochester N.Y.' (Radcliffe-Brown 1947). The barb was uncharitable but not entirely inaccurate. Thus Spencer in some of his writings, such as *Social Statics* (1899), described as 'social evolution' the progressive development of human society towards what he envisaged as its 'perfect form': a peaceful, egalitarian, altruistic and cooperative society characterised by industrialisation, minimal state regulation and a high level of division of labour (Spencer 1899; see Perrin 1976). Spencer draws on examples from history to construct a picture of the processes and transformations through which such an ideal society will be achieved. This grand vision of progressive development became characteristic of anthropological evolutionism, even as different authors came to disagree profoundly over its actual content. The evolutionists had impassioned arguments over whether, for instance, human societies had originally been patriarchal or matriarchal. They also proposed rather different accounts of what the endpoint might look like. Marx, not unlike Spencer, imagined a further utopian stage (in his case communism), whereas others, such as Morgan saw the apogee of human development in something much like their own bourgeois social arrangements. Evolutionists differed also over their accounts of the key stages in that process, and over the main engine of those transformations. But the difference between a unilineal, progressive and stage-bound 'social evolution' and the Darwinian picture of the evolution of species, could not be clearer. This core divergence is a clue to the otherwise puzzling fact that, as we shall see below, some of the anthropological evolutionists' later critics (the diffusionists and the functionalists) presented visions of society that were entirely congruous with Darwinian theory.

It must be said that I have slightly overdrawn, for clarity, the contrast between Spencerian and Darwinian evolutionisms. The two men in fact borrowed from each other's thinking much more that the above description suggests, and Spencer's work, which spanned over 60 years, was rich in complexities and reappraisals. As a number of scholars have pointed out, Spencer's sociology went well beyond the somewhat naive version of social evolutionism outlined above. For instance, Perrin (1976) detects no fewer than four distinct theories of social evolution in Spencer's work, of which the above is only the earliest and most simplistic, while the other three are much more in line with the Darwinian vision of complex, unguided and multi-stranded change – as Radcliffe-Brown himself recognised (Radcliffe-Brown 1947). Crucially also, Spencer's explicit and precise analogy of organisms and societies, and his focus on the ways in which different parts of society were mutually interdependent, announced much of what would later be rediscovered by anthropological functionalists and structural-functionalists (Peel 1971).[2] Other evolutionists, too, differed in the extent to which they imagined human history to have been unilineal or directed, and some of them, such as Tylor, also developed distinctly functionalist analyses.

However, Spencer's intellectual interactions with Darwin were relatively atypical, and it remains the case that anthropological evolutionism owed little in the main to Darwinism, as historians of the discipline such as Adam Kuper (2005: 15–36) and Thomas Trautman (1992a) have noted. As these scholars and others point out, key elements of the evolutionist view of history long predated Darwin. The recurrent trope of a contrast between 'savage' and 'civilised' humanity had been a fixture of European philosophical discourse since the 17th century. This was partly inspired by European encounters with previously unknown Amerindian populations which soon came to be imagined as stand-ins for the 'precursors' of European modernity. The notion of a world-history of humanity divided into 'stages' had been articulated by thinkers of the Scottish enlightenment and provided a frame through which a basic contrast between 'us' and 'them' could be developed into a stepwise history of human progress that doubled up as a typology of societies.

A more immediate element of context was what Trautmann calls 'the revolution in ethnological time' in the 1860s, prompted by archaeological discoveries that conclusively expanded the traditional biblical time frame of human history. As Trautmann puts it:

> Very suddenly the bottom dropped out of history and its beginnings disappeared into an abyss of time. The sudden, dramatic enlargement of the scale of human history demanded new content: and what rushed in to fill out the vast blank spaces in the newly enlarged frame was social evolutionism.
>
> (1992a: 380)

Another important conceptual building block was the flowering, from the early 19th century, of a range of comparative disciplines – comparative anatomy, comparative physiology, comparative grammar, geography (Goyet 2014) and, of particular relevance to anthropology, comparative law. A number of early anthropologists, such as Lewis Henry Morgan or Henry Sumner Maine, were themselves lawyers, steeped in the study of classical Roman law as well as modern law. This explains in part the rise of 'kinship' to prominence as an anthropological topic. It is not surprising that such lawyer–anthropologists would have taken as their focus the set of rules surrounding reproduction, family relations and the distribution of property. These seemed a direct analogue for the legal systems they knew (Trautmann 1992b; Kuper 2005).

At the heart of these various comparative disciplines lay the same insight, most prominently associated with philology: that one might reconstruct historical sequences, branching associations and progressions by comparing the state of affairs in a number of distinct contemporary cases. The increasingly close contact with non-European societies and ways of life through European imperial expansion and colonisation, and the multiplication of detailed accounts and reports by travellers, missionaries and others, provided the material for 19th-century evolutionists to envisage a comparative field of 'peoples' that had a richness and diversity akin to the comparative fields of other disciplines. Here was the rich material required to fill that enlarged temporal canvas. The following quote from E.B. Tylor, to whose work we return below, gives an explicit statement of this vision of comparison, and the assumptions it comes bundled with:

> By comparing the various stages of civilization among races known to history, with the aid of archaeological inference from the remains of pre-historic tribes, it seems possible to judge in a rough way of an early general condition of man . . . This hypothetical primitive condition corresponds in a considerable degree to that of modern savage tribes . . . [The] main tendency of culture from primeval up to modern times has been from savagery towards civilization . . . The European may find among the Greenlanders or Maoris many a trait for reconstructing the picture of his own primitive ancestors.
>
> (Tylor 1871: 1– 21)

There were serious intellectual flaws to the evolutionists' comparative vision. As Kuper forcefully puts it:

> The whole conception is fundamentally unsound. There is not even a sensible way in which one can specify what a 'primitive society' is. . . . Human societies cannot be traced back to a single point of origin. Nor is there any way of reconstituting prehistoric social forms, classifying them, and aligning them in a time series. There are no fossils of social organization.
>
> (2005: 5)

The politics of this vision have also been justly criticised. As anthropologists have recurrently pointed out, one key effect of this vision in anthropology was to leave a range of contemporary peoples 'out of time' (Thomas 1996), treating their societies as, in effect, living fossils, arrested at a 'previous stage' of a single progressive road that leads to European-style modernity. Evolutionism was the first (and sadly not the last – see below) anthropological paradigm to imagine non-Western peoples as stuck in an eternal present (Fabian 1983) and therefore to envision the contemporary world as made up of *Europe and the People Without History* (Wolf 1983 – see chapter 5, this volume, for a fuller discussion of history and anthropology).

Put like this, most present-day readers will easily recognise such a Eurocentric vision of history as politically and ethically unacceptable. And yet this sort of world-historical vision has been surprising resilient up to the present day. The narrative of global 'development' shares many of the same premises as 19th-century evolutionism (Ferguson 1996). One might say the same of contemporary references to 'social progress'. Think, for instance, of the ease with which particular practices and policies can be justified today by simply pointing to the fact that 'it's the 21st century'. Hidden within that self-evident assertion is an implicit story of the present as morally superior to the past (and occasionally, the West as morally superior to other places), that many 19th-century evolutionists would have found familiar.

For, conversely, many 19th-century evolutionists were on their own terms political progressives, even radicals.[3] 'Progress' was a key conceptual and political lynchpin of evolutionism. Many saw themselves as resisting the still common view that non-Western peoples were 'fallen', 'degenerate' or 'immoral', by insisting on the mental unity of mankind (we shall see this theme in Tylor below) and by replacing 'primitive peoples' within the grand tapestry of human 'progress'[4] – albeit in a supporting role. In so doing, as Trautmann put it 'Morgan rejected the permanent degradation and inferiority of the cultural other' (1992b: 171). And yet, this grand tapestry was woven through with patronising assumptions of unequal development, in some cases along racial lines.[5]

An evolutionist vision could also coherently imply that the present would be superseded by a better future. Engels, in *The Origins of the Family, Private Property and the State* (1972), famously turned Morgan's grand history of the development of human kinship systems into a radical critique of modern political-economic arrangements. What Morgan had presented as a triumphant rise towards civilisation under the watchful eye of God, Engels described more ambivalently as a move away from 'primitive communism', and, in the shift to the patriarchal family, as 'the world-historic defeat of the female sex' (1972: 67). Engels was not alone, as we have seen, in imagining that a further stage would resolve the obvious problems of 19th-century modernity. Whatever its intellectual inconsistencies, therefore, the device of connecting up different socio-political arrangements into a sequence of stages is, politically speaking, an inherently equivocal move – the sequence can be mined for very different political and moral lessons (see Bloch 1984). This explains the otherwise paradoxical reappearance of enthusiasm for Engels' evolutionist classic a century later under the impetus of Marxist and feminist scholars. In sum, the politics of evolutionism, then and now, are more complex than might first appear.

Evolutionism in practice: Tylor on avoidance

Evolutionism is often written about and contextualised in broad terms, as I have done above. Evolutionists themselves, however, are now seldom read, particularly by students. The effect is that much of the sophistication and complexity – and indeed diversity – of their arguments tends to be forgotten, and the continuities between their approaches and concerns and those of the anthropologists who critiqued and followed them, tend to be underestimated. As a counterpoint, I will give in this section a detailed example of an evolutionist argument from the work of E.B. Tylor, whose definition of the comparative method we saw above. I will describe this case in such detail because it exhibits some of the key failings of 19th-century evolutionist theorising, yet nevertheless displays the sophistication of some of the best thinkers of the period.

One of the works in which Tylor demonstrated his comparative method is a famous article in the *Journal of the Royal Anthropological Institute* entitled 'On a method of investigating the development of institutions; Applied to laws of marriage and descent' (1889). Tylor reported that he had tabulated the 'laws of marriage and descent' of between 300 and 400 peoples, as reported by missionaries, travellers and others. He then sought higher than average statistical correlations between seemingly unrelated customs. For instance, Tylor focuses on what he – in the characteristically dismissive and patronising language of his day – describes as

> a quaint and somewhat comic custom as to the barbaric etiquette between husbands and their wives' relatives, and vice versa: they may not look at one another much less speak, and they even avoid mentioning one another's names.
>
> (1889: 246)

Such 'avoidance' occurs in some cases between the husband and his in-laws; in others, avoidance occurs between the wife and her in-laws; in others again, avoidance goes in both directions. Tylor's first problem is to seek to explain this 'strange' set of customs. He does so by connecting this problem to a different one, concerning the residence of the married couple. In some societies in his sample, husbands went to live with their wives' families. In others, this arrangement is temporary; after which the couple move off to set up an independent household, or move to settle with the husband's family (Tylor terms this 'removal'). In others still, wives went to live with their husbands. Tylor focuses on the correlation between these two sets of customs, the set that relates to avoidance and the set that relates to residence patterns of the married couple. Note that in proceeding in this way Tylor is, like Spencer, deploying a basic insight that will be characteristic also of 20th-century functionalism: namely, that there is an interrelation between different rules, laws and customs in the same society – that they form a mutually supporting functional arrangement. Tylor's first problem – explaining an initially 'strange' set of customs in relation to other aspects of life in the same society – remains a classic one for anthropologists still today.

On the other hand, the second (and to him more fundamental) problem Tylor focuses on will strike most contemporary anthropologists as oddly formulated. Tylor is trying to reconstruct the *sequence* through which societies moved from one system of residence to the other. Did husbands first go to live with their in-laws, and only later this pattern was reversed such that wives would come to live with their husbands? Or was it the other way around? I hope I have written enough above to explain both why Tylor could see such a question as a meaningful one and why contemporary anthropologists see it as nonsensical. Contemporary anthropologists would look at the differences in Tylor's sample as evidence of just that – differences in customs between contemporary societies. By contrast, Tylor assumed that human society evolved in a single direction of travel, through a set of more or less fixed sequences or stages.[6] As a result, he thought that his sample of societies with different customs represented that set of stages. The difficulty, for him, was finding a way to put them back in their proper sequence. Hence his question: Which kind of residence, of husband-to-wife or wife-to-husband, came first? This question might seem to be of mainly 'antiquarian' interest. In fact, it spoke to a much broader debate that raged between evolutionist anthropologists of the 19th century. This debate turned on knowing whether human societies were originally matriarchal (as proposed for instance by Bachofen) or patriarchal (as argued by Maine. On this argument see Bamberger 1974; Kuper 2005).

Tylor seeks, through a sophisticated demonstration, to solve both of his problems simultaneously, by focusing on the correlation between his two sets of customs. Recall that there are three possibilities in each set: avoidance can be from husband to in-laws, from wife to in-laws, or mutual; residence can involve husband moving in with in-laws, wife moving in with in-laws or ("removal"). He shows that avoidance customs between the husband and his wife's family occur in all three types of residence, and are statistically more important where the husband is moving to live with his in-laws, either permanently or temporarily ("removal"). He shows that the other two kinds of avoidance (mutual and wife to her in-laws) only occur in cases of "removal" or where the wife moves in with her in-laws.

The first hypothesis Tylor draws will have already occurred to many readers: there is some relation between expressing a certain kind of formal distance with one's in-laws, and actually going to live with them. Tylor invokes this by comparison with his contemporary social norms, those of Victorian Britain:

> As the husband has intruded himself among a family which is not his own, and into a house where he has no right, it seems not difficult to understand their marking the difference between him and themselves by treating him formally as a stranger. So like is the working of the human mind in all stages of civilisation that our own language conveys in a familiar idiom the same train of thought ... we have only to say that they do not recognise him and we have condensed the whole proceeding into a single word.
> (Tylor 1889: 247–248)

The initially 'quaint and comic' custom, Tylor shows, turns out to have been perfectly logical and understandable to 'civilised' moderns. This is a good reminder that, despite their triumphalist reading of history as a progress towards Western modernity, many evolutionists like Tylor had a profoundly humanist assumption of the unity of humans in all 'stages' of civilisation.

This hypothesis matches some, but not all (or even most) of Tylor's data. True, in societies where husbands reside with their wife's kin (65 societies in his sample), there is husband to in-law avoidance (14 instances), and no wife to in-law avoidance. Wife to in-law avoidance occurs in societies in which wives go to live with their husband's kin (eight instances out of a total of 141 such societies). And it makes sense, in terms of his hypothesis, for both types of avoidance to be present in the cases of removal in which the couple lives first with one family then with the other (respectively 22 and five instances out of a total of 76 societies). But other significant data do not fit: both mutual avoidance and husband to in-law avoidance are present in societies in which wives come to live with their husband's kin. Indeed, the incidence of husband to in-law avoidance (9/141) is slightly higher than the converse in those cases (8/141).

Tylor explains these anomalies by relying on the image of social change in sequences. Note the diagram through which Tylor presents his correlations (Figure 1.1). The basic form of this diagram is a crossed table: types of residence on three horizontal 'rows', types of avoidance in three vertical 'columns', with the number of societies involved in

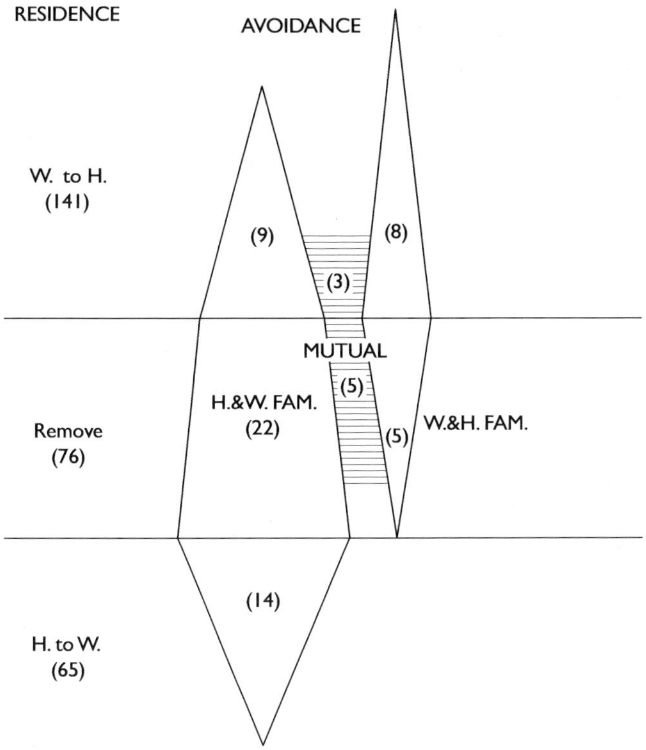

Figure 1.1 Residence and avoidance.
Source: Reproduced from Tylor (1889: 251).

each case. But Tylor implies a very different factor through his use of triangles and cross-hatching: namely, a sense of direction and transformation. The diagram is supposed to be read from the bottom up, as a sequence of stages. At the bottom is an 'initial' stage, represented by societies in which husbands go to live with their wives' kin. At the top is a 'final' stage, in which wives go to live with their husbands. In between is the 'removal' stage (husbands first go to live with their in-laws for a period then the couple moves to the husband's family) that Tylor presents as a transitional point between the other two stages. This narrative is used to explain the untidy data mentioned above. Husband to in-law avoidance is the earliest form of such avoidance. Wife to in-law avoidance develops later, in the transitional stage. This historical hypothesis allows Tylor to explain the incidence of husband to in-law avoidance in the final stage as a 'survival' from a previous state of affairs.

Through a clever twist, Tylor turns what initially seemed to be confounding data (the incidence of husband to in-law avoidance in societies in which the husbands do not live with their in-laws) into evidence for the 'direction of movement' of social evolution:

> From this distribution of the avoidance-customs, it appears that in the parts of the world open to the present inspection, the three stages of residence have tended to succeed one another in the upward order of the diagram. . . . For if it be supposed that

> the course of society was in the reverse direction, as would be represented by turning the diagram upside down, avoidance between the husband and the wife's family would be represented as arising in the stage when the husband lived away from it, while avoidance between the wife and the husband's family, which ought on this supposition to continue by survival into the stage of residence on the wife's side, is not found there. The avoidance-customs, though practically so trifling, are thus signals showing the direction of a movement, of which we shall more fully see the importance, namely, the shifting of habitual residence from the wife's family to the husband's.
>
> (1889: 252)

This is a 'direction of movement' that fits with Tylor's broader argument in the rest of the piece in favour of a general move in human society from matriarchy towards patriarchy.

In sum, this argument displays prominently the key conceptual errors outlined previously: assuming there was a single 'course of society', and therefore thinking that contemporary societies might be taken as examples of different stages within such a course. Even if one were to pass over these profound conceptual problems, the use and presentation of 'data' in Tylor's argument is also rather dubious. Some, as we shall see below, raised the crucial question of what constitutes a 'unit' for the purpose of this analysis (What is a custom? What is a people?), and of the validity of the second-hand material Tylor draws on. Scientifically minded readers might add that Tylor's attempt to draw 'statistical' inferences from such small samples is fairly unconvincing. Tylor's diagram is both impressively constructed and also somewhat disingenuous. His depiction of the top right-hand triangle as longer albeit narrower than the top left-hand triangle, in order to suggest that this is a actually smaller number is somehow more significant than the larger one, is a clever bit of visual manipulation.

And yet, for all these failings, this example shows the detail and thoroughness of some 19th-century evolutionist arguments. If anthropological evolutionists did ultimately rely on 'just-so-stories', these were not plucked out of thin air, but could be closely and coherently argued, based on impressive erudition and significant amounts of detail. This example counters the usual stereotypes of evolutionists as little more than crude Victorian ideologists. It also explains why the work of authors such as Tylor, Morgan or Spencer, while much berated and thoroughly excoriated by the next generation of anthropologists, nevertheless managed to set the terms of many of the problems that anthropologists thought through and reconceptualised throughout the 20th century: avoidance, marital locality, and more broadly, kinship systems, but also animism, magic or ritual. On avoidance, specifically, we shall see below how the problem would be reframed by structural-functionalist analysis. But the question of avoidance has continued to exercise anthropologists to this day in very different ways (Stasch 2002; Sanchez 2016). This is true more broadly – many of the puzzles that the evolutionists had set out to solve remain key foci of anthropological debate today.

The diffusion and cohesion of traits: Boas's double challenge

I noted above that, in the US as in the UK, 'modern anthropology' or 'anthropology proper' is often described as starting with critiques of the 19th-century evolutionist paradigm. However, this narrative takes a different form in each case. The critique of evolutionism was spearheaded in the US by Franz Boas, who played an equivalent foundational role in American narratives of disciplinary history, to that played by Bronislaw Malinowski in the British narratives.

A core element in Boas's critique of evolutionism had already been foreshadowed by a commentator on Tylor's argument above, Francis Galton. Galton, commenting from the audience when the paper was presented, had noted (in Tylor 1889: 270) that the procedure of comparing customs across different groups of peoples assumed that these customs could be treated as independent variables. But one group of people, Galton countered, might well borrow the customs of a nearby group. In which case, Tylor's elaborate hypotheses about survivals and sequences would hold little water, as the correlation of customs in any given setting might be merely a contingent effect of borrowing (for different comments on Galton's problem, see Handler 2009; Strathern 2004).

Boas developed this line of critique, noting that it was important to study the ways in which customs in fact travelled, changed, were borrowed and transformed. Such a study could not simply mix and match observations taken from all over the world. It would need to rely on the careful examination of the customs and material culture of a set of related peoples within a specific geographic area (Boas 1896: 1924). Boas himself had undertaken a number of field studies, first in Baffin Island, and then in the Pacific Northwest. From this material he drew insightful observations about the ways in which cultural traits (designs in pottery, myths, kinship arrangements) were copied and replicated from one culture to another. He integrated this line of critique with another: that what might seem superficially like the same trait in two different contexts might actually prove to be two very different traits. Together, these two lines of critique provided a devastating attack on the entire evolutionist enterprise, and opened the way to a new anthropological paradigm (Handler 2009).

Boas's double critical move was rooted in a particular intellectual context. Boas was born in Germany into a secular Jewish family. He received a PhD in physics from Heidelberg and then studied geography. This remained the broader framework in which he developed a particular approach to cultural transformation and variation. This, in turn, became the centrepiece of the anthropological tradition he founded when he moved to the USA. Boas had inherited his concern with the diffusion and cohesion of cultural traits from a debate amongst 19th-century German scholars such as Graebner, Schmidt and Bastian, who had sought to turn the romantic concern with *Kultur* – prominent in thinkers such as Herder – into historical science (Kuper 1999; 2005; Penny and Bunzl 2003).

If Boas was interested in the diffusion of traits, he also noted the ways in which cultural traits were transformed when they came to be integrated with the existing traits of a different culture. This attention to context also involved an attention to meaning and perspective. Boas's critique of anthropological evolutionism was also influenced by 19th-century German philosophical debates around the nature of science and the emergent distinction between a positivist approach to natural phenomena and a hermeneutic or interpretive approach better suited to distinctively human phenomena. In this view, society posed a categorical problem: Ought it to be studied as a natural phenomenon or as a human one? The two answers to this question are often exemplified by the contrast between the interpretivist German School of Max Weber and the positivist French sociology of Emile Durkheim – although as we shall see below, the picture was more complex. Boas was leaning towards the interpretive side: the study of cultures ought to be an interpretive, historical science rather than treating human societies merely as natural objects.

For instance, Boas noted that what looked – in a purely objective or behavioural sense – like 'the same' trait or practice in different cultural settings might in fact mean something profoundly different in those different cases or, as he put it, might stem from different 'psychological causes'. Boas takes 'murder' as a case in point: applying this Western juridical notion as a term for comparing killings in different cultures will necessarily miss the following fact:

> The person who slays an enemy in revenge for wrongs done, a youth who kills his father before he gets decrepit in order to enable him to continue a vigorous life in the world to come, a father who kills his child as a sacrifice for the welfare of his people, act from such entirely different motives, that psychologically a comparison of their actions does not seem permissible.
>
> (Boas 1911: 173)[7]

This presented a second, and more profound objection to Tylor's comparative method than Galton's problem of the potential diffusion of traits. Here, what Boas cast into doubt was the very procedure of identifying 'traits' for comparative purposes in the first place, unless and until a careful field study had shown these to truly derive from the same mental processes. This similarity of perspective was precisely what Tylor had (naively) assumed when imagining that the 'embarrassment' a Victorian might feel towards one's in-laws was a human universal. It is through observations such as these that Boas established the most distinctive feature of later American cultural anthropology: an attention to what came to be called cultural relativism (see chapters 8 and 14). Implied in this notion of cultures as contexts and viewpoints was a cultural holism – the image of cultures as integrated wholes, coherent above and beyond their disparate borrowed elements – that counterbalanced the diffusionist vision.

Both elements of Boas's anthropology – its diffusionism and its holism – made a nonsense of the sort of 'statistical' vision implied in the work of Tylor and other evolutionists: the thought that one might pick and choose disconnected customs to compare from a global library of people. They also made a nonsense of the grand historical just-so-stories of 19th-century evolutionism. The image of a single 'course of society' through time exploded into a vision of multiple micro-shifts, borrowings, transformations and reintegrations. At the same time, the image of a single 'human mind' devising solutions for changing historical circumstances exploded into a myriad questions about how differently situated humans might encounter the world from different cultural contexts.

Consistently with this cultural relativism, Boas was also a fervent critic of racism and antisemitism, and of attempts to find 'scientific' grounding for human inequality, including in some strands of social evolutionism. It is important to remember, however, that Boas's objection was not to biological explanations per se – he himself retained an abiding interest in human biology and how its study might inform cultural anthropology – but to their misapplication through racial science. Nor was Boas a critic of evolutionism of the Darwinian type, but rather of the – as we have seen – profoundly non-Darwinian vision of a single path of human history developing progressively towards a pinnacle of civilisation instantiated by 19th-century Euro-America. Boas's vision of traits diffusing and integrating across space and time was in many ways much more Darwinian than the 'evolutionism' of Spencer, Morgan or Tylor. It is therefore incorrect, however tempting, to draw a clear line between Boasian historical anthropology on the one hand, and on the other, those schools that like evolutionism and functionalism had sought to imagine anthropology as a biology of social 'organisms'.

True, the diffusionist 'arm' of Boas's thinking tended towards disaggregating cultures rather than envisioning them as organisms. As Boas's student Ruth Benedict wrote in an early work:

> It is, so far as we can see, an ultimate fact of human nature that man builds up his culture out of disparate elements, combining and recombining them; and until we have abandoned the superstition that the result is an organism functionally interrelated, we shall be unable to see our cultural life objectively.
>
> (quoted in Eggan 1954: 750)

And yet, the relativist 'arm' of Boas's approach implied an integration that brought the organic analogy back into view. Thus, Boas himself wrote that: 'Cultures differ like so many species, or perhaps genera, of animals' (1940: 254). Benedict was thus still following Boas when, in her later works, she took the opposite position to the one just quoted above. As Eggan notes, 'Benedict was soon engaged in the study of cultural patterns and configurations, and her *Patterns of Culture* (1934) represents a complete reversal of her earlier position – here superstition has become reality' (Eggan 1954: 750).

The two Boasian lines of attack (diffusionism and relativism) were complementary but also importantly in tension (Eggan 1954; Handler 2009). The first treated 'traits' as essentially detachable units that could move from one culture to the next – cultures in this view were aggregates of traits. The second treated cultures as integrated wholes in which individual traits formed a coherent totality. The first focused on relations and connections across history and geography. The second set the scene for a cultural relativism in which different cultures could be imagined as alternative viewpoints on the world.

In Boas's own work, the two visions – diffusionist and holist – were kept productively in tension. However, Boas's students and followers, who established American anthropology in the early 20th century, ended up torn and oscillating between these two visions of culture, as demonstrated in the example of Benedict above. In the short to medium term, the holist vision mostly won out. Boas's students, Benedict and Margaret Mead, in particular, became key exponents of a 'culture and personality' approach, in which cultures were treated as integrated contexts that inculcated into individuals particular moral, intellectual and aesthetic proclivities. This, in turn, fed into the interpretive turn spearheaded by Clifford Geertz (see chapter 8). In various forms, cultural relativism and holism was, for most of the 20th century, the major anthropological tradition in US anthropology.

However, diffusionism did not die out (nor indeed did evolutionism; a point I return to in the conclusion). The concern with connections, links and transformation across time and space remained a key element of the work of anthropologists such as Kroeber, and became important to a later generation of Marxist-inspired anthropologists who sought once again to ask questions on a global scale (see chapters 3 and 5). In a very different vein, the doubts raised against cultural holism by the 'writing culture' critique of the 1980s (see chapter 8), opened US anthropology up to a kind of neo-Boasian vision of 'flows', 'routes' and '-scapes' (Brightman 1995; Bashkow et al. 2004). Multi-sited ethnography, another child of the writing-culture critique (Marcus 1995; see chapter 6), turned these questions into a methodological vision that bore some resemblance to the earlier Boasian question of the diffusion and integration of 'traits'.

From evolution to function

Diffusionism held some appeal in the UK. Egyptologist Grafton Elliott Smith and anthropologist William James Perry attempted far-reaching hypotheses about the global diffusion of traits. In a less grand style, and closer to Boas's own methods and concerns, turn-of-the-20th-century anthropologists such as A.C. Haddon and W.H.R. Rivers asked questions about the diffusion of customs, tool types and kinship terminologies across regional areas. In the main, however, the challenge against evolutionism took a different form in the UK: namely, that of 'functionalism'. There were two main components to the paradigm shift from evolutionism to functionalism in British anthropology. The first was a sharpened theoretical interest in function, drawn in part from the work of French sociologists such as Durkheim. The second was the development of a particular fieldwork tradition often associated with Malinowski. This section examines both developments in turn.

The basic principle of functionalism is simple – it involves analysing a feature or object in terms of its function or purpose. Functionalism in biology explains the role of an organ or behaviour in terms of its value for the organism. The respiratory system, for instance, serves the organism by allowing the intake of oxygen and the exhalation of carbon dioxide. Functionalism in sociology transposes this argument to societies through the medium of the organic analogy – the idea that societies, like organisms, are to be envisaged as a set of interacting parts arranged in mutually supporting ways to form a functional totality. In this view, to explain social features such as customs or institutions is to point to their purpose – like organs in a body.

It is worth noting that functionalism introduces a different kind of teleology into sociological and biological explanation to the historical teleology we criticised above. The crucial flaw of evolutionist anthropological explanations was the assumption that the end-point of evolution could be known (that, broadly speaking, societies evolved 'towards' something like Western modernity), and that history could thus be reconstructed as a series of progressive stages towards that goal. But functionalist explanation is also teleological, in the sense that it explains entities by pointing to their end or purpose in relation to the organism. Here again, the point is that the total form of the organism has to be assumed in order for the purpose of the part to be known. And here again, this assumption would prove dubious – for what after all is a society? Is it really a total, organically integrated entity of the type the theory assumes? As we shall see, this would be one of the key critical points raised against functionalism in its most theoretically developed form: namely, the structural-functionalism of Radcliffe-Brown.

Durkheim's (r)evolution

It was Auguste Comte (1830) who originally introduced both the terms *positivism* and *sociology* in France. Durkheim, however, did more than any other thinker of the period, to establish sociology as a distinct positive science, irreducible to philosophy, psychology, biology or history. The key problem for Durkheim was to convince readers that explanations of social behaviour had a scientific consistency that distinguished them from mere historical interpretation or philosophical abstraction, and that sociological explanations could not be reduced either to individual psychological motives, or to biological propensities – in other words, social facts existed as an irreducible layer of reality, and thus deserved their own systematic science (see chapter 7 on this original distinction and its enduring afterlife). The key device in Durkheim's arsenal was functionalism, with its vision of society as an organic-like structure of mutually supporting parts.

This vision was not new. Elements of sociological functionalism were already important, as we saw above, in the works of some evolutionists such as Spencer and Tylor. For Spencer, the organic analogy was clear and explicit, and the question of the division of labour in society led to complex functionalist exegesis. When Tylor (as I described above), sought to explain the interrelation between avoidance practices and the customs of post-marital residence, he was performing something similar to that 'functionalist' analysis.[8]

For these authors, however, function and historical transformation were intimately interwoven questions. Durkheim's key contribution to this debate was to split the two questions asunder. Evolutionists, Durkheim charged, had too often confused *cause* and *function* (1895: 110–152). Durkheim claimed that in the teleological progressive vision of evolutionists, social institutions seemed to appear in order to serve the aims and purposes of a slowly improving humanity. For him, by contrast, the historical reason why innovations appeared was a separate question from the function they fulfilled at a particular point in time. Durkheim himself did not, however, do away with historical questions. His early work, such as *The Division of Labour in Society* (1984 [1893]) was straightforwardly evolutionist and while he later nuanced his position, Durkheim retained throughout his work the thought that societies had historically evolved (like organisms) from simple to increasingly complex forms – a view not unlike some of the more sophisticated visions of 'social evolution' proposed by Spencer. Conceptually, then, Durkheim's vision of human history was in the same general ball-park as that of the more sophisticated evolutionists. Methodologically, however, the distinction he introduced between history and function was revolutionary. It drove a profound wedge between the explanation of functional arrangements and the (often fanciful) retracing of human history.

A clear and subtle example of this (r)evolution in practice can be found in Durkheim's famous opus, *The Elementary Forms of Religious Life* (1915 [1912]). In that book, Durkheim returned to, and radically reframed, a question that had much exercised the evolutionists: namely, that of the origins of religion. For some, like Tylor or Spencer, at the root of primitive religion lay 'animism', a mistaken pre-scientific theory grounded in systematic attempts to speculate that had started in the wrong place, namely in mistaking dreams for reality. Others, like Frazer, saw in 'magic' an early form of rudimentary, practically oriented science, an attempt at mastery over the natural world. All of them drew these conclusions in part from accounts of contemporary societies they imagined as 'primitive'.

Durkheim's approach was superficially similar: he too turned to accounts of contemporary societies – those of Aboriginal Australia – to examine the question of religion. Durkheim also assumed that Aboriginal Australian societies lay near the simple end of his imagined simple–complex continuum, and contemporary Western societies near the complex end. Unlike Tylor, Morgan or Spencer, however, Durkheim's aim in comparing 'simple' and 'complex' societies was not to reconstruct a sequence or to tell a story of humanity's grand progression. Rather, Durkheim thought that by observing the workings of religion in 'simple' societies, where, as it were, 'less was going on', one might see more clearly what religion does – how it functions in relation to the rest of the 'organism'. By comparing this to the operation of religion in more complex societies, these insights could be tested and the variation of functional forms might be analysed. This is why Durkheim notes that he is not interested in the hoary historical question of the 'origins' of religion – 'Like every human institution, religion did not commence anywhere.

Therefore, all speculations of this sort are justly discredited' (1915: 8) – but in the 'ever-present causes' (p. 8) of religion in society, namely the functions religion fulfils.

Durkheim followed his question about function to two ingenious arguments. On the one hand, he argued that religion helped to maintain and shape the social order, binding individuals together through the performance of collective rituals. The resulting 'effervescence' periodically reloaded the feeling of collective belonging to a shared social entity. In worshipping God(s), Durkheim argued, people unknowingly worshipped society itself. On the other hand, religion provided an account of the world, a 'system of representations' and a powerful complement of moods and motivations, instilled in particular through rituals that shaped and aligned individual experience into a common and socially shared perspective.[9] These two strands were carefully interwoven into a single and powerful argument about the functional and conceptual significance of religion for society.

Aside from illustrating the subtle shift from evolutionary to functionalist concerns, the example of *The Elementary Forms* also helps to nuance the usual reading of Durkheim as a staunch positivist; the polar opposite to Weber's interpretivism. This is, of course, a not unreasonable simplification: Durkheim was as vocal in outlining and defending a positivist vision of sociology as Weber was for interpretivism. And yet, just as Weber recognised some value in functional explanations alongside interpretive ones (1978: 15 [1922]), Durkheim, particularly in his later work, became increasingly interested in the question of 'collective representations' – the influence of society on people's experience and understanding of the world. *Elementary Forms* exhibits this dualism. On the one hand, arguments about the role of religion in the maintenance of society took an external perspective on society as a whole, as a functionally interrelated organism – a classically positivist perspective. On the other hand, arguments about the way in which religion shaped mood, motivation and worldview launched the project of a sociology of knowledge. The problematic of the 'social construction of knowledge' (see chapter 13) can be read as an interpretivist vein running through Durkheim's otherwise positivist work (Handler 2009: 628; on interpretivism, see chapter 8). We shall find these two strands evolving into polarities within British functionalism.

Durkheim's nephew, collaborator and successor, Marcel Mauss, developed and interwove both of these strands in exquisitely sophisticated ways. One instance is his work on the gift (1970 [1925]), in which questions of interest, motivation and perspective on the one hand and socially binding arrangements and effects of gift-giving on the other, are worked together seamlessly. Another is Mauss's genealogy of the modern notion of the person (1985 [1938]), a historical sociology of a key concept that works by tracing the different kinds of personhood invoked and deployed in different types of social, ritual and political life (see chapter 15). In both works, and in another, co-written with Durkheim on 'primitive classification' (1963 [1903]), Mauss deployed a distinctive comparative method. These works begin with contemporary ethnographic accounts of 'primitive' societies, followed by accounts of classical Greece and Rome and ancient China and India, and ending with reflections on contemporary Western modernity. There are evident evolutionist undertones to this sequence with its move – again – from the 'simple' to the 'complex'. And yet a key point of the account was to unsettle Western self-conceptions. As Mauss put it, the aim of these works was to establish:

> the social history of the categories of the human mind ... We describe particular forms of them in certain civilisations and, by means of this comparison, try to discover in what consists their unstable nature, and their reasons for being as they are.
>
> (1985: 1)

The sophisticated project of a Durkheimian–Maussian sociology of knowledge, with its attention to people's own practices of categorising the world, has often been claimed as a precursor of the French structuralism of Lévi-Strauss (see chapter 2). Yet structuralism also involves a profound departure from this earlier work, most obviously perhaps in the rejection of functionalist arguments about social cohesion and the vision of society as an organism. By that point in the mid-20th century however – and indeed in some quarters today – those functionalist arguments were perceived by French anthropologists as not essentially Durkheimian, but primarily British in origin. How did Durkheim's functionalist sociology become a British anthropological phenomenon?

Fieldwork and Malinowskian functionalism

This question brings us to the second pillar of anthropological functionalism: namely, the rise of a particular fieldwork tradition in early 20th-century Britain. The particular form of long-term ethnographic fieldwork that arose in the early 20th century, and the new types of data it generated, necessarily shaped the sorts of questions anthropologists could ask and turned attention away from the evolutionists' earlier concerns. The functionalist paradigm shift was as much driven by this methodological revolution as by the theoretical one described in the subsection above.

If Malinowski plays such a foundational role in histories of British anthropology, it is to a great extent because of his success in presenting himself as the originator of this new type of ethnographic fieldwork, as well as his vocal attacks on the old order of anthropological evolutionism. Malinowski's monograph, *Argonauts of the Western Pacific* (1922), based on two years of ethnographic study in the Trobriand Islands, is often remembered as the first 'proper' ethnographic monograph. Its introductory chapter, in which Malinowski outlines the principles of ethnographic field study, the importance of living amongst 'the natives' for an extended period of time and conducting enquiries in their language, became a foundational reference point for ethnographers, and established Malinowski's credentials as the 'father' of long-term ethnographic fieldwork.

This technique of immersive fieldwork by participant observation came to be contrasted with the tendency of 19th-century evolutionist anthropologists to work 'from the armchair', merely collating and comparing data drawn from collations of native customs in a range of miscellaneous travelogues, missionary narratives and colonial officers reports. What came to be called 'Malinowskian fieldwork' was also contrasted with the first forms of dedicated ethnographic field research devised in the late 19th century by scholars such as Rivers and Haddon. These had taken the form of 'expeditions' in which various botanical, zoological and ethnological information was gathered while passing through a range of locations. The most famous of these was the Torres Straits Expedition, organised by Haddon, Rivers and Seligman (see chapter 7).

Like many origin myths, the narrative of Malinowski's single-handed invention of long-term immersive fieldwork is something of an overstatement. Malinowski was working in part to the methodological stipulations of Rivers and Haddon, and he was not alone at the time in doing so (Stocking 1983b).[10] Furthermore, his decision to stay for nearly two years in the Trobriand island of Kiriwina – rather than, as he had initially planned, to visit a number of Papua New Guinean locations in a more expeditionary style – was due to fortuitous and serendipitous circumstances, rather than being straightforwardly methodologically motivated (Young 1984). The image of Malinowski as a model fieldworker also became the subject of some controversy following the posthumous publication of his field diaries (see chapter 8). Nevertheless, Malinowski was an outstandingly skilled field-worker – even if 'intensive fieldwork' was in the air he, more than anyone, succeeded in exemplifying and popularising it. It helped that he was also a prolific and evocative writer, who perfected a new style of monographic writing, and a dedicated controversialist and self-publicist.[11]

The birth and spread of 'Malinowskian fieldwork' amongst early-to-mid-20th-century British anthropologists was conceptually revolutionary and in great part explained the paradigmatic shift away from evolutionism and diffusionism in the UK, for one simple reason: the type of data it provided. Worldwide compilations of (usually 'strange' or 'note-worthy') customs, studied from the armchair, had provided the ideal material for a sweeping comparative evolutionist vision of world history. Expedition-style data gathering of pre-defined sets of information (for instance about tools, kinship terms, etc.) provided the ideal material for a study of the diffusion and change of 'traits' across a geographic area, and for organising collections of artefacts in anthropology museums. Each of these methods added grist to the mill of their particular theoretical vision. Long-term fieldwork in a single location, on the other hand, produced a different kind of data: intricately con-textualised knowledge about everyday practices in any and all fields of activity, as well as stark and unusual moments of ritual or other intensity. This material provided the ideal support for raising the sorts of questions associated with a functionalist perspective. This was so for two reasons.

First, this type of data drew anthropologists' attention to the way in which different fields or domains of activity were interrelated. Malinowski made this holism into a funda-mental methodological injunction: 'An ethnographer who sets out to study only religion, or only technology, or only social organization cuts out an artificial field for inquiry, and he will be seriously handicapped in his work' (1922: 11). Thus, in *Argonauts*, in order to describe the economic cycle of inter-island exchange known as the *kula*, Malinowski had to draw together, amongst other topics: questions of value and exchange; accounts of dif-ferent forms of kinship relations and political authority that informed the proprieties of the *kula*; an enquiry into the technological, but also ritual practices involved in crafting the canoes for *kula* travel. The vision this encouraged fitted with the general functionalist view: a vision of societies as complex and coherent entities in which different activities make up an integrated and mutually supporting whole.

Second, this type of intensive fieldwork also brought centre stage the question of how the world made sense to the people living in this way. 'The final goal', as Malinowski famously wrote, 'is, briefly, to grasp the native's point of view, his relation to life, to realise *his* vision of *his* world' (Malinowski 1922: 25). Tylor, working from reported fragments, had thought he could derive universal psychological insights from Victorian jokes about in-laws. Malinowski, working from a rich experience of day-to-day decisions and evalua-tions he could not initially understand, sought to build a complex picture of the motiva-tions and meaning of practices such as the *kula* for the people involved in them.

The theoretical influences of Durkheimian sociology, combined with the new field-work tradition popularised by Malinowski, gave rise to a particular theoretical vision for early 20th-century British anthropologists, that one might think of as 'Malinowskian functionalism'. The personalism of that description is apposite: Malinowski was an institu-tion builder who trained a generation of anthropologists at the LSE, and he was not shy in assuming the mantle of intellectual leader. As Malinowski once wrote: 'the magnificent title of the Functional School of anthropology has been bestowed by myself, in a way on myself, and to a large extent out of my own sense of irresponsibility' (quoted in Radcliffe-Brown 1940a: 1). Often this work is described as functionalism *tout court*, by opposition to the later 'structural-functionalism' spear-headed by Radcliffe-Brown.

There are two key characteristics of this type of work. First, the importance of book-length monographs that examined the interrelation of different features of the same society, albeit usually focused on one main strand or institution – in opposition to the grand comparative endeavours of 19th-century evolutionism, and to the similarly ambitious if more conceptually sophisticated comparisons of the Durkheim–Mauss kind. Second, an attention to two types of questions and their interrelation: on the one hand, questions relating to how the different parts of society are integrated and function in relation to one another, on the other, questions about worldview and perspective. We saw these two strands emerge in Durkheim's work, and saw also how the type of field material produced by long-term participant observation lent itself to raising and answering those two types of questions.

The conceptual convergence was part of a broader intellectual exchange. Malinowski and anthropologists of his generation read Durkheim, of course, although they deployed his theories differently, as we shall see below. Malinowski, in particular, took little heed of Durkheim's injunction to separate sociological from psychological and biological explanations. Conversely, Mauss drew on Malinowski's *Argonauts*, for instance, in his argument on *The Gift*. Between functionalist theory and long-term fieldwork was a conceptual marriage made in heaven, for a while. The ethnographically derived concern with the native's point of view also rejoined in some respects the theoretical concern with cultural relativity and cultural coherence described above as part of the Boasian tradition. And, indeed, early-20th-century British and American perspectives were occasionally productively recombined, for instance, in the collaborations and conceptual intersections between the respective work of Cambridge-trained anthropologist Gregory Bateson and Boasians Mead (Bateson's first wife) and Benedict.

From function to structure: The fission of a paradigm

In early-20th-century British anthropology, the conceptual revolution ushered in by the functionalist paradigm was often described as so radical as to make earlier impassioned disputes between evolutionists and diffusionists seem like petty disagreements. The aspiration to reconstruct history – whether in the sweeping and approximate manner of the evolutionists, or in the careful and regionally circumscribed way proposed by the diffusionists, was coming to seem increasingly outlandish and uninteresting. The entire vision of human traits evolving and/or diffusing across 'peoples', painted on a grand canvas of space and time, was being replaced with a concern with the nuts and bolts of social and conceptual arrangements in particular locations, conceived as structured social systems and meaningful cultural wholes. And yet, as we saw previously, the shift had been only partial and in many ways a matter of method rather than profound outlook. From such gradual changes are radical paradigm shifts built.

New quarrels, however, were beginning to open up within the functionalist camp. Malinowski's position as the conceptual figurehead of British social anthropology soon came to face competition from a contemporary: Alfred Reginald Radcliffe-Brown. Radcliffe-Brown had been trained in Cambridge by Haddon and Rivers, just when

Malinowski was being trained at the LSE by Seligman. He was part of the generation of students sent out by those veterans of the Torres Straits Expedition to experiment with long-term fieldwork. Radcliffe-Brown wrote a monograph – *The Andaman Islanders* (1922) – that appeared the same year as Malinowski's *Argonauts*. Radcliffe-Brown's book had none of the writerly quality or pizzazz of *Argonauts*, nor did it foreground thick ethnographic knowledge or methodological innovation in the same way. The book's data evoked the fragmentary survey-style work of collecting myths and stories, rather than Malinowski's body-and-soul immersion and 'participant-observation'. The key weapon in Radcliffe-Brown's arsenal, however, was a close reading of French sociology and a theoretical self-consciousness that increasingly took the form of a conceptual purism, if not fundamentalism, in the analysis of social structure.

At the time their respective monographs were published in 1922, Radcliffe-Brown and Malinowski recognised each other as conceptual allies in the functionalist revolution. As George Stocking convincingly argues (1984a: 156–179), the shift in Radcliffe-Brown and Malinowski's relationship in the inter-war period was simultaneously personal, institutional and intellectual. Their public disagreements came to shape British social anthropology in profound ways. By the late 1930s, hard words had been exchanged, their two conceptual brands had been radically distinguished, and the theoretical limelight of the discipline was shifting from Malinowski's LSE to the Institute of Social Anthropology established in Oxford by Radcliffe-Brown. A radical theoretical polarity had emerged within the once-united conceptual field of British functionalism.

Radcliffe-Brown demarcated himself, in particular, from Malinowski's terminology in ways that radically inflected the development of functionalism. First, Malinowski had rather ecumenically focused on both 'society' (like Durkheim) and 'culture' (like Boas), later in his life even proposing a 'scientific theory of culture' (Malinowski 1944). Radcliffe-Brown made a point of sticking to 'society' as the focus of his work, explicitly critiquing the American anthropologist Leslie White's attempt to describe anthropology as a science of culture (Radcliffe-Brown 1949c) – the dig at Malinowski was nevertheless implicit in that argument. Anthropology for Radcliffe-Brown was quite simply 'comparative sociology'. The distinction was not just nominal but more profoundly conceptual, and harked back to the distinction between history and science, interpretivism and positivism, introduced above. Society, Radcliffe-Brown felt, was a proper object for a generalising science, that might seek to discover natural laws out of observed realities (1951). Culture by contrast, was a proper object for a historical study. Radcliffe-Brown insisted repeatedly on this division of labour. This left Boas and his American students on the side of history, tasked with the important, but not scientific, job of reconstructing the cultural history of particular peoples. It fell to (British) social anthropologists to undertake systematic comparisons of social systems in pursuit of general laws.

Second, Radcliffe-Brown pilloried Malinowski for his loose use of the term 'function' to refer 'to any and every relation of interdependence' and for failing to distinguish *social function* from a general concept of utility (quoted in Stocking 1984a: 173). In explicit opposition to Radcliffe-Brown, Malinowski (1939) had indeed made a point of arguing that his own 'brand' of functionalism rested on a rich analysis of the physiological and psychological needs of the individual and 'their satisfaction in culture'. This to Radcliffe-Brown suggested a profound confusion and misunderstanding. Like Durkheim, Radcliffe-Brown rejected as wrong-headed any attempt to derive sociological explanations from biological or psychological ones, or to explain social function in terms of its use for the individual. For Radcliffe-Brown, as for Durkheim, society was a *sui generis* level of reality and social facts required an explanation in terms of other social facts.

The result of this Durkheimian purism was, first, an abstraction from the human individual as a living being to the 'person' as a conjunction of social statuses and roles. Radcliffe-Brown had (in)famously written that:

> [t]he actual relations of Tom, Dick and Harry ... may go down in our field note-books and may provide illustrations for a general description. But what we need for scientific purposes is an account of the form of the structure.
>
> (1952: 192; see also chapter 6 for later critiques of this position)

Second, it involved a progressive narrowing of the notion of function. To argue that the function of a social fact can only be in relation to society as a whole (and not in relation to the individual's needs, be they biological or psychological) in effect meant that the only possible function of any social fact was the maintenance of society itself. In other words, the function of a social fact was explained by describing the way it contributed to maintaining a stable social structure. The fight between Malinowski and Radcliffe-Brown over the question of the role of the individual, and also of psychology, in anthropological analysis, established an enduring polarity in the discipline, with later authors returning to Malinowski to suggest a swing back from the abstractions of Radcliffe-Brownian 'structural-functionalism'.

What structure is not: Content and process

The moniker 'structural-functionalism' requires some explanation. It reflects the fact that while Radcliffe-Brown was still proposing, in essence, a species of functionalist analysis, 'structure' came to be the key term in these analyses, over and above 'function'. The shift from function to structure was partly conceptual (because function in the end was subsumed in an account of the role in maintaining a structure) and partly rhetorical – the mark of a demarcation from Malinowski who had appropriated the term 'functionalism'.[12]

What, then, was social structure to the structural-functionalists? As Siegfried Nadel wrote, it is easier to define structure negatively, by specifying what it is not[13] (1957: 7): namely function, content and process. First, as we saw above, structure is not function. Structure describes the shape of social arrangements, without necessarily commenting on their 'adequacy in regard to some stipulated effectiveness' (p. 7).

Second, structure is not content. The thing we call 'a society' might seem to be made up of individual human beings in interaction. But while this may be the empirical content of what anthropologists describe as 'a society', the 'structure' of society is something else – an arrangement that endures even when the individuals die or change their status. Imagine a set of lectures on a fundamental topic of a discipline, given year after year by a lecturer. The students will be different each year, the content of the lecture will change gradually, and the lecturer too will eventually change. On a larger time frame, perhaps, one of the students might eventually become the lecturer. The structure of the lecture as a social arrangement is what endures above and beyond these changes of personnel and empirical content. For Radcliffe-Brown (1940a: 3–4), social structure is thus built up, not out of individuals, but out of roles (student, lecturer, mother, daughter, king, postman, etc.), which endure as individuals pass through them. These roles could be tied together into dyadic relationships (parent–child, employer–employee), and also aggregated into broader 'social groups' (lineages, tribes, nations). Those groups, and lineages in particular ('unilineal descent groups'), became a particularly favoured object of analysis for structural-functionalists, as we shall see again below (Fortes 1953).[14] Nadel was rather isolated in seeking to build up the notion of 'roles' into a consistent focus for theory and analysis (1957) – a project that never really took off, but had important offshoots in sociology's social network analysis (see chapter 6). Radcliffe-Brown defined social structure as consisting of 'human beings considered not as organisms but as occupying positions in the social structure' (Nadel 1957: 5). This seemingly circular definition was in effect a strictly relational one: social structure is a thing only in the sense of being a relational arrangement of things – or rather, of people.

The definition of social structure as that which endures as individuals grow older, change status, are born and die, was also the way to name a conceptual problem: how does this structure manage to endure, despite these 'internal' changes, and pressures from outside? The core concern for structural-functionalism became that of explaining this 'stability' of social structure, the 'equilibrium of society'. Yet in a sense this stability itself was a theoretical artefact of the way social structure was defined. If you abstract what is shared, general and unchanging from the flow of actual human interaction and consider it as a separate object – if you separate structure from content – you simultaneously turn it into a problem to be explained.

The focus on balanced equilibrium relates to the second negative definition of structure: structure is not process. Radcliffe-Brown, following Durkheim's observations on the difference between cause and function, strictly separated the study of 'social statics' – the form of social structure at one particular point in time – from that of 'social dynamics' – the evolution of social structure through time, grounded in a careful study of actual historical changes and transformations. The distinction remained for Radcliffe-Brown, as for Durkheim, a methodological one. Radcliffe-Brown, like Malinowski and other anthropologists of the period, still firmly believed (obviously) that societies had a history, but, more specifically, that societies 'evolved', albeit now in the unguided, multilinear sense of the evolutionism of Darwin rather than in any 'developmental' or progressive sense (Radcliffe-Brown 1947 – on that distinction, see above). The functionalist framework envisaged societies as akin to organisms, involving functional arrangements to maintain their structure in the face of internal and external pressures (social statics), arrangements that had themselves diversified and transformed through time in the face of different historical and environmental circumstances (social dynamics).[15]

While both questions were important, Radcliffe-Brown shared with many mid-20th-century anthropologists the sense that the former was the only one that was empirically accessible for anthropologists. This was quite simply because most anthropologists at the time worked in societies without a written historical tradition, and whose past transformations were therefore considered to be unknown. Synchronic accounts of social structure, based on descriptions of actual practice of those societies today, had an aura of empirical validity. Diachronic reconstructions of those societies' pasts, in the absence of recorded history, became, as Radcliffe-Brown famously wrote of the evolutionists, merely 'pseudo-historical' (1952: 3) and to be strictly avoided.

Thus, in the main, functionalists in general, and structural-functionalists in particular, gave themselves the theoretical licence to write as if the societies they described were stuck in an eternal present, out of time. Of course, such synchronic structural accounts were abstractions: ultimately, the concrete reality which social anthropologists study is 'not any sort of entity but a process, the process of social life' (Radcliffe-Brown 1952: 4).

Nevertheless, the accumulation of these abstracted synchronic accounts of non-Western societies came to build up a sort of commonsense vision of a world divided into – as Eric Wolf would later sarcastically put it – 'Europe and the people without history' (1983; see chapter 5). Functionalism in all its variants, and structural-functionalism in particular, came to be seen as 'declarations of war against historical concerns' (Carneiro, quoted in chapter 5). A number of structural-functionalists tried to tackle the problem of social dynamics head-on (Fortes 1970; Goody 1971) but these efforts swam against the conceptual tide and were eventually overtaken by more radical efforts to rethink the entire statics/dynamics divide (see chapter 5).

A paradigmatic example: Nuer feuding

Edward Evan Evans-Pritchard, a former student of Malinowski, progressively aligned with Radcliffe-Brown's vision of anthropology against his former mentor's. His first monograph, *Witchcraft, Oracles and Magic amongst the Azande* (1937) examines questions of social structure, but its main focus is on the question of the rationality of witchcraft beliefs – a question of 'the native's point of view' if ever there was. By the time he came to write his second monograph, *The Nuer* (1940), Evans-Pritchard's structural-functionalist conversion was at its height. That book's argument about Nuer feuding stands as the most famous and convincing instance of a structural-functionalist argument about equilibrium, which sharply demonstrates the distinction between structure and content, and also between structure and process.[16]

The Nuer in the 1930s, according to Evans-Pritchard's description, were constantly engaged in blood feuds in which a murder would prompt retribution by the victim's kinsmen against the murderer and his kinsmen. This, in turn, would prompt further retribution (for a later account that contextualises Evans-Pritchard's study of the Nuer and revises some of its conclusions, see Hutchinson's *Nuer Dilemmas* [1996]). Evans-Pritchard examines feuding as an 'institution' that, while clearly violent and seemingly destructive, is in fact functional in maintaining the social structure. Rather as Durkheim had earlier focused on suicide (2002 [1897]) as a hard case for demonstrating the social motivation of action (what can be more personal, after all, than suicide?), Evans-Pritchard's analysis draws much of its power from the fact that feuding seems such a counter-intuitive example for demonstrating the maintenance balanced equilibrium. After all, what could be more anarchic, lawless and destructive than a context of perpetual feuding?

Evans-Pritchard's sophisticated argument presents not one, but two, functional aspects of the feud. First, feuding can be seen in its 'legal' aspect, as an institution for – however paradoxical this may seem – controlling and limiting violence between individuals. Evans-Pritchard notes, first of all, that feuding is not simply indiscriminate violence between individuals, but rather a type of violence within a particular social group (the tribe) that follows set patterns and prescribed rules. Between members of different tribes, or between Nuer and non-Nuer, there can be no feuding; just war. The first rule of feuding is that honour dictates retribution – this would in and of itself discourage would-be murderers, playing an equivalent dissuasive role to the trust in the police's prosecution of crimes in other societies. But other rules limit the destructiveness of the feud, by providing ways out of the escalation of murders and counter-murders. The leopard-skin chief, a figure of ritual but not political authority, can be appealed to by a murderer to act as an intermediary with the kin of the victim, in order to convince them that their honour can be satisfied by a payment in cattle in lieu of the death of the murderer. This outcome is still serious and thus dissuasive, but it provides a way to stop the feuding . . . for a time.

Indeed – and initially this is the most counter-intuitive aspect of this structural-functionalist vision – for feuding to serve this legal function or indeed any function at all, it must be an enduring phenomenon. If the resolution of feuds were successful in eradicating them for good, and feuds simply ceased because of effective arbitration, then their dissuasive function would fall away – feuding would have been an unfortunate historical episode rather than a functional social institution. It is thus part of Evans-Pritchard's functional scheme, that feuds rendered dormant by arbitration are always prone to being rekindled in the same or a later generation. The feuding may be paused, but it does not disappear. The institution is not therefore made up just of the rules to control feuding, but of the feuding itself. Or to put it otherwise, feuding is not evidence of the breakdown of social order, it is the solution for maintaining it, through the perpetual threat of violence and retribution, controlled by the possibility of arbitration.

To this 'legal' function of the feud, Evans-Pritchard adds a political function. This argument relies on an intricate description of Nuer political and kinship organisation that I will attempt to summarise briefly. In a very simplified way, one might imagine Evans-Pritchard's argument as in Figures 1.2 to 1.4. Nuer society as he describes it, is organised in a nested pattern of political groupings: 'tribes' consist of 'primary sections' that in turn are subdivided into 'secondary sections', that in turn consist of villages (Figure 1.2).

To this nested political system corresponds a genealogically organised kinship structure in which a 'clan' that traces its ancestry to one, presumably mythical, elder is subdivided into 'maximal lineages', each associated with one of that elder's descendants, in turn

Figure 1.2 Political system.

Source: Author.

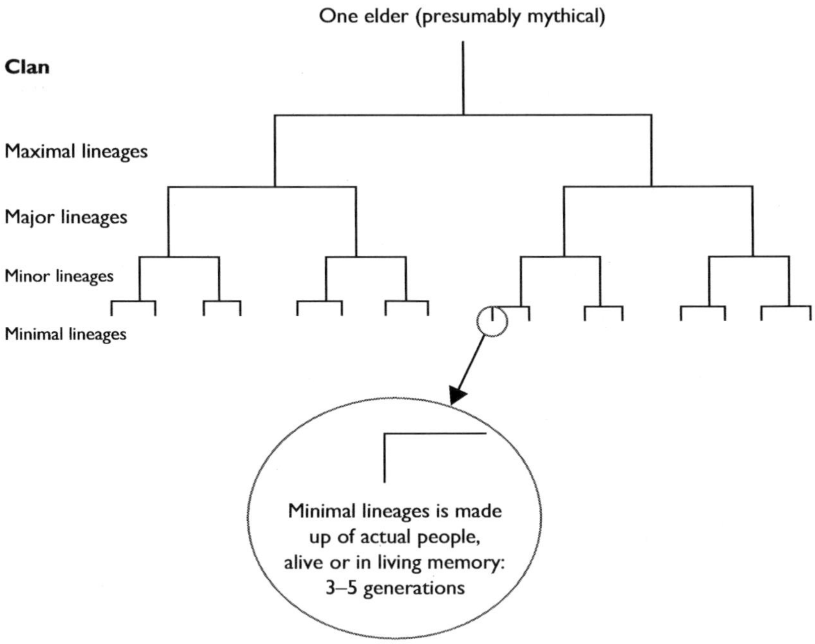

Figure 1.3 Kinship system.

Source: Author.

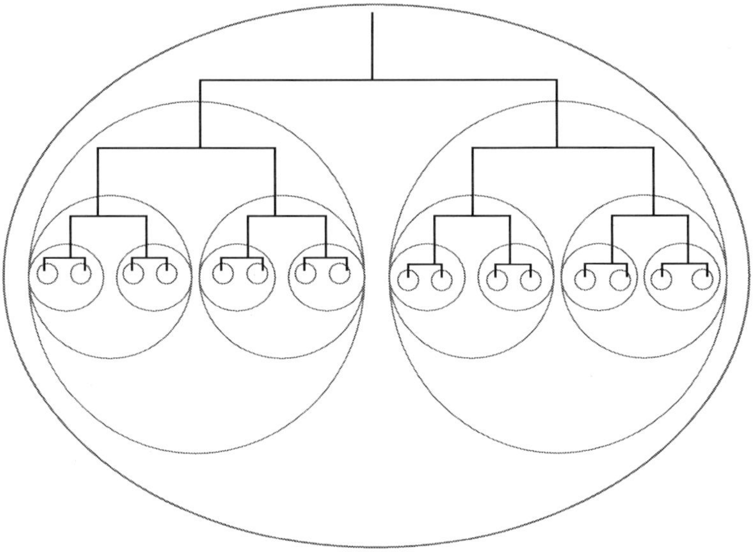

Figure 1.4 A conceptual skeleton.

Source: Author.

subdivided into 'major', 'minor' and finally 'minimal' lineages along the same pattern of genealogical branching (Figure 1.3). These minimal lineages, consisting of about three generations, are the actual corporate kin groups, on the ground, as it were, each of them forming the local elite of each particular village.

The branching pattern of lineages, all the way up to the clan, maps onto the nested pattern of sections, all the way up to the tribe, a conceptual architecture that enables each village, around its minimal lineage, to reckon its genealogical proximity or distance from other villages (Figure 1.4). As Evans-Pritchard puts it, demonstrating the deep-rooted metaphor of society as an organism, 'The system of lineages ... is a conceptual skeleton, on which the local communities are built up into a system of related parts' (1940: 212).

This nested political organisation, in which communities reckon their distance from one another as segments of a shared genealogy, became famous in anthropology as the paradigm of a 'segmentary system' (1940: 4). This segmentary system is then the canvas upon which the political function of the feud operates. Feuds between members of different villages tend to pit these villages against each other. These hostilities can, however, be superseded by an eruption of feuding between a member of one of these villages and one belonging to a different primary section, at which point the formerly hostile villages join in their opposition to their more distantly related enemy. This enmity can, in turn, be superseded by a feud erupting with a member of another primary section. When that feud is resolved and the higher order enmity subsides, the temporarily allied secondary sections might once again resume their earlier hostility. The effect on the structure as a whole is that of a perpetual movement of 'fission' and 'fusion' as segments by turns unite together in opposition to others and fall apart again around their own internal tensions – 'The Nuer Tribe and its divisions are to be understood as an equilibrium between these two contradictory, yet complementary tendencies' (Evans-Pritchard 1940: 148).

Once again, this perpetual motion, driven by hostility and antagonism, might seem a poor candidate for an instance of social stability. And yet this is precisely Evans-Pritchard's claim, neatly summarised in the following quote:

> The feud is a political institution, being an approved and regulated mode of behaviour between communities within a tribe. The balanced opposition between tribal segments and their complementary tendencies towards fission and fusion, which we have seen to be a structural principle, is evident in the institution of the feud, which on the one hand, gives expression to the hostility by occasional and violent action that helps to keep the sections apart, and, on the other hand, by the means provided for settlement, prevents opposition developing into complete fission.
>
> (1940: 161)

In this vision, Nuer society as a whole retains its carefully balanced equilibrium between complete fragmentation and compete centralisation, even as its form is constantly shifting.

The Nuer retained some of the broader range of concerns characteristic of Malinowskian functionalism, including the question of 'the social construction of knowledge' that had also been so central to Evans-Pritchard's earlier work on Azande witchcraft. One chapter, for instance, is devoted to Nuer perceptions of space and time. The book also has a holistic aspect, demonstrating the integration of economy, space and time, cattle symbolism, political organisation and kinship organisation. Nevertheless, the sharpness of the structural-functionalist argument above demonstrates clearly the partial move away from

Malinowski's interest in individual perspective and experience. After all, one can only see the feud as a functional institution that maintains social equilibrium by abstracting the experiential perspective of those caught up in it, which on Evans-Pritchard's own account would have been one of perpetual – and timeless – uncertainty of living in a minefield of dormant violence on the verge of being unleashed. The main problem of *Witchcraft, Oracles and Magic* was to ask how those beliefs make sense to the Azande; the main problem of the Nuer is to give 'an account of the form of the structure' (Radcliffe-Brown 1952: 192).

Back to comparison

Radcliffe-Brown's structural-functionalism, and its continuities and breaks from evolutionism, is neatly illustrated by his rereading of the question of avoidance relationships (not mentioning the name or acknowledging the presence of one's in-laws), introduced into anthropology by Tylor. Radcliffe-Brown returned to this problem a number of times (1940b; 1949a). Radcliffe-Brown proceeds in a similar way to Tylor, comparing instances of avoidance behaviours between different societies. Rather than map these customs, as Tylor had done, against patterns of marital residence, to try to deduce some sequence in which one form of avoidance might have turned into another, Radcliffe-Brown considers avoidance relationships in relation to another set of behaviours, that he terms 'joking relationships'. These are relationships of 'permitted disrespect' (Radcliffe-Brown 1940b: 196) between in-laws, in which one or both parties are expected and entitled to play practical jokes on the other, or mock or insult each other, without any resulting animosity. Radcliffe-Brown notes that, whereas avoidance relationships tend to be between in-laws of different generations (e.g. the husband and his wife's parents), joking relationships tend to be between in-laws of the same generation (e.g. brothers-in-law). He relates avoidance and joking behaviours to the 'normal' social expected behaviours, within and between generations of the same kin group: respect and deference between adjoining generations (parents–children), friendship and camaraderie within generations (brothers and sisters). Avoidance and joking between in-laws thus emerge as distorted or exaggerated versions of those respective kinds of normal behaviours within kin groups.

These comparisons and contrasts set the scene for Radcliffe-Brown's explanatory move (1940b: 196–197 for all quotes in this paragraph): marriage in a sociological sense is not simply a union between two individuals but rather 'a readjustment of the social structure' by which the wife and husband are brought into new relations with their respective in-laws. The vision is one of a partial merger between two families, lineages or clans, imagined in effect as micro-societies. Between these families, one would expect 'social disjunction' – a difference in interests that can lead to hostility. Without erasing this fundamental social disjunction, marriage creates new ties of 'social conjunction', because the wife's family retains 'a continued interest in her and in her children'. This recombination of two social units into a new arrangement, and the combination of social conjunction and social disjunction, Radcliffe-Brown argues, will necessarily put structural strains on both. The exaggerated distance and respect of intergenerational in-law avoidance, and the ritualised and joking hostility between in-laws of the same generation, both address those tensions in the social structure by avoiding occasions for conflict, or making conflict into a joke.

The comparison between Radcliffe-Brown's argument and Tylor's, as outlined above, is instructive. On the one hand, we see the radical breaks between the functionalist (including structural-functionalist) and the evolutionist paradigms. Tylor sought to reconstruct a sequence – proximally of the shift in patterns of post-marital residence, and ultimately, of the grand evolution of human society from matriarchy to patriarchy. By contrast, such historical questions are entirely absent from Radcliffe-Brown's vision, in which societies and groups face perennial structural problems and find functional solutions in an eternal present. Tylor drew quite explicitly on an appeal to the point of view of the individual, albeit in a naively universalising way (what does it feel like to intrude into someone else's family?), a type of concern with individual perspective that emerges in a more sophisticated form in Malinowski's ethnographically rich functionalism. Radcliffe-Brown's structural-functionalism, by contrast, leaves aside the perspective of 'Tom, Dick and Harry' and focuses on the problems faced by abstract entities called kin groups. Incidentally, there is a vision here, characteristic of structural-functionalism, of kin groups as essentially micro-societies made up of people related by descent (archetypally, fathers and sons), who have to face the 'problem' of relations by marriage: in order to perpetuate the group, alliance with members of a different group is (unfortunately) necessary and (naturally) problematic. The limitations and assumptions of this 'descent' theory of kinship were later thrown into stark relief by Lévi-Strauss's contrasting 'alliance' theory (see chapter 2).

On the other hand, the comparison between Radcliffe-Brown and Tylor also shows up one important methodological similarity: both scholars relied on the comparison of partly decontextualised social forms. Whereas Malinowskian functionalism had tended to focus on the monographic study of single societies, examining the interrelations of different aspects and elements, Radcliffe-Brown spearheaded the reintroduction of a more comparative view. Indeed, Radcliffe-Brown explicitly argued that functionalist anthropologists, in their enthusiasm for long-term fieldwork, had too quickly rejected the virtues of a view from the 'armchair' (1951). As societies were like organisms, one could envisage, alongside the study of the functional interrelation of their different parts, a kind of broader 'comparative morphology' (Radcliffe-Brown 1940a: 6), in which different kinds of kinship systems or political systems might be excised from their organic context and laid alongside each other, rather as biologists might compare the functional differences and similarities amongst types of digestive systems, or respiratory systems in different species – another vision that Radcliffe-Brown inherited from Durkheim (1964). This approach was exemplified in the two perhaps most characteristic works of structural-functionalism: the edited volumes *African Political Systems* (Fortes and Evans-Pritchard 1940) and *African Systems of Kinship and Marriage* (Radcliffe-Brown 1950). Each book was a collection of chapters by anthropologists who had also written monographs on 'their' respective societies, and who were here treating one single aspect – respectively, politics and kinship – introduced by an essay that sought to bring out key differences and similarities between the cases, in view of a taxonomic exercise.

We can at this point pause to give a schematic opposition between the Malinowskian functionalism examined above and Radcliffe-Brownian structural-functionalism: the two varieties of functionalism that dominated British anthropology in the early-to-mid-20th century. The paradigmatic instance of the former is the full-length holistic monograph, showing the interrelation of diverse aspects of society and intricately interweaving questions of social function and the perspective and experience of being a member of that society. The paradigmatic form of structural-functionalism, by contrast, is the edited volume, in which one structural aspect is singled out, extracted from its broader context and compared across multiple societies (this genre would come to have a long life: see, for instance,

Middleton and Tait 1958; Middleton and Winter 1963; Goody 1966: 1971; Richards 1971; Fortes 1972). This polarity between the thickly contextualised, multi-layered monograph, and the theoretically driven edited volume maps onto the classic caricatures of Malinowski as a poor theorist and Radcliffe-Brown as a poor fieldworker. However crude and unfair these depictions may have been, they set up a polarity that remains with the discipline to this day (see introduction).

Over the course of the early to mid-20th century, Radcliffe-Brown's version of the functionalist paradigm slowly gained theoretical ground over Malinowski's in British anthropology. The question of social equilibrium and social structure became, for a time, the core concern of the discipline; a 'framework' that, it was hoped, might finally make anthropology into a science. But the monographic form that Malinowski had established persisted, and with it a concern with holism that necessarily pushed the boundaries of strictly 'structural-functionalist' concerns and opened up adjacent questions. Most of the works of the period – despite the occasionally strident positioning of their authors – fall along a continuum between these two polarities, rather than in one of two 'camps'.

Long after the demise of both Malinowskian functionalism and structural-functionalism as paradigms, the tensions between the two schools left anthropology with a set of foundational oppositions – between the field and the armchair (although we now rather think of it as 'the desk'), between a holistic description of single cases and comparative explorations of single problems, between rich ethnography and sharp theory, between an attempt to grasp individual motivation and the 'point of view' from a particular sociocultural position and the conceptual vision of patterns and dynamics. These oppositions are still occasionally used to draw battle lines between authors as they once were between Malinowski and Radcliffe-Brown. Mostly, however, they operate as productive tensions within single anthropological works.

Critiques and new departures

As I noted in the introduction, much of later 20th-century anthropology can be described as a reaction to and critique of (structural)-functionalism. Alternatively – as so many of the initial critiques had at least one foot inside structural-functionalism themselves – one might think of these critiques and departures as the structural-functionalist paradigm fissioning through internal debates. A third way of thinking about this (and these three ways are not incompatible) is as U-turns towards earlier themes evoked in this chapter, that structural-functionalism had sidelined: history, individual persons, culture and meaning. These critiques are recurrent in and spread throughout this book. I will therefore deal with the main axes of critique more synthetically here than I might otherwise.

One might draw together critiques of (structural)-functionalism into three main lines. The first line of critique is essentially philosophical: it charges that the purported object of (structural-)functionalism, namely functioning social institutions articulated into stable social structures, simply does not exist in the way required for the theory to make sense. The second line of critique focuses on exclusions – what the (structural-)functionalist view leaves out; the theory is egregiously selective, runs the charge. The third line of critique is more specifically historical and relates the exclusions and selections of (structural-) functionalism to the specific historical context in which the theory developed, namely early 20th-century British colonialism.

The battle over structure

Evans-Pritchard's intellectual trajectory is not subsumed in the shift described above, from his allegiance to Malinowski to his fealty to Radcliffe-Brown. There was a third phase to his work, whose inception was marked by a resounding lecture given in 1950: the year of the publication of *African Systems of Kinship and Marriage*. This lecture proceeded to savage the structural-functionalist paradigm in unambiguous terms:

> The functional or organismic theory of society ... reigns in social anthropology in England today ... it has been most clearly and consistently stated by Professor Radcliffe-Brown. Human societies are natural systems in which all the parts are interdependent, each serving in a complex of necessary relations to maintain the whole. The aim of social anthropology is to reduce all social life to laws which allow prediction ... a society can be understood satisfactorily without reference to its past. But ... there are many objections to the functional theory. It is no more than an assumption that human societies are systems of the kind they are alleged to be ... The theory assumes, moreover, that in the given circumstances no part of social life can be other than what it is and that every custom has social value, thus adding to a naïve determinism a crude teleology and pragmatism ... for these and other reasons I cannot accept, without many qualifications, the functional theory dominant in English anthropology today.
>
> (1950: 120)

Evans-Pritchard's alternative was to argue that social anthropology ought to leave behind its pretension to being a science on the model of biology, and understand itself as a humanity; more specifically, a genre of interpretive history. Evans-Pritchard returned to the fundamental tension between the positivist vision of human societies as natural systems and the interpretive tradition in which societies

> are seen as systems only because social life must have a pattern of some kind, inasmuch as man, being a reasonable creature, has to live in a world in which his relations with those around him are ordered and intelligible.
>
> (1950: 123–124)

We saw this double concern with positivist objectivism and with the interpretive question of perspective in the later work of Durkheim, and in that of the early functionalists. We saw the radical narrowing of focus away from interpretive questions in structural-functionalism, including in the shift between Evans-Pritchard's Azande and his Nuer work. Now, perspective came back with a vengeance. Evans-Pritchard's later work exemplified different facets of his vision: the sophisticated interpretive work of *Nuer Religion* (1956), in which (structural-)functionalist aspects were thoroughly muted in the search for understanding, and most of all the sadly under-read historical ethnography *The Sanusi of Cyrenaica* (Evans-Pritchard 1949; see chapter 5). Anthropology's historical turn took some time to materialise, but when it came, as Evans-Pritchard predicted it would, it swept away many of the assumptions and concerns of (structural-)functionalism. Evans-Pritchard's turn towards interpretation in his later work also began to sow the seeds of a reconciliation between British social anthropology and American cultural anthropology, much later in the 20th century.

Although Evans-Pritchard does not cite it, his Marret lecture bore many conceptual resonances with a paper published the previous year by Claude Lévi-Strauss on the relationship between anthropology and history (1949). The connection is telling. Lévi-Straussian structuralism became the main European challenger and alternative to structural-functionalism in the mid-20th century. The path it takes is certainly not that of interpretive history, and as it is examined in chapter 2, I will not dwell on it here. However, Lévi-Straussian structuralism, like Evans-Pritchard's later work, starts from a radical critique of the whole attempt to construct anthropology on the model of a comparative biology of human societies. The use of both 'function' and 'social structure' implied, in the structuralist view, naive misconceptions. As Lévi-Strauss famously and rather unfairly wrote, 'to say that a society functions is a truism; to say that everything in society functions is an absurdity' (1949: 376). The vision of social structures on the model of biological ones was, to him, no less naive. Rather, Lévi-Strauss argued, the observed regularities and patterns of social life – be they in terms of kinship, or ritual arrangements or myth – stemmed from elsewhere: from systematic, subconscious and often implicit rules located in the mind of culturally shaped individuals. Structure was (as its name implies) paramount in structuralism, but it was a very different phenomenon indeed from 'social structure'.[17]

Evans-Pritchard's and Lévi-Strauss' critiques of structural-functionalism were echoed, and in some respects sharpened, by British structuralist authors such as Edmund Leach and Rodney Needham, who unpicked the entire structural-functionalist enterprise of systematic generalising comparison in a series of forensic and devastating critiques (Needham 1962; 1975; Leach 1966). If social structures were not real objects on the ground, related to each other through evolutionary relationships of functional differentiation, then the whole project of seeking to derive typologies and rules from their comparison was a nonsense – structural-functionalist comparisons were thus, as Leach famously put it, mere 'butterfly collecting': picking up on and arranging pretty yet inherently meaningless patterns of similarity and difference.

What goes unsaid

Alongside and in conjunction with these fundamental philosophical objections, a number of critiques focused on the important things a (structural-)functionalist paradigm left out. Key amongst these were, as we have seen, history, transformation and change. (Structural-) functionalism, it was charged, had over-reacted to the 'pseudo-history' of the evolutionists, and in contrast produced pictures of societies stuck in an eternally stable present (Leach 1964; see chapter 5).

A second, related, line of attack focused on the fact that conflict in structural-functionalism tended to be presented as in some way functional. Tensions and even violence within society were described as contributing somehow to the maintenance of the social order. Differences in status, wealth, power, gender or ability to act were characterised as forming a well-ordered system of 'roles' adding up to a functional division of labour. The logic of the account left no room for questions of inequality, oppression, or indeed revolution or intentional social transformation. These first two lines of critique were particularly associated with the increasing take-up of Marxist social theory in anthropology and cognate disciplines. Early instances of these critiques, for instance in the work of Max Gluckman and the Manchester School, did not directly cite Marx, but imported a concern with transformation, conflict and inequality without entirely breaking with earlier structural-functionalist lines of analysis (see chapter 6). Later neo-Marxist anthropologies involved a more head-on attack on the whole (structural-)functionalist enterprise (see chapter 3). And yet it remained the case, to quote one of my undergraduate mentors, the regretted Susan Benson, that much anthropological Marxist analysis was effectively 'structural-functionalism with an outboard motor'.

Another set of elisions was picked up by the turn to feminist and gender anthropology (see chapter 12). This was centrally premised on challenging the classic model of a functional and complementary division of labour between men and women, in which men inhabited a public political sphere and women inhabited a private domestic sphere. A double erasure was at play here – first, an erasure of the relationships of power and domination between women and men in these purportedly functional and complementary arrangements. Second, and more profoundly, structural-functionalism was charged with the erasure of women's wider role in society, as so much of the structural-functionalist perspective took politics to be the essence of society, and the business of men. This charge was well founded in many cases. Evans-Pritchard, for instance, casually notes in the introduction to *The Nuer* that

> The Nuer, like all other peoples, are also socially differentiated according to sex. This dichotomy has a very limited, and negative, significance for the structural relations which form the subject of this book. Its importance is domestic rather than political and little attention is paid to it.
>
> (1940: 7)

There were, however, some accounts of the period that, while steeped in functionalist concerns, nevertheless paid close attention to the role and experience of women, such as Audrey Richards' *Chisungu* (1956).

A third classic line of critique focuses on the (structural-)functionalist elision of individual differences and human agency. This critique was carried most prominently by Leach, Raymond Firth and 'transactionalist' authors such as Fredrik Barth (see chapter 4). This critique represents in a sense the posthumous revenge of Malinowski against Radcliffe-Brown: both Leach and Firth drew directly and self-consciously on Malinowski's concern with individual persons, their motivations and activity as a counterpoint to structural-functionalist elisions of individual agency.

To many of these critiques (although not perhaps quite so straightforwardly in the case of gender), a structural-functionalist might find a simple and fairly convincing reply: they do not describe things that structural-functionalists somehow 'forgot', or aspects of the world that they were strangely impeded from seeing. Who could miss the fact that there are differences and tensions within societies? Or that there is such a thing as history? Or that individuals have a hand in what they do and that their actions are not wholly determined by social constraints? Rather, these features – history, individual agency and the forces that make society unstable and pit it against itself – were *intentionally* excluded from the framework, in order to provide a clear model, and because their stated aim was to discover structural regularities. The structural-functionalists' contention was never, they might say, that societies were fully functional, perfectly stable, a-temporal entities that entirely constrained their members' actions. Rather, the point was that by modelling this ideal state of synchronic stability one might then precisely have a standpoint from which to gauge exceptions, historical change and dysfunction or disaggregation. We saw how this point was made in relation to structure and process by Durkheim and Radcliffe-Brown: process was a fascinating and key question, but it had to be disaggregated from structure heuristically, for the purpose of analysis. Durkheim made the same point of agency and human freedom, in a frequently overlooked passage in his book *Suicide*:

> Without wishing to raise a question of metaphysics outside our province, we must note that this theory of statistics does not deny men every sort of freedom. On the contrary, it leaves the question of free will much more untouched than if one made the individual the source of social phenomena . . . the stability of demographic data results from a force external to the individual . . . It may be granted that some people resist the force and that it has its way with others.
>
> (2002: 289–290)

In other words, for many of the points raised above, structural-functionalists could charge their critics with missing the crucial difference between forgetting something and leaving it out for the purposes of analysis. They might be comforted by the fact that many of their critics, from Marxists to transactionalists, implicitly retained many more of the basic premises of the study of functioning structures than they openly acknowledged. For, indeed, a structural-functionalist might ask: How, without an account of a stable structure, might one ever *see* what change transformed, what individual agency shaped or came up against, what particular aspects of social life were innovative, destructive or exceptional?

History and responsibility

This is a convincing retort. It comes up short, however, in the face of a different line of critique, that did not simply pick out 'gaps' in structural-functionalism, but related these gaps to the political and historical context in which the theory had developed. One of these relevant elements of context has already been hinted at and forms an important part of the feminist critique: namely, the fact that gendered assumptions amongst (mostly male) anthropologists of the period played a role in their tendency to leave women out of the picture (see chapter 12).

The other key element of context that became retrospectively indissociable from an understanding of structural-functionalism was that of British colonial rule. The key salvo here was Talal Asad's edited volume *Anthropology and the Colonial Encounter* (1973a). This volume was the first systematic assessment of the relationship between social anthropology and British colonialism. The starting point of the volume is the obvious yet rarely discussed fact that European anthropology in the first half of the 20th century was 'devoted to a description and analysis – carried out by Europeans, for a European audience – of non-European societies dominated by European power' (1973a: 15). What this meant, most obviously, was that anthropologists were dependent for their access to the so-called 'primitive peoples' they studied, upon the institutional structures of British colonialism, as indeed they looked to funding from the British government to support their research. In so doing, (structural-)functionalist anthropologists often tended to portray their work as useful to the colonial enterprise, both in the general sense of educating future colonial officers to the realities of the societies they were going out to 'administer', and occasionally in the more precise sense of providing specific models and explanations that might guide colonial policy.[18] The extent to which such anthropological advice was actually useful or interesting to colonial officers, however, and the personal political proclivities of individual anthropologists, their often distant attitudes towards the colonial project and their frequent sympathies with the colonised people they studied have been the subject of much discussion (see, for instance, Asad 1973a; Kuper 1973; Stocking 1991).

The more profound point of Asad's critique, however, bore on the implicit conceptual interactions between anthropological structural-functionalism and its political context. Asad (1973b) compared, in this respect, two instances of European scholarship about non-European peoples and their respective relationship to European power over these peoples: 20th-century structural-functionalism and 19th-century orientalism. There was a striking match, in each case, between the content of the descriptions provided and the strategic interests and perspectives of the European powers at the time. Thus, on the eve of massive imperial expansion into the Middle East, orientalists presented a picture of Islamic societies as perennially dysfunctional, swinging wildly between tyrannical power and violent, mindless insurrection. Orientalists in other words painted a picture of societies seemingly 'in need' of Western intervention. By contrast, writing in the context of a settled and fairly stable colonial presence in Africa, the structural-functionalists painted, as we saw above, pictures of stable, well-ordered societies organised through natural balances of power, or ruled in effect by popular consent. No problem here, went the subliminal and implicitly conservative message: all's well under His Majesty's rule. In each account, there were important exclusions. The orientalists ignored the various processes of mutual accommodation between rulers and ruled in the history of the societies they described, and also the impact of commercial relations with Europe and Asia, treating these societies as self-contained. The structural-functionalists presented – as we saw above – a picture of society from which internal conflict and historical change were missing. Seen from Asad's historical perspective, these respective exclusions no longer seem so innocent.

There are some limits to Asad's analysis – for one, functionalist approaches existed in other contexts than British social anthropology of African societies, including in 19th-century evolutionist accounts, as we have seen above. Second, there is a subtle irony to the fact that Asad's analysis is itself not untinged by functionalism. Nevertheless, it is certainly true that structural-functionalist visions of African societies laid out side-by-side as distinct units, ripe for inspection and comparison, echoed the 'problem space' created by British colonial administration, through which anthropologists gained access to those societies in the first place (see also Kuper 1973). Societies that might otherwise have been encountered as messy, shifting, overlapping, were encountered through the frames of administration as inherently separate and clearly labelled. In a later paper, Asad concludes that 'The role of anthropologists in maintaining structures of imperial domination has, despite slogans to the contrary, usually been trivial ... but if the role of anthropology for colonialism was unimportant, the reverse proposition does not hold' (1991: 315). Colonial relations of power were part of the intellectual, and not merely of the practical context of early 20th-century functionalist anthropology, and this fact took some time to come to light.

This in turn means that the question of the 'heuristic exclusions' of structural-functionalism – as of any other theory – has to be treated with some care. The problem in sum is not simply that, in the abstract, structural-functionalism 'left some things out' (history, inequality, or the individual) – observing this is only the beginning of a critique. Every theoretical perspective, every account, every question asked leaves some things out. The question is what these exclusions, in turn, do. As philosopher of science, Karen Barad put it, 'one can't simply bracket (or ignore) certain issues without taking responsibility for the constitutive effects of these exclusions' (Barad 2007: 58; see also chapter 13).

Of course, the question of what these exclusions do, politically speaking, remains a complex one. In some respects, early 20th-century accounts and analytics were inherently challenging to widespread colonial viewpoints, notably on the purported irrationality of 'natives', and therefore on the need for enlightened interference and administration (Kuper 1973). More specifically, anthropologists such as Evans-Pritchard drew on their conceptual work in order to explicitly critique colonial policies on witchcraft (James 1973). Similarly, one can read Evans-Pritchard's account of the Nuer, for instance, precisely as a counter-example to Asad's contrast between orientalism and functionalism. Where colonial administration saw Nuer feuding as an irrational and violent practice that justified forceful intervention, Evans-Pritchard redescribed it as a finely tuned system whose logic was intensely rational and in danger of being mangled by colonial interference. That is also an important message in the introduction to *African Political Systems* (Fortes and Evans-Pritchard 1940). In sum, an organismic view of societies as stable systems of social practice and social thought, from which exceptions, tensions and anomalies were heuristically excluded, could also be read as working against the grain of colonialist assumptions demonstrating the rationality of seemingly alien systems of belief and the validity – and precarious balance – of non-Western forms of social organisation. In sum, the politics of (structural-)functionalism were are at least as complicated as the politics of evolutionism (see above).

Conclusion: What old theories are good for

A key aim of this overview of the beginnings of modern anthropological theory has been to show how theoretical perspectives that are in many ways profoundly different, even opposed (such as evolutionism, diffusionism, and [structural]-functionalism) emerged out of an extended set of conversations and arguments between scholars in at least four countries. This conversation laid the ground for much of what was to follow in 20th and 21st-century anthropological theory, if only by providing key exemplars and ancestors *against* whom new theoretical departures could be articulated.

This procedure is merely a replication of the cut that functionalists or Boasians sought to make between themselves and evolutionism. The rejection of evolutionism was the gambit of an up-and-coming generation of scholars in the UK, US and France seeking to position themselves as the vibrant modern alternative to a set of established assumptions and methods. Later, the evolutionists descended into near-oblivion, remembered as precursors of anthropology rather than anthropologists proper, while the functionalists (and structural-functionalists) in the UK and the Boasians in the US came to play for the next generation the role that the evolutionists had played for them. Witness the number of chapters in this book that begin by contrasting the school or theorist they examine with the errors and failings of functionalism. Rupert Stasch writes in chapter 2 that: '[a]n account of the different ways anthropologists have departed from structuralism would be a description of almost the entire discipline'. Much the same could be said for functionalism and structural-functionalism – and the first chapter in that story would be, precisely, the rise of structuralism itself.

This opens up the broader question of what 'old theories' – schools and styles that have now been abandoned as systematic attempts to explain the society and culture – are good for. As I noted in the introduction it is reasonable to recall such theories as reminders of mistakes one ought to avoid repeating. While the organic analogy is still going strong in popular discourse, as it probably has in some form since ancient Greek times, professional anthropology's love story with that particular metaphor is mostly at an end. There are few anthropologists today who would take seriously the key trope that informed all of the schools and styles examined above, and this is probably all to the good. More generally, putting, in hindsight, old theories 'in their historical context' is a powerful move. From that somewhat contrived theory–context relation (structural-functionalism and British colonialism in Africa, for instance), one can draw important lessons for the present. Lessons, precisely, about the very notion of context, and the power and danger of leaving certain aspects out of the frame, as all theoretical perspectives – including those of critical postcolonial scholars – necessarily do.

But recalling old theories merely as caveats is distinctly narrow minded. A fair and generous account of now-abandoned theories should also be able to recapture some of the rigour and sophistication, some of the excitement and promise of those old theories when they were new. This is, of course, another caveat: a salutary reminder to relativise our own present theoretical enthusiasms. But this view also opens up the more exciting possibility that old theories might be mined for new insights, as indeed many of the theories represented in the following chapters do, be it implicitly or explicitly. Thus, many evolutionist questions garnered a new lease of life when anthropologists turned once more to history from the 1960s onwards (see chapter 5); and some functionalist concerns continued to animate their immediate critics (see chapter 6). Essentially, diffusionist concerns with how things move and how they are transformed and reintegrated have recently been mined, as I noted above, for insights into the postmodern condition of global flows and networks. And the key question functionalists and structural-functionalists sought to address is still very much with us. It concerns the strange existence of any kind of social stability and regular social behaviour despite the constant turnover of individual members of any social assemblage.

Yet there is another, in some ways more profound, legacy of the theories examined in this chapter. This takes the form of the series of contrasts that I have evoked along the way in this discussion: contrasts between interpretation and generalisation, between the field and the armchair, between structure and process, between a study of the integration of traits and their decontextualisation and comparison. These contrasts began their life as demarcation lines between these various schools: boundaries drawn by individual scholars in the process of marking oneself off from conceptual opponents. They have proved enduring, but they have changed their form and import. While they still operate in some quarters as facile resources for critique and name-calling, in the main, these contrasts have become productive alternatives within specific arguments and positions, the resource for arguments in a both/and, rather than either/or mode. More than any single concept or approach, it is those tensions and their occasional temporary synthesis that characterise the enduring and productive legacy of the schools we have examined in this chapter.[19]

Notes

1 The term 'evolution' appears for the first time in the sixth edition of *The Origin* (Darwin 1872), even though one single instance of the word 'evolved' had been present (in the book's final sentence) since the first edition.

2 This leads Peel, for instance, to note that in Spencer's book '*The Principles of Sociology* there is so much functional analysis of an entirely modern kind that it is hard to understand why so many

latter-day functionalists (though not Radcliffe-Brown) should have supposed that they were rebelling against Spencer the evolutionary-functionalist' (Peel 1971: 183).

3 Morgan, for instance, had advocated for the rights of the Iroquois.

4 'We owe our present condition, with its multiplied means of happiness, to the struggles, the sufferings, the heroic exertions and the patient toil of our barbarous, and more remotely, of our savage ancestors. Their labors, their trials and their successes were part of the plan of the Supreme Intelligence to develop a barbarian out of the savage and a civilized man out of this barbarian' (Morgan 1877: 554, cited in Ferguson 1996: 155).

5 'The Aryan family represents the central stream of human progress, because it produced the highest type of mankind, and because it has proved its inherent superiority by gradually assuming the control of the earth' (Morgan, quoted in Kuper 2005: 62).

6 'Even the diagrams of this paper may suffice to show that the institutions of man are as distinctly stratified as the earth upon which he lives. They succeed each other in series substantially uniform over the globe, independent of what seem the comparatively superficial differences of race and language, but shaped by similar human nature acting through successively changed conditions in savage, barbaric, and civilised life' (Tylor 1889: 269). See Figure 1.1 below for an instance of one of these all-important diagrams.

7 I am grateful to Rupert Stasch for bringing this passage to my attention and for invaluable comments on this section.

8 Functionalism was a key aspect, also, of Darwinian evolutionism – while, as we saw, the word 'evolution' does not appear in the first edition of *The Origin of the Species*, function(s) or functional(ly) appear 30 times (Darwin 1859). The vision of biological change driven by a process of selection and adaptation is inseparable from a functional account of the role of organs, and Darwin spends much time tracing in particular the ways in which organs may serve different functions at the same point in time or come to develop new functions. This point was, and has remained, crucial to Darwinian rejection of creationist teleology: it allows for random variations to give rise to new and unexpected functional possibilities. Indeed, one might say, somewhat paradoxically, that it was their functionalism and not their evolutionism that made some evolutionist thinkers like Spencer quasi-'Darwinian'.

9 Note that, as Irvine (see chapter 7) points out, Durkheim's account of the collective effervescence of ritual broke his own methodological rule of treating psychological phenomena and sociological ones as strictly separate.

10 As Stocking notes, Rivers had recommended, as early as 1913, a programme for intensive work 'in which the worker lives for a year or more among a community of perhaps four or five hundred people and studies every detail of their life and culture; in which he comes to know every member of the community personally; in which he is not content with generalized information, but studies every feature of life and custom in concrete detail and by means of the vernacular language' (Rivers, quoted in Stocking 1983b: 92).

11 Author, for instance, of books with titles such as *The Sexual Life of Savages* (1932).

12 To the eternal confusion of contemporary anthropology students, Radcliffe-Brown, who evidently was profoundly concerned with the function of social facts, nevertheless wrote that '[a]s a consistent opponent of Malinowski's functionalism, I may be called an anti-functionalist' (1949b: 321). His own description of his work was, as we saw above, comparative sociology. Increasingly, however, the approach of Radcliffe-Brown and his followers came to be described as 'structuralism' – not to be confused, as we shall see, with the French structuralism spearheaded by Lévi-Strauss, in explicit opposition to the work of the British 'functionalists' (including Radcliffe Brown)!

13 Not unlike 'actor-network theory' (see chapter 13).

14 The structural-functionalists had inherited their concern with descent groups from the evolutionists, for whom they formed the building block of 'primitive society' (Kuper 2005: 5). However, through the notion of social structure they made something quite distinctive of that notion.

15 This implicitly evolutionist underpinning of functionalism explains an otherwise puzzling feature of this work. Despite their forceful critiques of 19th century evolutionism, most anthropological functionalists of the early-to-mid-20th century – and indeed the American Boasians of

the same period, too – continued to describe the societies they studied as 'primitive', as opposed to the 'modern' ones they lived in. Thus, even though functionalists turned away from historical questions and concerns, they continued to implicitly imagine a grand historical canvas in which the social forms they studied featured as 'earlier' or 'less advanced' versions of European ones (Kuper 2005).

16 For an exceedingly clear and systematic account of this argument, which informed my own reading, see Kuper (1973).

17 I am slightly overstating this point for clarity, because the confusion between the term structure as deployed by structuralists and structural-functionalists is such a common problem. However, once one has integrated the profound philosophical differences at stake in those two uses of 'structure', one can also begin to consider some of the tentative conceptual bridges between the very late writings of Radcliffe-Brown (on totemism, for instance), and those of Lévi-Strauss – a point the latter himself acknowledged despite his critiques of many points of the former's work (Lévi-Strauss 1963).

18 Thus for instance, the editors of *African Political Systems* noted: 'We hope this book will be of interest and of use to those who have the task of administering African peoples' (Fortes and Evans-Pritchard 1940: vii).

19 For a fascinating exploration of how such recurrent contrasts operate as resources within the theoretical life of the social sciences more broadly, see Abbot (2001). As if to prove my point about enduring legacies and reinventions, Abbott notes that – for all its characteristically late-20th century reference to chaos theory and fractals – his theoretical perspective is grounded in conceptual points developed by Evans-Pritchard in his study of political segmentation amongst the Nuer.

Bibliography

Abbott, Andrew 2001. *Chaos of Disciplines*. Chicago, IL: University of Chicago Press.

Alfred Reginald Radcliffe-Brown 1922. *The Andaman Islanders*. New York, NY: Free Press.

Ansell-Pearson, Keith, Paul-Antoine Miquel and Michael Vaughan 2010. Responses to evolution: Spencer's evolutionism, Bergsonism, and contemporary biology. In Keith Ansell-Pearson and Alan Schrift (eds.), *The History of Continental Philosophy: Bergsonism, Phenomenology, and Responses to Modern Science. The New Century. Volume 3*, p. 347. Chicago, IL: University of Chicago Press.

Asad, Talal (ed.) 1973a. *Two European Images of Non-European Rule, Anthropology and the Colonial Encounter*. New York, NY: Humanity Books.

——— 1973b. Two European images of non-European rule. In Talal Asad (ed.), *Two European Images of Non-European Rule, Anthropology and the Colonial Encounter*, pp. 103–118. New York, NY: Humanity Books.

——— 1991. Afterword: From the history of colonial anthropology to the anthropology of western hegemony. In George W. Stocking (ed.), *Colonial Situations*. Madison, WI: University of Wisconsin Press.

Bamberger, Joan 1974. The myth of matriarchy: Why men rule in primitive society. In Michelle Zimbalist Rosaldo, Louise Lamphere and Joan Bamberger (eds.), *Woman, Culture, and Society*, pp. 263–280. Stanford, CA: Stanford University Press.

Barad, Karen Michelle 2007. *Meeting the Universe Halfway: Quantum Physics and the Entanglement of Matter and Meaning*. Durham, NC: Duke University Press.

Bashkow, Ira, Matti Bunzl, Richard Handler, Andrew Orta and Daniel Rosenblatt 2004. Introduction. *American Anthropologist* 106 (3): 433–434.

Benedict, Ruth 1934. *Patterns of Culture*. Boston, MA: Houghton Mifflin Harcourt.

Bloch, Maurice 1984. *Marxism and Anthropology: The History of a Relationship*. Oxford: Oxford University Press.

Boas, Franz 1896. The limitations of the comparative method of anthropology. *Science* 4 (103): 901–908.

——— 1911. *The Mind of Primitive Man: A Course of Lectures Delivered Before the Lowell Institute, Boston, Mass., and the National University of Mexico, 1910–1911*. London: Macmillan.
——— 1924. Evolution or diffusion. *American Anthropologist* 26 (3): 340–344.
——— 1940. *Race, Language, and Culture*. Chicago, IL: University of Chicago Press.
Brightman, Robert 1995. Forget culture: Replacement, transcendence, relexification. *Cultural Anthropology* 10 (4): 509–546.
Comte, Auguste 1830. *Cours de philosophie positive*. Paris: Rouen frères (Bachelier).
Darwin, Charles 1859. *The Origin of the Species by Means of Natural Selection, or the Preservation of Favoured Races in the Struggle for Life*. London: John Murray.
——— 1872. *The Origin of the Species by Means of Natural Selection, or the Preservation of Favoured Races in the Struggle for Life; Sixth Edition, with Additions and Corrections*. New York, NY: D. Appleton and Company.
Durkheim, Émile 1895. *Les règles de la méthode sociologique*. Paris: F. Alcan.
——— 1915. *The Elementary Forms of the Religious Life*, translated by Joseph Ward Swain. London: George Allen & Unwin.
——— 1964. *The Rules of Sociological Method*. New York, NY: Free Press.
——— 1984. *The Division of Labour in Society*. London: The Macmillan Press.
——— 2002. *Suicide: A Study in Sociology*. 2nd edition. London: Routledge.
Durkheim, Emile, and Marcel Mauss 1963. *Primitive Classification*. Vol. 273. Chicago, IL: University of Chicago Press.
Eggan, Fred 1954. Social anthropology and the method of controlled comparison. *American Anthropologist* 56 (5): 743–763.
Engels, Friedrich 1972. *The Origin of the Family, Private Property, and the State*. New York, NY: Pathfinder Press.
Evans-Pritchard, Edward Evan 1937. *Witchcraft, Oracles and Magic among the Azande*. Oxford: The Clarendon Press.
——— 1940. *The Nuer: A Description of the Modes of Livelihood and Political Institutions of a Nilotic People*. Oxford: Oxford University Press.
——— 1949. *The Sanusi of Cyrenaica*. Oxford: Clarendon Press.
——— 1950. Social anthropology: Past and present; the Marett lecture, 1950. *Man* 50: 118–124.
——— 1956. *Nuer Religion*. Oxford: Clarendon Press.
'Evolution, N.' 2017. *OED Online*. Oxford: Oxford University Press. Accessed 1 July 2017. www.oed.com/view/Entry/65447.
Fabian, Johannes 1983. *Time and the Other, How Anthropology Makes Its Object*. New York, NY: Columbia University Press.
Ferguson, James 1996. Development. In Alan Barnard and Jonathan Spencer (eds.), *Encyclopedia of Social and Cultural Anthropology*, pp. 155–160. London: Routledge.
Fortes, Meyer 1953. The structure of unilineal descent groups. *American Anthropologist* 55 (1): 17–41.
——— 1970. *Time and Social Structure and Other Essays*. London: Berg Publishers.
——— 1972. *Marriage in Tribal Societies*. Cambridge: Cambridge: Cambridge University Press.
Fortes, Meyer, and E.E. Evans-Pritchard (eds.) 1940. *African Political Systems*. Oxford: Oxford University Press.
Goody, Jack 1966. *Succession to High Office*. Cambridge: Cambridge University Press.
——— 1971. *The Developmental Cycle in Domestic Groups*. Cambridge: Cambridge University Press.
Goyet, Francis 2014. Comparison. In Barbara Cassin, Emily Apter, Jacques Lezra, and Michael Wood (eds.), *Dictionary of Untranslatables: A Philosophical Lexicon*, pp. 159–164. Princeton, NJ: Princeton University Press.
Handler, Richard 2009. The uses of incommensurability in anthropology. *New Literary History* 40: 627–647.
Hutchinson, Sharon Elaine 1996. *Nuer Dilemmas: Coping with Money, War, and the State*. London: University of California Press.

James, Wendy 1973. The anthropologist as reluctant imperialist. In Talal Asad (ed.), *The Anthropologist as Reluctant Imperialist, Anthropology and the Colonial Encounter*, pp. 41–69. New York, NY: Humanity Books.
Kuper, Adam 1973. *Anthropologists and Anthropology: The British School 1922–1972*. London: Allen Lane.
——— 1999. *Culture: The Anthropologists' Account*. Cambridge, MA: Harvard University Press.
——— 2005. *The Reinvention of Primitive Society: Transformations of a Myth*. London: Routledge.
Leach, Edmund Ronald 1964. *Political Systems of Highland Burma*. London: G. Bell and Sons.
——— 1966. *Rethinking Anthropology*. Vol. Monographs on social anthropology, no. 22. London, New York, NY: Athlone Press & Humanities Press.
Lear, Jonathan 2006. Allegory and myth in Plato's *Republic*. In Gerasimos Xenophon Santas (ed.), *The Blackwell Guide to Plato's Republic*. London: Blackwell.
Lévi-Strauss, Claude 1949. Histoire et ethnologie. *Revue de métaphysique et de morale* 54 (3/4): 363–391.
——— 1963. *Totemism*. Boston, MA: Beacon Press.
Malinowski, Bronislaw 1922. *Argonauts of the Western Pacific*. London: G. Routledge and Sons, Ltd.
——— 1932. *The Sexual Life of Savages in North-Western Melanesia; an Ethnographic Account of Courtship, Marriage, and Family Life among the Natives of the Trobriand Islands*. London: G. Routledge and Sons, Ltd.
——— 1939. The group and the individual in functional analysis. *American Journal of Sociology* 44 (6): 938–964.
——— 1944. *A Scientific Theory of Culture*. Chapel Hill, NC: The University of North Carolina Press.
Marcus, G.E. 1995. Ethnography in/of the world system: The emergence of multi-sited ethnography. *Annual Review of Anthropology* 24 (January): 95–117.
Mauss, Marcel 1970. *The Gift: Form and Functions of Exchange in Archaic Societies*. London: Routledge.
——— 1985. A category of the human mind: The notion of person; the notion of self. In Michael Carrithers, Steven Collins, and Steven Lukes (eds.), *The Category of the Person: Anthropology, Philosophy, History*, pp. 1–25. Translated by W.D. Halls. Cambridge: Cambridge University Press.
Menand, Louis 2011. *The Metaphysical Club: A Story of Ideas in America*. New edition. London: Flamingo.
Middleton, John, and David Tait 1958. *Tribes Without Rulers: Studies in African Segmentary Systems*. London: Routledge.
Middleton, John, and E.H. Winter 1963. *Witchcraft and Sorcery in East Africa*. London: Routledge.
Morgan, Lewis Henry 1877. *Ancient Society*. New York, NY: H. Holt and company.
Nadel, Siegfried Frederick 1957. *The Theory of Social Structure*. London: Cohen and West.
Needham, Rodney 1962. *Structure and Sentiment: A Test Case for Social Anthropology*. Chicago, IL: University of Chicago Press.
——— 1975. Polythetic classification: Convergence and consequences. *Man* 10 (3): 349.
Peel, J.D.Y. 1971. *Herbert Spencer: The Evolution of a Sociologist*. London: Heinemann Educational Books.
Penny, H. Glenn, and Matti Bunzl 2003. *Worldly Provincialism: German Anthropology in the Age of Empire*. Ann Arbor, MI: University of Michigan Press.
Perrin, Robert G. 1976. Herbert Spencer's four theories of social evolution. *American Journal of Sociology* 81 (6): 1339–1359.
Radcliffe-Brown, A.R. 1940a. On social structure. *The Journal of the Royal Anthropological Institute of Great Britain and Ireland* 70 (1): 1–12.
——— 1940b. On joking relationships. *Africa: Journal of the International African Institute* 13 (3): 195–210.
——— 1947. Evolution, social or cultural. *American AnthropologistNew Series* 49 (1): 78–83.
——— 1949a. A further note on joking relationships. *Africa: Journal of the International African Institute* 19 (2): 133–140.

——— 1949b. Functionalism: A protest. *American AnthropologistNew Series* 51 (2): 320–323.
——— 1949c. White's view of a science of culture. *American Anthropologist* 51 (3): 503–512.
——— (ed.) 1950. *African Systems of Kinship and Marriage*. Oxford: Oxford University Press.
——— 1951. The comparative method in social anthropology. *Journal of the Anthropological Institute of Great Britain and Ireland*, 15–22.
——— 1952. *Structure and Function in Primitive Society*. London: Cohen and West.
Richards, Audrey Isabel 1956. *Chisungu: A Girl's Initiation Ceremony Among the Bemba of Zambia*. London: Psychology Press.
——— 1971. *Councils in Action*. Cambridge: Cambridge University Press.
Salmon, Gildas 2013. *Les Structures de l'esprit: Lévi-Strauss et Les Mythes*. Paris: Presses universitaires de France.
Sanchez, Andrew 2016. Profane relations: The irony of offensive jokes in India. *History and Anthropology* 27 (3): 1–17.
Spencer, Herbert 1867. *First Principles*. 2nd edition. London: Williams and Norgate.
——— 1899. *Social Statics: Abridged and Revised: Together with The Man versus the State*. London: D. Appleton and company.
Stasch, Rupert 2002. Joking avoidance: A Korowai pragmatics of being two. *American Ethnologist* 29 (2): 335–365.
Stocking, George W. 1983a. *Observers Observed: Essays on Ethnographic Fieldwork. History of Anthropology; Vol 1*. Madison, WI: University of Wisconsin Press.
——— 1983b. The ethnographer's magic: Fieldwork in British anthropology from Tylor to Malinowski. In George W. Stocking (ed.), *Observers Observed: Essays on Ethnographic Fieldwork*, pp. 70–120. Madison, WI: University of Wisconsin Press.
——— 1984a. Radcliffe-Brown and British social anthropology. In George W. Stocking, *Functionalism Historicized: Essays on British Social Anthropology*, pp. 131–191. Madison, WI: University of Wisconsin Press.
——— 1984b. *Functionalism Historicized: Essays on British Social Anthopology*. Madison, WI: University of Wisconsin Press.
——— 1991. *Colonial Situations: Essays on the Contextualization of Ethnographic Knowledge*. Madison, WI: University of Wisconsin Press.
Strathern, Marilyn 2004. *Partial Connections*. Lanham, MD: Rowman and Littlefield.
Thomas, Nicholas 1996. *Out of Time: History and Evolution in Anthropological Discourse*. Ann Arbor, MI: University of Michigan Press.
Trautmann, Thomas R 1992a. The revolution in ethnological time. *Man* 27 (2): 379.
——— 1992b. *Lewis Henry Morgan and the Invention of Kinship*. Reprinted edition. Berkeley, CA: University of California Press.
Tylor, Edward Burnett 1871. *Primitive Culture: Researches into the Development of Mythology, Philosophy, Religion, Art, and Custom*. Vol. 1. London: Murray.
——— 1889. On a method of investigating the development of institutions; Applied to laws of marriage and descent. *The Journal of the Anthropological Institute of Great Britain and Ireland* 18 (January): 245–272.
Weber, Max 1978. *Economy and Society: An Outline of Interpretive Sociology*. Berkeley, CA: University of California Press.
Wolf, Eric 1983. *Europe and the People without History*. Berkeley, CA: University of California Press.
Young, Michael W. 1984. The intensive study of a restricted area, or, why did Malinowski go to the Trobriand Islands? *Oceania* 55 (1): 1–26.

Chapter 2

Structuralism

Rupert Stasch

Structuralism is a school of thought centered on the idea that relations between sociocultural elements are more basic than the elements themselves. To understand an aspect of people's lives, structuralists look to how it stands positionally in a network of surrounding other elements.

Anthropological structuralism is associated with Claude Lévi-Strauss, who in the 1940s learned of recent breakthroughs in linguistics premised on the primacy of relations, and began applying this idea to culture. Through discussion of Lévi-Strauss's writings, in the 1960s, structuralism became the most influential intellectual movement of its time. It remains historically the theory of greatest cross-disciplinary impact ever produced by anthropology. Most anthropologists now explicitly describe themselves as rejecting structuralist ideas and methods. Yet we still have many quiet debts to Lévi-Strauss's innovations, and much to learn from structuralism about working with ethnographic data. Additionally, the theory provides a helpful reference point by which to understand and evaluate other analytic frameworks.

Engaging seriously with the core idea of structuralism requires more than summing it up abstractly, as I did above. Across this chapter, I develop this idea in a more substantial way, first by retracing the way it took shape in linguistics, and then considering how Lévi-Strauss metaphorically adapted it to new subjects. I also apply the idea to certain ethnographic data from Robert Lowie's book *The Crow* (1935). Finally, I close with brief comments on the large subject of structuralism's afterlives and its limiting assumptions.

Saussure's concept of *langue* or 'linguistic structure'

Lévi-Strauss's exposure to modern linguistics began in 1942 through lecture courses given by Roman Jakobson about phonology and about Ferdinand de Saussure.[1] Lévi-Strauss and Jakobson were both living in New York and teaching at a Francophone university-in-exile staffed by Jewish intellectuals who had fled Nazi Europe. Jakobson himself had contributed to the major scientific advances that had been made in the 1920s and 1930s in phonology, the linguistic subfield concerned with the sound systems of languages. Concepts from phonology were the linguistic ideas that Lévi-Strauss then most directly adapted to analysis of anthropological subjects. But Ferdinand de Saussure's earlier work on linguistic order more generally provided important groundwork for the innovations in phonology. Later scholars would see Saussure's *Course in General Linguistics* as structuralism's charter text. This was published posthumously by his students in 1916, based on their own notes from lectures he had delivered.

In these lectures, Saussure argued for fundamentally reorienting linguistics. He offers a new vision of the discipline by offering a new vision of what language is, which he introduces by proposing two dichotomies that help separate off parts of the total phenomenon of human language from his main object of study. The first dichotomy is between *langue* and *parole*, French terms that one English translator of Saussure's book renders as 'language' versus 'speaking' (Saussure 1959). As Saussure develops these categories, *langue* could also be glossed as 'grammar, linguistic structure, linguistic system, code', and *parole* could be glossed as 'speech, use, performance, discourse, practice'. Saussure argues that linguistics should be concerned with *langue*: the code or structure, rather than events of speech use. He aligns *langue* versus *parole* with social versus individual. Thus, individual intention is something he factors out of the proper object of linguistic study. Grammatical rules are something speakers enact unconsciously, as the collective conventions of their social community.

A second dichotomy is between diachrony and synchrony, or study of language as it changes through time versus linguistic code as it exists 'statically' in a slice of time. Prior to Saussure, a dominant idea about how to approach analysis of any given linguistic element was to look at its history. Linguistics was philology, the analysis of changes of linguistic structure across time, and the reconstruction of how different languages branched off from parent languages and from each other. Even today, it is a common popular ideology about language that to understand the truth of words, we can look to their etymological origins. Saussure was steeped in careful practice of historical linguistics, and his theoretical innovations were all built on, and oriented toward, concern with doing historical linguistics accurately. He discovered that, for this purpose, it was crucial to study synchronic linguistic structure as a complex reality of its own, connected with but different from processes of diachronic linguistic change.

Having hived off speech and diachrony as *not* his new object of study, Saussure forwards a model of the makeup of that new object in its own right, *langue* as a synchronic system. What *langue* or 'linguistic structure' consists of is relations. A linguistic code is a system of elements all interlocked with each other in a vast network of relations of similarity or contrast, and relations of the elements' possible and prohibited ways of being combined with each other to form other levels of elements. This system of relations is immanent or virtual, rather than existing in the flow of time. All different elements of the system are present to each other at once.

Saussure sometimes emphasizes *difference* as the paramount type of relation making up *langue*. He states that 'In language [*langue*] there are only differences', and that the 'phonic elements' of a language should not be considered as 'sounds having an absolute value but with a value purely oppositive, relative, and negative'.[2]

Phonology and the system-relativity of sound categories

From among all different levels of linguistic structure, the level of patterning of sounds turned out to be the one where linguistic elements *do* preeminently consist of relations of difference from other elements in a phonological system. The relational entity Saussure refers to as a 'phonic element' was being newly described by other linguists of his time as a 'phoneme'. Subsequent advances in analyzing systems of phonemes were what Roman Jakobson summarized in his New York lectures in 1942.

A basic fact of sound systems of languages, central to the theory of phonemes, is that different languages differently divide up the continuum of possible sounds humans can make with our articulatory organs. Often, sounds recognized as different in one language are considered the same in another.

beat	boot
bit	put
bait	boat
bet	bought
bat	

Figure 2.1 Contrasts between selected English vowel phonemes.

Source: Author.

Figure 2.1 lists nine English words consisting of three successive sound segments. (That the words are written with up to six letters is a result of the idiosyncratic history of English writing conventions and sound changes. The actual sound segments making up a word are not the same as letters in the writing system.) The nine words differ from each other only in the vowel that falls between the two consonants, except for 'put', which also involves a change in initial consonant. The words are arrayed according to differences in tongue position. A speaker makes the vowels in the left column with the tongue forward, and those in the right column with the tongue back. Reading down a column, from each word to the next the tongue moves from higher positions to lower ones.[3]

These vowels are crucial parts of the structure of English, yet all they consist of is their distinctness from other elements in the sound system. They signify pure difference of form. Once existing as distinct empty markers, they are available as building blocks for larger units on which the language actually hangs semantic meanings, such as words. Each language has a sound system of its own, the units of which are not natural but are defined by the linguistic system itself. Languages lump and split all the potential sounds humans can make or hear differently.

Figure 2.2 is the top half of a diagram Saussure offers to express the idea that linguistic categories exist through the introduction of cuts of difference into an undifferentiated continuum. With vowel phonemes, for example, the substrate of what I will call undifferentiated 'muck' includes the infinite gradations of slightly different positions a speaker's tongue could be in, and the continuum of slightly different physical sounds so produced. The dividing lines are conventions adhered to in a specific language for recognizing some vowel sounds as different and others as the same. Some languages divide the space of vowel production into just two categories, while others divide that space (and other articulatory variables in how a vowel can be produced) into about thirty. Languages with similar numbers and kinds of vowel categories may locate the boundaries between them differently.

A phoneme is thus not a specific physical sound, but an abstraction grouping together many physical sounds, and separating that grouping from other abstract phonemes and the physical sounds grouped in them. A phoneme is an abstract category, within a specific language system, defined by its difference from the other phonemes in the system. The sound units of a language are not objective but relative.

Another example is the pattern that in English, the sounds [l] and [r] are separate phonemes, while in many other languages, the physical sounds [l] and [r] are merely different versions of a single phoneme. In the Korowai language of Indonesian Papua that I have studied, for example, [rar] and [lal] are two ways of saying the same word, and Korowai tend to hear those two pronunciations as the same sound.

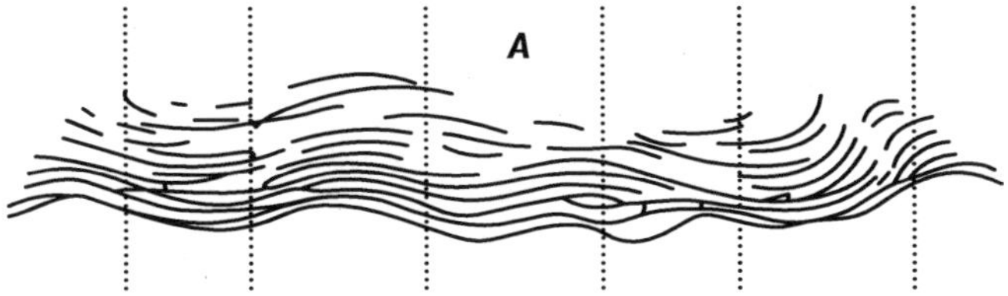

Figure 2.2 Linguistic structure as differentiation of 'muck' (detail from Saussure 1916: 156).
Source: Public domain, Wikimedia Commons.

What people experience as someone speaking a language with a 'foreign accent' is the pattern of which sound differences matter or not in the speaker's mother tongue, interfering with production of speech in another language acquired later in life. Speakers carry their early, language-specific training in how to say and hear sounds into other languages they learn.

Saussure on *langue* as a relational network of signs

There is a further major layer to the ideas from linguistics to which Lévi-Strauss was exposed. So far, I have discussed the makeup of *langue* or 'linguistic structure' by using words like 'element', 'category', or 'phoneme'. But Saussure proposes more specifically that the elements of language should be thought of as *signs*, which he defines as a unity of a 'signifier' and a 'signified' (following a longer tradition of thought). On his account, for example, a word consists of a conjoining of a sound form, the signifier, and a concept or semantic meaning, the signified. One property of linguistic signs is that the relation between signifiers and signifieds is arbitrary: there is no intrinsic reason a specific signifier would be linked to a specific signified. The reasons for the link are in the larger system surrounding that sign. We encounter this whenever we think about different words used to designate the same referents in different languages. We also encounter it if we alienate ourselves from a word in any language by repeating it continuously. Droning repetition leads to heightened awareness of the word's physical form, and loss of the illusion of that form's immediate link to a semantic meaning.

Faced with anything that seems like a 'sign', it can be useful to ask what is its signifier and what is its signified, drawing on Saussure's vocabulary. More profoundly, this aspect of Saussure's theory offers an alternative to the popular assumption that meanings exist first, and language or symbolic systems then secondarily assign symbols to represent them. Instead of signified meanings preceding signifiers, Saussure holds that distinct categories in the order of signifieds come into existence in tandem with the creation of distinct categories in the order of signifiers. Figure 2.3 juxtaposes two of Saussure's visual models of this idea.[4] The left diagram depicts introduction of cuts into two bodies of muck, coinciding with creation of unities between specific territories in muck 'A' and muck 'B'. The right diagram portrays the same idea of *langue* as made of a network of signs, each a unity of signifier and signified. The crux of Saussure's model is the system-constituting primacy of the double-pointed arrows *between* signs in the right diagram, or in the left diagram the dotted lines that *cut* each body of muck and *join* portions of one with portions of the other.

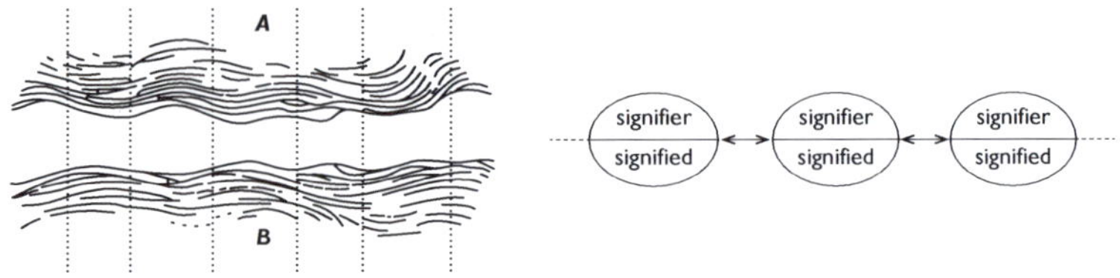

Figure 2.3 Two portrayals of the primacy of relations in the makeup of a linguistic system (from Saussure 1916: 156 and 159).

Source: Public domain, left diagram from Wikimedia Commons, right diagram redrawn by author after Saussure's version.

In all layers of this process – signifiers, signifieds, and signs joining the two – the elements are defined not by their physical properties or other intrinsic contents, but by their position in the network of other elements.

Lévi-Strauss's 'phonological' theory of totemism

Levi-Strauss metaphorically adapted these ideas to analysis of culture in general. In certain writings on kinship and myth starting from the mid-1940s, he explicitly used linguistic vocabulary encountered via Jakobson. But the easiest example to grasp of how Lévi-Strauss transposed the above-summarized ideas into cultural analysis is his theory of 'totemism' (Lévi-Strauss 1963).

Totemism is a worldwide pattern of people linking social groups to animal species or other natural objects. For example, Korowai people of Indonesian Papua belong to small named patriclan groups, and each patriclan has an animal species that the group is said to have originated from, such as the sulphur-crested cockatoo (*Cacatua galerita*), the red-bellied river turtle (*Emydura subglobosa*), or a fork-tailed catfish (*Arius sp.*). Variations on this pattern are widespread in kinship-based societies, but also occur in mass metropolitan societies in systems of sports team names, social club emblems, and the like. For a century prior to Lévi-Strauss's work, totemism was a major topic of anthropological investigation. Prominent evolutionists such as John McLennan, Edward Tylor, and James Frazer had devoted much exoticizing attention to the pattern. Emile Durkheim's *Elementary Forms of Religious Life* (1995 [1912]) was subtitled *Totemic Systems of Central Australia*. But as Lévi-Strauss documents across much of his own short book on the topic, the great variety of different *ways* that people link animals and social groups meant that no satisfactory comparative definition or theory of totemism had yet been produced.

The theory Lévi-Strauss proposes begins with this definition: 'The term totemism covers relations, posed ideologically, between two series, one *natural*, the other *cultural*' (Lévi-Strauss 1963: 16, emphasis in original). A set of human social groups, such as Korowai clans, would be an example of what he means by a cultural 'series', and animal species would be an example of what he means by a natural 'series'. Prior theories of totemism had focused on the relation between human group and animal: to understand totemism, think about the relation between members of the clan and the specific totem animal. Lévi-Strauss, inspired by the linguistic idea of phonological signs as 'purely oppositive, relative, and negative', argues that this approach was looking in the wrong place. We should look instead at the relation of difference *between* animal species. Here are two passages where he states this theory:

> To say that clan A is 'descended' from the bear and that clan B is 'descended' from the eagle is nothing more than a concrete and abbreviated way of stating the relationship between A and B [is] analogous to a relationship between species.
>
> (p. 31)

> The animal world and that of plant life are not utilized merely because they are there, but because they suggest a mode of thought . . . [The mind] postulates a homology between differential features existing, on the one hand, between species *x* and *y*, and on the other, between clan *a* and clan *b*.
>
> (p. 13)

It is not that people use a bird species such as sulphur-crested cockatoo as a metaphor for a social group, but that they use the *differences* between cockatoos, turtles, catfish, and other species as a metaphor for differentiation of human social units (see also Lévi-Strauss 1963: 77). An animal totem is a way of thinking about differentiation itself. Like a phoneme in a language's sound system, the primary meaning of any given animal species is pure difference from other categories.

This account is a transposition into ethnographic data of the Saussurean idea depicted in the two diagrams of Figure 2.3. The 'natural series' and 'cultural series' in Lévi-Strauss's definition of totemism correspond to the two respective layers of 'muck' in the left diagram of that figure. Another way of expressing the account is a symbolic analogy of the form 'A is to B as X is to Y'. Like the diagrams in Figure 2.3, this structuralist formula expresses the idea of order as consisting not of *elements* but of *relations*. The symbolic analogy states that the relation between elements A and B stands in a relation of likeness to the relation between elements X and Y. Those qualities of relationality are basic to the existence, force, or significance of the elements, rather than the elements existing first. Prior analysts had focused on properties of animal A, properties of human group X, and possible reasons for linking animal A and human group X. Lévi-Strauss accords primacy instead to the '. . . is to . . . as . . . is to . . .' relations in the analogic formula, rather than the elements.[5]

For illustrations of Lévi-Strauss's model possibly familiar to readers' own experience, consider some fragments of popular thought about gender in the United States. In recent decades there, it has become common to talk about gender difference by talking about the hormones 'testosterone' and 'estrogen'. Saying, 'there was a lot of testosterone in the room', could be a way of describing that a meeting had consisted mainly of men, or been dominated by masculine interactional styles. In experience of domestic pets, there is a partial tendency for cats to be experienced as embodying human femininity, and dogs to be experienced as embodying human masculinity, independently of the actual gender of the pets. John Gray's self-help book, *Men are from Mars, Women are from Venus*, owed its runaway success partly to the poetics of its gender–totemic title. In each of these cases, difference in a 'natural series' (hormones, animals, or planets) is used to talk about difference in a 'cultural series' (genders). As Saussure's model schematized in Figure 2.3 would have it, masculinity and femininity do not exist from the outset as distinct and easily understood categories, but are ambiguous, undifferentiated, and chaotic. People use difference in an external series of categories to call gender difference into clear existence and keep a grip on it.

'Pig and cassowary are uncle and nephew'

A more elaborate ethnographic illustration of Lévi-Strauss's model of totemism can be found in a link that Korowai of Indonesian Papua make between animals and a specific kinship relation. In Korowai people's forest environment, the two largest animal species are pig and cassowary, a large flightless bird similar to an ostrich. Korowai routinely say, 'Pig and cassowary are uncle and nephew'. Why?

In the Korowai language, only mother's brothers are called uncles (*mom*). Father's brothers are called 'father' (*ate*). So the category 'uncle' has a different 'value' in Saussure's sense in Korowai than in English. Also, Korowai patriclans are exogamous, so marriages are between men and women of different clans. Thus, while fathers and their children are members of the same clan, mothers and their children are members of different clans (Figure 2.4). This means children and their mother's *brothers* are also from different clans. The word 'clan' itself is also the word for 'species' (*gun*). Uncles and nephews are, in a sense, wholly different species of humans. Yet, in Korowai kinship, the relation between uncles and nephews or nieces is especially close. When people have trouble at home or difficulty in life generally, they seek help from maternal uncles. (Radcliffe-Brown and others promoted a Latin-derived special term, the 'avunculate', to talk about this pattern common in many societies.)

Concerning pig and cassowary, meanwhile, what these animals have in common is not only that they are the largest and most desired game in the environment, but also that they are both freaks of nature. They are so different from other animals in size and other qualities that pigs are not even included in the general Korowai category of 'mammal' (*nduo*), that only includes smaller terrestrial mammals; and cassowaries are not included in the general term for 'bird' (*del*), that only includes smaller winged flying animals, including

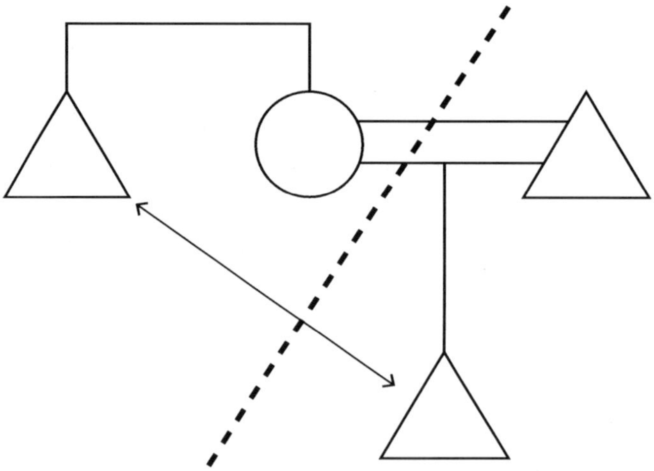

Figure 2.4 Maternal uncle and sister's son (dashed line indicates difference of clan).
Source: Author.

bats. Instead, Korowai lump pig and cassowary together as a single category of their own, 'pig–cassowary' (*gol–küal*), meaning, in effect, 'big game'.

Thus the saying 'pig and cassowary are uncle and nephew' deals with two relatives, uncle and nephew, who are fundamentally different from each other but also mutually close in a special relation different from the rest of kinship. And the saying deals with two animals, pig and cassowary, who are fundamentally different but also close, in their anthropomorphically large size and other anomalous features. This common quality of being profoundly close and profoundly different is probably why it makes sense for people to identify this human kinship relation and animal species relation with each other.

A detail exceptionally congenial to Levi-Strauss's model is that when discussing this bit of received wisdom, Korowai do not care which animal is which relative. When I asked this question, people would say they did not know, or they would give an ad hoc answer that would change across different conversations. What the statement affirms is that 'the pig–cassowary *relation* is an uncle–nephew *relation*'. It is the relations between the animals, and between the kin, that are the focus of thought, not the pig, cassowary, uncle, or nephew as single elements.

Korowai also say it is because pig and cassowary are uncle and nephew that a person cannot eat the two animals' meat on the same day. This seems to parallel a taboo on nephews or nieces marrying spouses from their mother's clan. Korowai say that if a person did mix pork and cassowary meat, his or her rectum would fall out. If you mix the wrong things in your belly then as a consequence you lose control also of the boundary separating your body's insides from what is outside of it. The taboo concerns maintaining boundaries between categories that are already in a volatile situation of being extremely different but also extremely alike.

Generalizing structuralism to wider ethnographic subjects

This style of analysis can be extended beyond ethnographic topics that might neatly consist of alignments between two 'series' of differentiated elements. The methodological spirit behind the specific formalisms of the diagrams in Figure 2.3 is something more flexible: a practice of looking for analogies, looking for what is being put in the same position as what, and asking what more general principles are implied by those identifications.

This is a framework that emphasizes the *autonomy* of 'structure', 'code', or sign system. The system is an independent entity unto itself. Thus, a structuralist does not try to explain a human phenomenon, or the larger system of which it is a part, by reference to something outside or underneath that system itself. For example, a person applying this approach does not explain what the element means by its intrinsic properties. Nor do structuralists explain something by its function or its history. To say the practice fills a function is to say there is something outside the system of interconnected elements that determines that system.

A structuralist also does not try to explain an element primarily by appeal to people's conscious intentions, or their own explanations of what they do. We saw above that separation of structural order from consciousness is one strand of Saussure's broader separation of *langue* 'linguistic structure' and *parole* 'speaking'. Speakers are not conscious of the grammatical rules that they follow, in the sense of being able to state those rules accurately in propositional form. So too, very controversially, applying a structuralist sensibility to ethnographic materials means not being interpretively limited by what people explicitly say about their lives. Lévi-Strauss's structuralism pushed anthropology much further than it had previously gone in presenting an explicit alternative to the commonsense ideology that conscious individual psychology is the basic determinant of cultural practices.

What a structuralist analyst *can* do is make arguments on the basis of formal patterning alone. Patterns of the combination and position of elements in relation to each other are a kind of evidence in their own right, which need to be sifted to arrive at principles that seem to account for them. Such patterning of details is good evidence in its own right to support interpretive generalizations, even if it is not backed up by people's statements of conscious awareness and intention.

Crow 'taking another's wife' in the 1860s

To understand and evaluate this overall methodological program requires applying it to data more complex than those I have cited so far. Here I will model how such an application can proceed, using the example of a specific social institution in the lives of Crow Indians in the mid-19th century. Crow are a Native American group who lived in the Yellowstone River basin in what is today the U.S. states of Montana and Wyoming. Robert Lowie did fieldwork with Crow Indians during summer trips between 1907 and 1931. Like other students of Franz Boas, Lowie's ethnographic writings were descriptive compilations of data with no strong theoretical agenda, beyond the then revolutionary agenda of accurately describing the complexity of Native American life practices without reducing them to fantasized evolutionary schemes. Also like other Boasians, Lowie studied memory culture. He talked with Crow consultants who lived on reservations, were practicing agriculture and wage labor, and were subordinated to structures of United States government rule and the industrial market economy. But the books he wrote were less about those current conditions of Crow life than remembered conditions of one or two generations earlier, when in the late 1800s, Crow were living on or outside the fringes of U.S. government rule, in an economy centered on hunting of buffalo by horseback.

In his book *The Crow* (1935), Lowie describes a social practice referred to in the Crow language as 'taking another's wife', and that Lowie terms 'wife-kidnapping' (p. 186; see also Bauerle et al. 2003: 44). Crow men belonged to clubs, and in the 1860s there were two main such groups the men chose between, named Lumpwoods and Foxes. Competition in 'wife-kidnapping' was a central aspect of the clubs' existence. This competition took place for half a month in the spring, in the seasonal transition between a snowy winter period when people camped at regular sites on rivers in their largest annual aggregations, and summer months spent in pursuit of buffalo herds (see also Bauerle et al. 2003: 85). It was also in the summer that men engaged in long distance raids against neighboring non-Crow people like Lakota and Blackfeet, whose horses they stole. During the brief spring 'wife-kidnapping' period, a Fox man would approach the wife of a Lumpwood man, tell her to come with him, and she would, thereby permanently ending her previous marriage. Conversely, a Lumpwood man would 'kidnap' the wife of a Fox man, permanently ending those spouses' marriage. But the new marriage made by kidnapping was only temporary: the couple would be lovers for a brief period, and then separate (Lowie 1935: 56).

Lowie and later scholars nowhere interpret this institution in any general way. When I have asked students to read Lowie's description and interpret the institution, their first ideas are typically to try to explain it by functions it fulfills (see chapter 1), by the desires of individual Crow persons, or by how it originated.[6] A more satisfactory basic understanding of the institution can be gained, however, by asking the structuralist methodological question: 'What is relational to what?' One concrete rule-of-thumb by which a structuralist analysis can be pursued, for example, is to look for things that in some sense occur in the same 'position', hypothesize that they are actually the same thing, and ask what that more abstract single thing might be.

For example, one pattern of literal *spatial* positioning is that, after a wife-kidnapping, the captured woman is paraded by the club whose member has captured her, riding a horse behind a man who has previously rescued another Crow from enemies by retrieving that Crow man from where he has fallen to the ground and galloping away with him behind. Additionally, the man and woman specifically ride on a horse stolen from an enemy camp (Lowie 1935: 190; Bauerle et al. 2003: 70, 108). In other words, the woman is put in the same position as the Crow man who rode behind, and put in the same position as a horse stolen from non-Crow enemies, by riding *on* such a horse. But if the abducted wife of an opposite club man is likened to a horse stolen from enemies and to a Crow man rescued from enemies, then the relation between Foxes and the Lumpwoods is being likened to that between Crow and non-Crow enemies. Rivalry between the two clubs is a metaphor *within* Crow society of the relation of Crow society as a whole with those enemies.[7]

If the clubs compete over different activities, those activities are also in a sense occurring in the same 'position'. Besides competing over how many of each other's wives they kidnap, the clubs also compete over which club *touches* enemies first in battle later in the summer (Lowie 1935: 186). Whoever manages to touch an enemy fighter gets more honour, whether that touch is smashing the enemy or just tapping them lightly (pp. 216, 228). What wife-kidnapping and touching the enemy can be hypothesized to exemplify in common is a certain quality of transaction in relation to others: taking from them against their wills, in violation of their sense of control over themselves, their personal space, and what is dear to them. The enemy who is touched may be killed or merely outmaneuvered, but in either case, a Crow man managed to approach within reach of that person and 'touch' him, despite the other's efforts to harm the attacker.

In a similar vein, the horse on which a captured wife was paraded was not just any stolen horse, but one stolen from a tether alongside an enemy's tent. The raider did not just stop outside the enemy's camp where horses were milling around, but went brazenly under the nose of the sleeping owner, and snuck away with the horse. Nothing about a horse stolen in this way is intrinsically better than one stolen from outside a camp, with respect to its power of carrying a rider. But such a horse is better as a signifier of *appropriation* as a quality of action, that is also central to wife-kidnapping: the startling disruption of others' sense of secure belonging in relation to possessions or persons they hold close.

On a broader level, it is also useful to ask what the *whole institution* of wife-kidnapping occurs in the same position as. One answer is that it is very much in the same position as regular marriage. During the parading of a captured woman, she wears the same ornaments and dress as a new bride (pp. 190, 214; Bauerle et al. 2003: 108).[8] But while the union between kidnapper and kidnapped is like a marriage, wife-kidnapping transgressively disrupts regular marriage and violates its norms. It is regular marriage with a twist.

One rule of wife-kidnapping that defines it as contrasting with regular marriage is the already-mentioned stipulation that the original spouses can never reunite. Another such rule concerns which men can kidnap which women. A kidnapper cannot approach just any woman married to a member of the opposite club, but can only approach a woman with whom he was previously a lover (Lowie 1935: 186). 'Taking another's wife' harkens back to histories of dating prior to when people settled down with long-term spouses. Wife-kidnapping draws earlier modes of transient romance and sexuality back into the present, overwriting stable marriage.

stable marriage	transient dating; romance as 'hunting' and moving on
love, kinship	war, impingement, appropriation
domesticity, childbearing	fighting enemies, killing buffalo
attachment to life, kin	recklessness toward death
winter	summer

Figure 2.5 A value polarity of Crow life in the 1860s.

Source: Author.

Running through these different details, there is a more abstract pattern, a polarity of contrasting values and visions of social relations, that I try to schematize in Figure 2.5. Across Crow life of this period, there is a conflict between an ideal of stable marital attachment and valuing life with kin, and an ideal of transient appropriation. The rule against earlier spouses reuniting affirms that non-attachment and living for the transient moment of appropriation is the higher good. So too the more general organization of Crow society around summertime competition between the two clubs affirms that transient appropriation in war, hunting, and seduction is a more valued mode of social action than stable domestic unions and other forms of kinship attachment that the society also revolves around. In the emotional experience of men in particular, the hierarchical superiority of the right-hand values in Figure 2.5 over the left ones was enforced by a socially required response of a man whose wife was being abducted. Even though in many cases such a man was actually feeling severe emotional distress, he was required not to object, but instead to stand by, performing a demeanor of impassive non-attachment (pp. 55, 186, 188, 191). So, too, men elected early in the spring to each club's 'officer' positions were expected to die that summer, as a consequence of their role's requirement that they run to the front of a confrontation and remain there (pp. 177–178, 183–185, 228). Also linking club solidarity to right-column values in Figure 2.5 is the pattern that a widower's bereavement was belittled by club-mates with graveside jests like, 'You will not have a wife today, shall you?' (p. 179). In other forms of verbal artistry, club-supported values of reckless masculine detachment from one's own life, and unhesitating brave appropriation of the lives of others, were contrasted pejoratively with reproduction-linked bodily processes of feminine menstruation and parturition (e.g. Lowie 1935: 116, 184).

While wife-kidnapping is in the same position as marriage and is an altered form of it, we saw earlier that it is also in the same position as stealing horses, warfare, stealing a fellow Crow out from under an enemy blow, touching an enemy in battle, and facing enemies with the expectation of dying. Wife-kidnapping is a kind of love and romance centered not on stable attachment to each other across time, but on the 'hunt': on the pleasure and love of first crossing the boundary from distance to romantic and sexual closeness. Lowie reports, for example:

> [E]xcess of virtue marked out a man for raillery. If he lived too long with one wife, his joking relative might say, 'You are as though next to a dead thing'. . . For, Gray-bull explained, women are like a herd of buffalo, and a husband who cleaves to one wife

> is like a hunter who has killed the last of the fugitive animals and stays by the carcass because he lacks spirit to pursue others.
>
> (p. 48)

Evidence of what is schematized in Figure 2.5 also occurs in the sanction followed when a couple *did* break the rule against reuniting after separation by a wife-kidnapping episode, which is that men of the other club cut up the blankets of not only the married couple but of all men in the husband's club (p. 189). If a Fox man reunited with his former wife, all Fox men might thus run away with their buffalo-skin blankets, to try to prevent Lumpwoods from cutting them up. It is likely that the sanction takes this form because what occurs in the same position as blankets is spouses' bodily closeness (p. 188). The sanction attacks exactly what the couple affirmed by reuniting, and what wife-kidnapping is pitched against in the first place. The couple asserts the value of long-term spousal relations as a space apart from the wider society, while the attack on blankets reasserts that club life is superior after all.

Part of the 'position' of cultural elements is where they fall in time. 'Taking another's wife' occurred in the spring, right after the clubs had returned to organizational life through election of new officers. It was after the brief wife-capture interlude that Crow then started raiding enemies or being attacked by them, with bravery in these external fights also being an arena of club competition. That wife-kidnapping falls *next to* fighting enemies in time suggests that wife-kidnapping and raiding are alternative forms of a single category, the value of transient appropriation. This quality of relational action is ascendant across the summer period. When the snow comes, clubs go dormant and domestic family life is dominant.

Concerning women's subjectivity, Lowie gives examples of specific women who hid away from the camp to avoid being kidnapped and thereby separated from their husbands, and he gives examples of spouses whose marriages were ended against their desires, to their severe distress across later life (1935: 53–54, 59, 186–188). But he also reports that when the clubs were coming back to life early each spring, it was regularly a married woman who would signal to her ex-lover in the other club that he should kidnap her in the upcoming season (p. 191). The wife-kidnapping period itself would end, Lowie reports, 'when all wives amenable to capture had changed husbands', and he quotes the gloating paraders of a captured woman as announcing, 'One of the Lumpwood girls has married one of us Foxes of her own accord!' (pp. 189–190).[9] He also reports that women were usually – though not always – able to rebuff an undesired abduction by emotional pleading (p. 189). These tendencies for women to be partial agents of wife-kidnapping events suggest that women in their own subjective experience of the institution not only lived out the tensions schematized in Figure 2.5, but also sometimes shared with men the hierarchical ranking of the right-column values as being superior to the left ones.

Overinterpretation versus the ethnographic craft of tracing connections

It would be instructive to consider additional Crow ethnography, and to develop more nuanced generalizations than I have so far proposed. The above discussion also raises questions about finer details of different Crow experiences and practices 150 years ago that are not answerable today, perhaps limiting the possibility of ever producing a satisfactory interpretation of the men's club activities I have discussed. But hopefully my account was substantial enough to illustrate how structuralist analysis can be carried out in practice. I also hope that this account has provoked in readers some feelings of being surprised and convinced by certain connections I drew, as well as some feelings of resistance and skepticism toward those connections.

One common response to this form of analysis is discomfort with the possibility of overinterpretation. For example, is it not just accidental that members of one Crow men's club cut up the blankets of the other club? In connecting that pattern to the bodily intimacy of spouses, am I making statements that Crow of the period would not themselves have recognized as meaningful or true, and so falsely imposing my thought into other people's lives?

This is a reasonable response, and the issues raised in it do not have simple solutions. There are a few broad points I would offer in turn, though, for thinking about this response. First, the hypothesis, 'There are no accidents' is itself a helpful working summary of structuralism's methodological assumptions. Second, structuralist interpretations are not arbitrary, but rather there is a robust criterion for evaluating them as well supported or poorly supported. This criterion is that the more ethnographic details can be fitted into a posited pattern, the stronger the evidence that the analyst is on the right track. Conversely, it is routine that a broad generalizing model would be continuously revised into more complex and nuanced forms, in the course of attempting to fit an increasingly finer and more varied range of empirical details to its terms. Third, while it is good to wish to avoid making untrue interpretations, *refraining* from interpretation is also itself a form of overinterpretation. Like other overinterpretations, a stance of the form 'I as analyst am not obligated to engage with certain details of people's lives, or should not put them into any explanatory framing other than accident or what practitioners themselves say', involves assimilation of empirical details to a general pattern from the analyst's own mental universe, whether it is a universalizing theory of accident, a universalizing theory of one's own entitlement to avoid intellectual engagement with difficult material, or a universalizing theory that individuals are sovereigns of our own thought and action.

A more positive form of these points is the historical transformation of anthropological practice I alluded to at the beginning of this chapter, when I suggested that anthropologists today have many quiet debts to structuralism even while often overtly disavowing it as an explicit framework. Lévi-Strauss's most lasting contributions probably were not his overt theoretical ideas, such as the ones I have touched on above. Rather, his most enduring influence came through how he worked with ethnography. What many anthropologists today unconsciously owe to Lévi-Strauss is methodological appreciation for interconnections across sociocultural practices. They owe to him the idea that 'the truth is in the details', so to speak: that in working with sociocultural materials, the basis of an original and accurate insight is close sifting of data, and imaginative recognition of non-obvious patterns of articulation between them. In the wake of structuralism, it has become routine to trace links across narratives, rituals, architectural forms, practices of food production and consumption, practices of bodily adornment, geography, cosmology, formal institutions of political order, and many other areas of life. Structuralism gave anthropology new reach in seeing abstract categories or principles in ethnographic particulars.

Conclusion: Unwinding the separation of *langue* and *parole*

If many schools of anthropological thought do share with structuralism a commitment to tracing non-obvious ethnographic 'connections' in the broad sense just outlined, this would mean they also implicitly share with structuralism an understanding of their objects of study as in some sense patterned, ordered, or systematic. Yet there are good reasons why adherents of most subsequent schools of thought view themselves as breaking fundamentally with structuralist orientations about the nature or location of this order.

An account of the different ways anthropologists have departed from structuralism would be a description of almost the entire discipline, which I cannot attempt. Yet it is remarkable how many of the departures made across the last half-century can be traced to the rolling back of assumptions contained within Saussure's original separation of *langue* versus *parole*. 'Practice' (*parole*) is itself a multilayered category, and so there are many interconnected layers to what 'system' (*langue*) is understood to be autonomous from, when practice is hived off from it. I now end by briefly charting a few aspects of human life that are marginalized by Saussure's dichotomy, and that most anthropologists since Lévi-Strauss have instead been at pains to recognize as *central* to their objects of study. There is a close link between separating 'practice' from 'system', and the structuralist insight that 'system' exists as pure 'static' relationality of elements to each other. Unwinding the distinction between *langue* and *parole,* while remaining committed to close analytic engagement with the relational interconnectedness of sociocultural elements, is not simple.

One area that has seemed particularly urgent to integrate with accounts of the systematic order of people's lives is consciousness or subjectivity. While structuralism offers a valuable corrective against the idea that humans transparently choose and express the shape of their own lives, there are strong grounds for recognizing that processes of consciousness play major roles in creating or remaking that shape. Anthropological inquiry is poorly served by a stark dichotomy of transparent self-knowledge versus unconscious order, a polarization of possibilities that some structuralist rhetoric seems to encourage. Rather, anthropological work requires instead an understanding of conscious as often multilayered or oblique. Many interpretive generalizations I set forth above about Crow 'wife-kidnapping' do seem to have been expressed in Crow discourse, although often as semi-tacit implications and entailments on the edges of stated propositions, rather than at those propositions' center. Perhaps a Crow person of the time *would* actually recognize some of those interpretations, if put in the right form. One of the most valuable benefits of structuralist modes of reasoning for fieldworkers is exactly that they can lead toward questions and propositions that *do* make sense to research interlocutors, bringing the fieldworker to deepened engagement with nuances of their thought.

One theoretical approach to integrating consciousness and system that emerged from the high-structuralist theoretical milieu was the Marxist philosopher Althusser's model of 'interpellation' (1971), that preserved a strong version of the autonomy of 'system' but merely described personal consciousness as an internalization of categories of that wider order (see chapter 3). Anthropologists of recent decades, by contrast, have generally eschewed models of the interface between subjectivity and wider order that depict this interface as so seamless and unidimensional. Linguistic anthropologists, for example, have accorded positive and complex roles to 'reflexivity' or 'ideology' on the part of language users, in shaping their languages and other aspects of their systems of life (Schieffelin, Woolard, and Kroskrity 1998; Agha 2007). The later work of Foucault has been widely taken up by many authors now working in the 'anthropology of ethics' tradition, as offering an ethnographically fruitful framework for exploring comparative variation in what a thinking and acting subject is, while also considering it possible that people may reflect on their own structural conditions of life, engage critically with alternative possibilities within those structures, and thus make their world in addition to being made by it (Foucault 1978; Faubion 2001; Laidlaw 2014). Countless forms of anthropological concern with 'agency' also fundamentally strive to square the circle of the Saussurean dichotomy (Ahearn 2001; Ortner 2006).

Closely related to the theoretical and ethnographic status of the acting subject are a whole further range of features of 'practice' as something that unfolds in a particular time and place, between particular persons, who live in bodies as well as minds, and are *aware of* practice itself as something they are doing. Bourdieu (1977), in a critique of Lévi-Strauss that closely followed Voloshinov's (1972 [1929]) very early critique of Saussure, drew attention to actors' own stances toward the here-and-now performative contingency of their practices, as itself a source of the value and efficacy of those practices and a level of their orderly organization. Even in the example of language, it turns out that there are many aspects of language structure that *encode* situations of practice. For example, the pronouns *you* and *I* encode a situation of language use, where one person is encoded in the grammar of the language as having the role of 'the one talking', and another person present on the scene is encoded as having the role of 'the one being talked to' (Benveniste 1971 [1956]). In other words, the linguistic structure itself contains a picture of a social scene of action or use, in a here-and-now of discourse performance at a specific time between specific people; *parole* is inside *langue,* so to speak, against the terms of Saussure's premises. So too it turns out that a lot of culture and society at large consists of general models or structure of how to *use* signs: conventions of genres of speech use and genres of interaction, like holding a lecture, or holding a sporting match, or holding an election. The *practice* or *action* cannot be distilled away from the *structure*, because the structure is a model of and in action. It is a reflexive sense of doing things at a here-and-now. My quasi-structuralist reading of Crow wife-kidnapping, above, led very quickly toward a model of contrasting, hierarchized values that were qualities *of action*: impinging transactionally on some person at a moment and place, versus enduringly partnering with someone in loving solidarity over extremely long successions of different moments. Structure is structure *of* and *about* activity, and vice versa.

Questions of temporality are particularly complex and urgent. It takes just the tiniest reflection to realize that people do not generally live in one time at once, but live through modes of understanding that are themselves about time's flow and about other times than the present. Phenomena like memory, bereavement, historical consciousness, or expectations of the apocalypse, utopia, career success, or uncertainty are structural forms. They are forms of synchronic order, so to speak, that stand in an internal relation with diachrony. Processes of change and reproduction across time are 'inside' of order, in culturally organized ways, so there is strictly speaking no such thing as *synchronic* order. There is rather order and patterning that is *of* temporality, constituted of temporal processes and people's cultural understandings of life in time, all the way down. It is in this vein as well as for other theoretical reasons that, since the 1980s, anthropologists have shifted to strongly endorsing the incorporation of historical perspectives and historical research at the center of ethnographic work. This overall shift has often been promoted in explicit opposition to structuralism, which most scholars stereotype as having been an ahistorical or anti-historical framework (though see, for example, Sahlins 1985; Gow 2001).

Also intertwined with these other dimensions of practical life are value and power. In the brief 'wife-kidnapping' example, we saw that the order of categories complexly in play across Crow life of the time was not a matter just of difference, but of hierarchy: of what people understand as being more important than what. Even greater anthropological attention in the wake of structuralism has focused on 'who benefits?': the question of how specific orderings of categories favor the well-being of distinct sets of people within a society or network, or might only be held or promoted by specific people and institutions.

Also implicated in these same lines of complexity is the question of a system's contradictions, heterogeneity, plurality, or openness. To some extent, there is a logical correlation between the idea of a sociocultural element being constituted by its relational interdependence with surrounding elements, and an assumption that sociocultural orders have a degree of unity and boundedness. For an element to be analyzable as having a stable relational makeup, there would need to be a stable and limited relational field that it sits within (Merlan 2005: 173). Yet actual human lives and even single actions are generally lived across a plurality of orders. The stability or significance of what is heuristically identifiable as one 'system' often seems to rest in its contrasts and articulations with what is 'outside' it or 'other' to it (an area of anthropological inquiry sometimes inspired by Bakhtin, e.g. 1981).

There are many further areas of theoretical concern that anthropologists have opened up extensively in the aftermath of structuralism. Concerning the few just touched on here, my purpose has not been to chart the actual concepts that have been developed in these areas, nor to chart the theoretical and empirical weaknesses that these concepts resolve or perpetuate. Rather, my main goal has been to give a sense of the ways the different layers of 'practice' are connected with each other. When one of these issues that was bracketed off by Saussure's conceptual framework (and by Lévi-Strauss's metaphoric adaptions of it) is brought back into consideration, as an integral part of order or system, all the others come rushing back in as well.

Dealing seriously with these theoretical and empirical complications does not necessarily mean an end to looking to how sociocultural elements are defined by their relational positions. It does, however, mean not locating those relationalities in a transcendental sphere unto themselves apart from practice and the heterogeneity of different logics of system, which makes an approach of tracing relational connections more difficult and complex.

Notes

1 For some of the content, see Jakobson (1978; 1990) and at slightly greater length (1984). Jakobson and Lévi-Strauss's lecture schedules are recorded in annual catalogs of the École Libre des Hautes Études, available at digitalarchives.library.newschool.edu/index.php/Detail/collections/NS050101.04. For example, titles and conflicting timing of their courses in the school's first term are shown on p. 16 of the February 1942 catalog.

2 For the first quotation, see Saussure (1959: 120), as well as the longer and more interesting passage in Saussure (1993: 141). For the second quotation, see Saussure (1996: 116), as well as Saussure (1959: 119) and (1972: 164, 433–434n111, 466n236). On the editorial process by which the original French 1916 edition of Saussure's book was produced, see for example Joseph (2012: 632–635) and Engler (2004).

3 Difference in vowel systems is the single most prominent way that English dialects vary from each other, so reading this figure will produce different results by speaker. Although I stated that the words in Figure 2.1 all consist of only three segments, the vowels in *bait* and *boat* are actually diphthongs: the tongue moves from one position to another in the course of uttering the segment.

4 See also the somewhat different renderings in Saussure (1993: 96, 135, 138–140).

5 Another anthropological area in which Lévi-Strauss argued for shifting attention from entities to relations was the study of kinship. As Candea discussed in chapter 1, Radcliffe-Brown and others in the tradition that came to be known as 'structural-functionalism' held that political and social

order in most societies studied by anthropologists was maintained primarily through the existence of matrilineal or patrilineal 'descent groups'. Lévi-Strauss, in *The Elementary Structures of Kinship* (1969 [1949]) and many other works, argued instead that marriage and associated exchange processes between in-laws (or 'affines') across generational time were the actual crux of kinship. This contrast between the structural-functionalist paradigm and Lévi-Strauss's approach was later sometimes described as a debate between 'descent theory' and 'alliance theory'. There are many nuances to Lévi-Strauss's work on kinship and marriage, and many nuances to structural-functionalist work. But the broad force of Lévi-Strauss's 'alliance theory' was to suggest that relations between groups were more basic than the groups themselves, again metaphorically akin to the idea about language expressed in Saussure's diagrams reproduced in Figure 2.3.

6 I owe use of this area of historical Crow ethnography in university teaching to Robert Brightman.

7 In a compilation of oral historical narratives initially collected and translated in the 1950s, Bauerle et al. (2003: 17–18) recount a remarkable specific instance of further creative elaboration, in a reverse direction, of the likening of captured female wives to rescued male comrades. A Fox man had taken the wife of a Lumpwood. The same Fox man, coincidentally, was rescued in battle by that Lumpwood. The next spring, the Lumpwood man rode into a circle of Foxes, and said 'One I have shared a woman with, I have come for you. Come with me to my club', to which the other man responded 'So it is' and mounted the horse behind his rescuer. As they rode away to where Lumpwood club members were gathered, the front rider called back, 'That is the way I [rescued him]. If you want him back, get him back as I did [rescue him in battle], and you can have him back'. In this way, the rescued man's club membership changed from Fox to Lumpwood, modeled on a woman's change of marital union.

8 Bauerle et al. (2003: 112) recount a complex case of a man whose dramatic fall in prestige followed in part from transgressively reuniting with his wife after they had been separated by another man 'taking' her. When the earlier kidnapper unexpectedly encounters the remarried pair again later, the fallen man, addressing his wife, refers to the former lover as 'your husband'.

9 Bauerle et al. similarly translate the Crow idioms by which celebrating men regularly described women's volitional subjectivity as 'she came on her own volition and free will', 'she made up her own mind', 'all on her own free will', or in a sung celebration 'She came on her own free will and on her own' (2003: 70, 105–107). In one specific event of 'taking another's wife' that was recollected in oral histories in particularly close detail, a Lumpwood interloper moved against the marriage of a Fox man who had taken four Lumpwood wives, an excessive humiliation. In this instance, the interloper made his attempt based on an extended principle that 'one could lay claim even if they had played man and wife or pretended to be lovers as children', rather than having ever been sexually involved. The interloper's overture to the Fox man's wife is quoted as: 'That guy [the addressee's current husband] keeps doing some things until it is probably too much. Let us, you and me, try something ourselves'. To this the Fox man's wife replied, 'So it will be' (2003: 106).

References

Agha, Asif 2007. *Language and Social Relations*. Cambridge: Cambridge University Press.

Ahearn, Laura M. 2001. Language and agency. *Annual Review of Anthropology* 30: 109–137.

Althusser, Louis 1971. Ideology and ideological state apparatuses. In *Lenin and Philosophy and Other Essays* pp. 123–173. London: Monthly Review Press.

Bakhtin, Mikhail M 1981. *The Dialogic Imagination: Four Essays*. Austin, TX: University of Texas Press.

Bauerle, Phenocia (ed.), compiled and translated by Henry Old Coyote and Barney Old Coyote Jr. 2003. *The Way of the Warrior: Stories of the Crow People*. Lincoln, NE: University of Nebraska Press.

Benveniste, Emile 1971 [1956]. The nature of pronouns. In *Problems in General Linguistics*, pp. 217–222. Miami, FL: University of Miami Press.

Bourdieu, Pierre 1977. *Outline of a Theory of Practice*, translated by Richard Nice. Cambridge: Cambridge University Press.

Durkheim, Emile 1995 [1912]. *The Elementary Forms of Religious Life*, translated by Karen Fields. New York, NY: The Free Press.

Engler, Rudolf 2004. The making of the Cours de linguistique generale. In Carol Sanders (ed.), *The Cambridge Companion to Saussure* pp. 47–58. Cambridge: Cambridge University Press.

Faubion, James D. 2001. Toward an anthropology of ethics: Foucault and the pedagogies of autopoiesis. *Representations* 74 (1): 83–104.

Foucault, Michel 1978. *The History of Sexuality, Volume I: An Introduction*, translated by Robert Hurley. New York, NY: Vintage.

Gow, Peter 2001. *An Amazonian Myth and its History*. Oxford: Oxford University Press.

Jakobson, Roman 1978. *Six Lectures on Sound and Meaning*, translated by John Mepham. Hassocks: Harvester Press.

——— 1984. La théorie saussurienne en rétrospection. *Linguistics* 22: 161–196.

——— 1990. Langue and parole: Code and message. In Linda R. Waugh and Monique Monville-Burston (eds.), *On Language*, pp. 80–109. Cambridge, MA: Harvard University Press.

Joseph, John E. 2012. *Saussure*. Oxford: Oxford University Press.

Laidlaw, James 2014. *The Subject of Virtue: An Anthropology of Ethics and Freedom*. Cambridge: Cambridge University Press.

Lévi-Strauss, Claude 1963. *Totemism*, translated by Rodney Needham. Boston, MA: Beacon Press.

——— 1969 [1949]. *The Elementary Structures of Kinship*, translated by J.H. Bell, J.R. Sturmer, and R. Needham. Boston, MA: Beacon Press.

Lowie, Robert Harry 1935. *The Crow Indians*. New York, NY: Farrar and Rinehart.

Merlan, Francesca 2005. Explorations towards intercultural accounts of socio-cultural reproduction and change. *Oceania* 75 (3): 167–182.

Ortner, Sherry B. 2006. *Anthropology and Social Theory: Culture, Power, and the Acting Subject*. Durham, NC: Duke University Press.

Sahlins, Marshall 1985. *Islands of History*. Chicago, IL: University of Chicago Press.

Saussure, Ferdinand de 1916. *Cours de linguistique générale*. Edited by Charles Bally and Albert Sechehaye, assisted by Albert Riedlinger. Lausanne: Payot.

——— 1959. *Course in General Linguistics*, translated by Wade Baskin. New York, NY: Philosophical Library.

——— 1972. *Cours de linguistique générale. Édition critique préparée par Tullio de Mauro*. Paris: Payot.

——— 1993. *Troisième Cours De Linguistique Générale (1910–1911): D'après Les Cahiers D'Emile Constantin = Saussure's Third Course of Lectures on General Linguistics (1910–1911): From the Notebooks of Emile Constantin*. Eisuke Komatsu (ed.), translated by Roy Harris. Oxford: Pergamon.

——— 1996. *Premier Cours de Linguistique Generale (1907): d'apres les cahiers d'Albert Riedlinger = Saussure's First Course of Lectures on General Linguistics from the Notebooks of Albert Riedlinger (1907)*. Eisuke Komatsu (ed.), translated by George Wolf. Oxford: Pergamon.

Schieffelin, Bambi B., Katherine A. Woolard, and Paul V. Kroskrity (eds.). 1998. *Language Ideologies: Practice and Theory*. Oxford: Oxford University Press.

Voloshinov, V.N. 1972[1929]. *Marxism and the Philosophy of Language*. The Hague: Mouton.

Further reading

Douglas, Mary 1966. *Purity and Danger: An Analysis of the Concepts of Pollution and Taboo*. London: Routledge and Kegan Paul.

Dumont, Louis 2006. *An Introduction to Two Theories of Social Anthropology: Descent Groups and Marriage Alliance*. Oxford: Berghahn Books.

Holdcroft, David 1991. *Saussure: Signs, System and Arbitrariness*. Cambridge: Cambridge University Press.

Lévi-Strauss, Claude 1963. *Structural Anthropology*, translated by Claire Jacobson and Brooke Grundfest Schoepf. New York, NY: Basic Books.

——— 1966. *The Savage Mind*. Chicago, IL: University of Chicago Press.
——— 1969. *The Raw and the Cooked*, translated by John and Doreen Weightman. New York, NY: Harper and Row.
Lowie, Robert Harry 1959. Back to the Crow. In *Robert H. Lowie, Ethnologist: A Personal Record*, pp. 41–66. Berkeley, CA: University of California Press.
Ricoeur, Paul 1974. Structure and hermeneutics. In Don Ihde (ed.), *The Conflict of Interpretations*, pp. 27–61. Evanston, IL: Northwestern University Press.
Robbins, Joel 2004. *Becoming Sinners: Christianity and Moral Torment in a Papua New Guinea Society*. Berkeley, CA: University of California Press.
Sahlins, Marshall 1976. *Culture and Practical Reason*. Chicago, IL: University of Chicago Press.
Shapiro, Judith 1988. Gender totemism. In Richard Randolph, David Schneider, and May Diaz (eds.), *Dialectics and Gender: Anthropological Approaches*, pp. 1–19. Boulder, CO: Westview.
Stasch, Rupert 2009. *Society of Others: Kinship and Mourning in a West Papuan Place*. Berkeley, CA: University of California Press.
Valeri, Valerio 2000. *The Forest of Taboos: Morality, Hunting, and Identity among the Huaulu of the Moluccas*. Madison, WI: University of Wisconsin Press.

Chapter 3

Marxism and neo-Marxism

Caroline Humphrey

In the mid-1960s, anthropology in Britain and the USA was bubbling along, preoccupied with its own disciplinary concerns. Functionalism (see chapter 1) and structuralism and semiotics (see chapter 2) were being critically questioned in Britain and replaced with new ideas from Edmund Leach, Mary Douglas, Frederick Barth (see chapter 4) and many others, while in the USA, an 'explosion of paradigms' (Geertz 2002) was starting to erupt, including interpretative anthropology (see chapter 8), socio-biology, cognitive anthropology (see chapter 7) and the 'ethnography of speaking'. But the events of 1968 – revolts across Europe, the Prague Spring and civil rights protests in America – changed the mood. Many anthropologists turned their attention to economic and political problems in the world. Marxism, with its world historical purview and promise of liberation, seemed the way out from debates that now seemed parochial. 'Neo-Marxism', the term for the theoretical movement that was born at this time, was a series of radical reinterpretations of Marx inspired by two mainsprings: first, the need to adapt Marxist thought to address contemporary issues such as de-colonisation, rampant capitalism and the inequality and exploitation afflicting small non-capitalist societies, and second, the desirability of rethinking straight Marxism by incorporating elements from other intellectual traditions, such as critical theory, psychoanalysis or existentialism. The movement proper, that is, the occupation of the fashionable forefront by a particular analytical tool-kit and vocabulary, was relatively short-lived. Keith Hart has claimed that it suddenly 'dropped out of anthropology' with the election of Margaret Thatcher and Ronald Reagan in 1979 to 1981.[1] However, this was not a full disappearance and it was not before neo-Marxism had made a radical transformation of anthropology's horizons. Whole fields as they are understood today are unthinkable without the theoretical impetus provided by the neo-Marxists – economic anthropology, feminism, gender and kinship, colonialism/post-colonial studies, anthropology of the city and historical anthropology. Furthermore, as I shall suggest at the end of this chapter, anthropologists continue to draw insights from Marx both directly and *indirectly* – via outlying disciplines that have pushed Marxist thinking in new directions.

'Marxism'

Karl Marx (1818–1883) said towards the end of his life that he was not a Marxist, thus recognising that a separate body of work had grown up around his work (including simplifications with which he did not entirely identify). 'Marxism' has many strands, but the version of it that reached anthropology was an economic interpretation of history that emphasised four central theoretical elements:

a the physical reality of human life, the centrality of labour and the constant dynamic of practices;
b the systemic organisation of social relations of production;
c the conflict between different forms of such organisation; and
d the historical evolution of different stages of development, to culminate in the overthrow of capitalism by communism.

Before neo-Marxism appeared on the scene, American anthropologists based at Columbia University had taken up Marx's concern with the conditions faced by labouring people and then applied a broadly Marxist approach using the massively greater amount of information that had become available since his day. Leslie White in *The Evolution of Culture* (1959) proposed a theory about the ways in which the technical base gives rise to social structure that gives rise to ideology. Students called this the 'cultural layer cake' (Sahlins 2000: 515). Sidney Mintz had begun writing on sugar-cane plantation workers, slavery and Caribbean peasantries even earlier. His life-time concern with colonialism and the capitalist world system resulted in a series of works, the best known being *Sweetness and Power: The Place of Sugar in World History* (1986). Eric Wolf, a colleague of Mintz at Columbia, likewise focused on peasantries from the 1950s onwards, publishing *Peasant Wars of the 20th Century* in 1969. Wolf later turned to address anthropological methodology and its background history in imperialism and colonialism. His *Europe and the People Without History* (1989) was written to redress the bias of a 'first world' perspective, focusing on the fur trade in Russia, the silver trade and the Atlantic slave trade. Broadly evolutionary, Wolf's argument – that isolated and largely forgotten historical actors are part of world history – was inspired by his own ruminations on the relation between power and ideology and the ways in which different kinds of Marxist theory address both general laws and situated ethnography (see also chapter 5).

In Britain, meanwhile, the historian and life-long communist, Eric Hobsbawm, had been addressing other non-mainstream subjects ('primitive rebels'/'social bandits', industrial workers under imperialism, or problems faced by working-class labour movements). His work was little known in the separate disciplinary enclave of anthropology and most anthropologists were sceptical of Marxist theory (Firth 1972). This is perhaps why, when in 1965 Hobsbawm published his seminal 'Introduction' to the English translation of Marx's little known *Grundrisse*, titled *Pre-Capitalist Economic Formations*, the anthropological world at first largely ignored it. 'Neo-Marxism' had not yet hit the scene. Yet a few years later, as I well remember, many students pored over it. The ostensible subject of the book, the delineation of a series of stages or epochs of primitive communal, slave-owning, Roman, German, feudal, bourgeois, and so on, societies, was not what interested us, and Hobsbawm himself warned against accepting any of these categories or their sequence uncritically (1965: 19). Rather, what was theoretically innovative was a number of other widely applicable ideas that Hobsbawm had extracted from Marx: the role of exchange in the historical process and the 'individualisation' of groups and persons; the notion of 'appropriation' (of 'bits' of bodies, land, shelter, food, labour, technologies, etc.) and its relation to the idea of property; the focus on craft skill, pride and organisation and its transformation into a type of property with industrialisation; the notion of 'internal contradictions' and the distinction between systems that resist and those that favour historical transformation; the theory that a breakdown in one form of society opens not one but *alternative* paths in subsequent history; and indeed, the cautionary thought that, contrary to crude historicist interpretations of Marx, 'there is nothing in Marx to authorise us to look for some "general law" of development that might explain its tendency to develop into capitalism' (p. 43). One grand idea emerges from Hobsbawm's version of Marx: to draw a distinction between the modes of production made up of specific relations observed in one context and the theorisation of the abstract 'forms' of such relations, that can exist in a variety of periods or socio-economic settings (p. 59). This last idea, not necessarily derived only from Hobsbawm (see below), was one of the hallmarks of neo-Marxism.

Neo-Marxism

The neo-Marxist anthropologists derived certain key ideas from Marx's writings[2] and used them as a hermeneutic to analyse (or critically reanalyse) ethnography in various parts of the contemporary world. These key concepts include:[3]

- 'Forces of production' (also known as the 'base') – the materials, technologies and energies, including raw human labour, that go into production of the means of life. The 'infrastructure' is the organised material basis of production, but does not include labour.
- 'Relations of production' – the social organisation whereby the forces, and specifically labour, are systematised via rights, property, etc.
- 'Mode of production' – the systematised unification of the forces and the relations, a totality that has a historically specific ideology (the latter, including law and religion, is called the 'superstructure').
- 'Use value' – the practical value that items have in daily life and consumption.
- 'Exchange value' – the value acquired when objects are exchanged for one another; the quantity of other commodities an object is worth; in complex societies, exchange value is expressed in money/price.
- 'Labour theory of value' – the idea that true value of a product is created by the labour (along with the materials) that go into its making. The capitalist does not reward the labourer with the full value of the work and thus is able to extract a profit ('surplus value').

A word of caution is in order before I discuss how these ideas were put to work to discuss contemporary problems. Most of the impetus for neo-Marxism in anthropology came to Britain from Paris. It is important to realise that by the mid-1960s, although the Communist Party was strong in France and certain philosophers were political supporters of the USSR and/or Maoist China, the anthropological elaboration of Marxism was divorced from the state ideologies employed in those countries. In other words, you could be a 'Marxist anthropologist' without reference to the ways – often turbulent, destructive and hidebound – in which those Communist Parties were trying to put versions of the very same ideas into practice. Neo-Marxism was not just academic theory, however, since it arose as a kind of moral imperative towards the injustices of the world, and specifically as a reaction to the very school from which many of its practitioners emerged: Lévi-Straussian structuralism (see chapter 2). By moving away from the relatively contained fields (mythology, ritual, aesthetics) in which structuralism had seemed to settle, the neo-Marxists aimed to address questions of political economy and the historical progress of diverse social forms. Nevertheless, in doing so, the arsenal of ideas the anthropologists used with which to think about change was in many cases refracted through their prior use by Lévi-Strauss (Nugent 2007). And thrusting into the convoluted Parisian anthropological scene was the critique proposed by the philosopher Louis Althusser, author of the influential *Reading Marx* (1965, English translation 1970). Althusser argued that vast acres of Marx's writings can be left aside (those afflicted by idealism, humanism, evolutionism and a teleology of history) in favour of 'symptomatic reading' of the later work. This reading would reveal theory that Marx himself was hardly aware of and able to express only obliquely, and thus a space would be freed for 'theoretical practice' – innovative interpretations that were able to do service in the realm of thought, released from their moorings in either the earlier writings or the empirical world.

Many French anthropologists grasped this freedom to play with what we might call 'epi-Marxist ideas', but in their case, they were interested in their application in real situations. Translations into English were followed by publication of textbook-like edited volumes of neo-Marxist theory and debate (e.g. Hindess and Hirst 1975; Clammer 1978; Seddon 1978). The following is a guide to some of the concepts they employed, these being formulations with a neo-Marxist twist and reflecting new concerns. Hence they are at more of a remove from Marx's own writings than the list of concepts given earlier.

- 'Dialectics'– no longer seen abstractly as the productive clash of conceptual 'opposites', dialectics was turned (or returned) to explain the conflict within a 'mode of production' between the 'forces' and the constricting 'relations', resulting in the emergence of a new mode. 'Dialectics' might refer to class conflict in the hands of some practitioners, yet many took pains not to reduce it to 'dialectical materialism' (the formula beloved of the Soviets), that presumes the decisive role of economics in these struggles, rather than politics, law or ideology.
- 'Praxis' – not just the process by which an idea or skill is enacted or realised, in neo-Marxism, the word tends to be used to mean action that shapes and changes social relations.
- 'Reproduction' – not only reproduction of biological life, but its incorporation into the socio-political reproduction of whole forms of domination.
- 'Social formation' – an identifiable socio-economic whole in which more than one mode of production coexists.
- 'Articulation' – the way in which the interests of one social group (or mode of production) connects with those of a different other group or mode, most often used to analyse the relation between capitalism and some non-capitalist mode of production; more abstractly used to refer to the form taken by the production of unity out of different elements.
- 'In the last instance' – the idea, prominent in Althusser but not universally followed by the anthropologists, that economic practice determines the relations between other spheres, thus being the underlying causal element in determining that other aspect of a given society (the political, religious, etc.) that dominates the society as a whole.

Modes of production

'Mode of production' was a key idea and provides a privileged way into understanding other terms of the debate. In the hands of neo-Marxist anthropologists, 'mode of production' was a radical idea. It offered a way of thinking about political–economic organisation that was not tied to the conventional quasi-evolutionary and/or functionalist schemes (e.g. 'tribes', 'chieftains', 'states'). Above all, it rejected the implicitly European perspective on the linear progress of history – even though that was clearly present in Marx's own writings, such as the *Grundrisse*. Thus, the 'slave-owning mode of production' need not be thought of as a stage between 'primitive communal' and 'feudal' society and could be investigated anywhere, at any time. This freedom to rethink was taken up by Emmanuel Terray (1969; 1974) in his analysis of pre-colonial West African social formations. It was already recognised that many of these societies combined farmers organising their labour communally and warrior aristocracies whose wealth was based on their control of long-distance trade. Terray argued that this picture underestimated the role of slaves: the captives in Gyaman and Ashanti who produced the surplus from which the aristocracy's means of domination were obtained. Long-distance trade enabled the aristocracy to 'cash in' this surplus. That was done through their exclusive acquisition of exotic prestige goods (paid for by products of slave labour) and their control of the distribution of such high value objects. The social formation as a whole was therefore constituted by a conjunction ('articulation') of a kin-based mode of production with a mode of production based on exploitation of slave labour. This analysis, it was argued, was both more systematic and more inclusive than earlier accounts of pre-colonial West Africa.

More influential than Terray, at least in British anthropology, was Claude Meillassoux's *Maidens, Meal and Money* (1981 [1975]), as he took on the structural-functionalist heartland of West African lineage society. Re-imagined, this became the notion of the agricultural self-sustaining community, an idea based on detailed fieldwork among the Gouro of the Ivory Coast. 'La communité d'auto-subsistence', also known as the 'domestic mode of production', was based on equal access to the means of production, household self-sufficiency and forms of circulation tied to seniority and prestige. Exogamous marriage was essential to its functioning, because the circulation of women was the necessary means of reproducing the conditions under which production could take place. The other fundamental relation was the seniority of elders over juniors, which gave them control of the circulation of prestige goods that counterbalanced the movement of women. The logic of kinship terminology and genealogy was not an independent moral sphere (as in Fortes 1949) but the ideological effect of this way of organising domestic labour and the role of elders in acquiring and distributing valuable goods. It is not difficult to see the influence of Lévi-Strauss's *Elementary Structures of Kinship* in this formulation. Yet Meillassoux's neo-Marxist approach brought about a conceptual leap: most notably in the idea of 'reproduction'. The 'maidens' in the title of his book did not just happen to exist. They (and other kinds of people) had to be produced socially, along with grains and money. The anthropological task was to discover and specify the circuits of reproduction whereby the system as a whole was reproduced. Note, however, that in Meillassoux, the notion of reproduction does not imply stasis; tensions in the functioning of the lineage system, for example, led to the latter's division into classes and the emergence of state-like structures. Even more radical was his argument that there is a logical connection between the kinship structures that control reproduction and the forms of exploitation found in capitalism.

The work of Maurice Godelier struck even more widely than that of Meillassoux, as he was interested not so much in the characterisation of a particular kind of society – in his case, the Baruya of Papua New Guinea – as in how anthropology could help answer wide theoretical questions, such as 'Is there a hidden logic to economic systems?', 'How do material realities, natural and man-made, impact on human thought?' and 'What are the conditions of our understanding of social reality?' In the 1960s and 1970s, Marxism, *very* broadly construed, gave Godelier a way to provide some answers. In later decades, as his work increasingly addressed gender inequality and sexuality, he abandoned Marxism and he now calls himself an ex-Marxist. Here, I briefly describe his earlier writing, especially his seminal collections of essays, *Perspectives in Marxist Anthropology* (1977 [1973]. This book makes central the hypothesis of the determinant role of the social relations of production. Thinking about this idea in relation to non-capitalist societies led Godelier to take up and shake around the conventional Marxist understanding that the base/infrastructure has a determining role in respect of the superstructure.

The question anthropology posed to Marxism, Godelier observed, was: what is the role of the economy in societies where kinship and not economics is the dominant organising principle of social organisation? The same issue, more broadly construed, plagued Marxist analysis of all pre-capitalist societies. As Kahn observes in his review of Godelier's book (1978: 486), several 20th-century authors, including Godelier, used the following quotation from Marx as a way of framing the problem:

> [T]hat the mode of production determines the character of the social, political, and intellectual life generally, all this is very true for our own times, in which material interests preponderate, but not for the middle ages, in which it is Catholicism, nor for Athens and Rome, where politics reign supreme . . . This much, however, is clear, that the middle ages could not live on Catholicism, nor the ancient world on politics. On the contrary, it is the mode in which they gained a livelihood that explains why here politics, and there Catholicism, played the chief part.
>
> (Marx 1887 [1867]: 82n)

Godelier's answer in the case of kinship-dominated societies like the Baruya is that kinship functions as relations of production, political relations and as ideology. In other words, it acts as both superstructure and infrastructure (see the first list of Marxist terms above). The idea that kinship organises the gaining of a livelihood did not, however, invalidate the labour theory of value. One of Godelier's most original chapters in *Perspectives* was his discussion of notions of value not as an abstract idea but through ethnography. The salt the Baruya produced, having an initial value determined by the work households put into acquiring, refining and transporting it, acquired a different 'exchange value' when traded externally for other objects, and due to inter-tribal demand also became a medium of exchange as 'salt money'. Needless to say, these values were manipulated by the elders, the bogey figures in so many analyses of domestic modes.

Godelier's reinterpretation of the Asiatic mode of production (AMP) resonated in a different way, because the debate occurred largely outside the realms of anthropology. In the 1960s, any mention of the AMP had become politically taboo in the world's dominant Communist Parties. Consequently, to revive it was seen as an act of aggression in the tense confrontation of ideologies of the Cold War. Marx himself, thinking it seems mainly of India, had described the AMP as a stagnant backwater in social evolution. Its combination of closed self-sustaining communal production, absence of landed property and despotic government was resistant to evolution. The various rulers, who came and went like storm clouds in the sky, took tribute for projects that glorified the unity of a given social formation (extravagant ceremonies, irrigation systems, pyramids, wars), but they did not use the surplus extracted in such a way that would change or promote evolution of social relations. Essentially, the communal units could carry on perfectly well on their own and were left to do so (Hobsbawm 1964: 33–34). By the 1950s, the horrifying thought that this stagnant system, commonly associated with 'Oriental Despotism', might apply to the Soviet Union made the AMP unmentionable in Russia. Godelier retrieved the AMP and gave it life by making the issue a theoretical one: systems that seem to stay the same for centuries need as much explanation as ones that change, he wrote – so what in fact are the factors that actively maintain stasis? Again, he used specific ethnography and history to address this question, looking in this case at a classic site of the AMP, nomadic empires of Inner Asia. His conclusion, that in the long run they were not so stagnant after all and evolved in various different ways, did little to endear him to the Soviet ideologues. In their view they had rescued the nomads from backwardness, whether characterised as immemorial torpor or not, and this very fact demonstrated that their system could have nothing to do with the AMP.

The neo-Marxist historian Perry Anderson dismissed the AMP as an unworkable idea in 1974. Nevertheless, it lives on to the present as a recurrent framework for discussion of a host of themes, such as hierarchy, difference, caste, colonialism, the state and Eurocentrism. It is germane to this chapter that debates about the AMP are especially lively in India (Jal 2014) because this underlines that such Marxist categories are not merely academic: they are ways in which people understand their own history – and the insulting history imputed to them by others. Nowhere is this point more apt than in the case of the peasant mode of production (PMP). Study of 'peasants' took off in an engulfing wave in the 1970s, sweeping though anthropology, sociology, agrarian studies, history, geography and social theory. Yet, even at the time, to me the title of the *Journal of Peasant Studies*, founded in the period, had a funny ring to it – and this must have been because after fieldwork in Soviet collective farms I had met people categorised legally as 'peasants' who knew too much history to identify themselves in that way.[4]

The first formulation in the 1920s of what came to be known as the peasant mode of production was as much political dynamite as the AMP was to become later. Its author, the agronomist Alexander Chayanov, was arrested in 1930 and shot in 1937. His crime was that his theory went against the writings of Lenin. Chayanov argued on the basis of detailed empirical studies that the PMP rested on the balance between labour and consumption in the peasant household. The higher the ratio of dependents to workers in the household, the more labour the workers had to give in order to feed them. This situation operated as a cycle, in which the birth of more children and survival of aged dependents would be succeeded by periods with many able-bodied adults or the formation of new small households. People with many dependents would expand their farms and work as hard as was necessary to supply the household, but when the members were satisfied they would not perpetuate the drudgery in order to produce a surplus. Because the mode was divorced from market pricing and without precise accounting, the mode could be described as 'irrational', that is, 'self-exploitation' consisting of long hours despite declining rates of return (see the discussion in Donham 1981). Lenin, on the other hand, while admitting that independent Chayanov-type peasants existed in Russia, argued that they were a fast-disappearing minority. The rural scene should rather be understood as a battlefield in which rapacious proto-capitalist landlords were creating a mass agrarian proletariat, the latter being ripe for revolution. The publication of Chayanov's *The Theory of Peasant Economy* in English in 1966 coincided with Eric Wolf's *Peasants* (1966). Suddenly, the anthropological world woke up to the existence of previously disregarded swathes of boringly non-exotic farmers. The issue of 'agrarian transition', the process referred to by Lenin, suddenly became interesting. The PMP, if only as something to argue against, ignited passionate debates about the various instances of family farming around the world, questioning whether or not they were obstacles to 'development', were or not a potential revolutionary force and the ways in which they did, or did not, engage with political movements, markets, entrepreneurialism and capitalism.

Neo-Marxist critiques: For and against

Much of the impact of neo-Marxism lay in its 'critique' (itself a newly modish word in the vocabulary) of other positions. Meanwhile, Marxism in general and neo-Marxism in particular came in for some heavy counter blows. As an example of neo-Marxist positive critique let me cite Jonathan Friedman's reworking of Edmund Leach's *Political Systems of Highland Burma*. Friedman argued that the oscillation described by Leach between egalitarian and ranked social forms was only 'apparent'; if the Kachin are considered in terms of a larger theoretical model, this superficial oscillation was only part of a multilinear development generated by a particular structure of social reproduction, one that tended in a longer history towards the emergence of a centralised hierarchy (1975: 161–163). Using the full panoply of the Marxist apparatus, Friedman attempted to show that a number of necessary technical activities (involving a techno-ecology of field sizes, distances, population density, forest regeneration, etc.) are organised socially, rather than the social organisation being determined by these activities. These 'relations of production' were manifest in a unified structure that was both vertical (community, chiefs, spirits) and horizontal (alliance and exchange). The contradictions in the system derived from increased demand for surplus by the chiefs as against the constraints of technology and difficulty of territorial expansion, resulting in the egalitarian revolts mentioned by Leach. However, the changes in patterns of dominance could only properly be explained by taking into account the whole social formation and its transformations over the long term. Friedman's (and others') well-argued critiques of *Political Systems* did not destroy the significance of the book, but on the contrary added to the enduring interest in the ambiguities of Leach's 'structural oscillation', that continues to raise new empirical and theoretical questions (Robinne and Sadan 2007).

Marshall Sahlins occupies a strangely interstitial place in the battle of critiques. He was in Paris during 1968 and imbibed much French structuralism and Marxism. Yet his own versions of 'generalised exchange' and 'the domestic mode of production' (1972) differ markedly from the respective theories of Lévi-Strauss and Meillassoux. Sahlins does not argue against the French theories but simply proposes his own eclectic ideas. The domestic mode of production (DMP) in his hands consists of self-sustaining household production characterised by limited wants, with no capitalist-style market but exchange between the producing units. Seemingly inoffensive, the idea was in fact provocative. Meillassoux weighed in with an attack: Sahlins's DMP is too unspecific, never indicates whether the households are hunter-gatherers, fishermen, pastoralists or farmers, and does not indicate the type of society or historical period (1981 [1975]: 6–7). Donald Donham (1981) argued that despite the 'mode of production' tag, Sahlins was inspired not by Marxism but by Chayanov, and furthermore that the work of both writers was, in certain crucial respects, consistent with neo-classical economics. Both worked from the neo-classical basis of analysis – individual decision-making under structural constraints. Where Sahlins differed from Chayanov was that he attributed the domestic producer's lack of interest in accumulation/ profit to culture, whereas Chayanov never wrote that peasants do not want more and better goods, only that they could not obtain them since they had no other resource than their own labour (Donham 1981: 529). As for exchange between households, Donham argued that Sahlins' DMP failed to specify the overall political–economic structure in that it happened and did not take into account that debts between households could involve labour and not just goods (p. 531). In another comment on Sahlins' DMP, I suggested (Humphrey 1998) that domestic agricultural production would require special conditions for households to be 'autonomous' and 'centrifugal' as Sahlins assumes, and that the existence of communal activities and nested hierarchies of rights requires the model to be rethought.

Sahlins gave as good as he got. His book, *Culture and Practical Reason* (1976), starts out with a diatribe against the economic determinism of Peter Worsley's extraordinarily detailed Marx-inspired critique of Fortes' work on the Tallensi (1956).[5] Inveighing against 'praxis theory' and the determining role of practice, Sahlins insists that understanding of economic life has to start with culture. But 'culture' for Sahlins is not a matter of the symbolic/ intellectual versus the material. Culture is understood comprehensively as *including* social relations and indeed the 'rational-cultural logics of the concrete' (2000: 18). The language of superstructure and economic base was not helpful, since the two seemed to be tangled together. Alluding to the problem that the superstructure might have to include the economic base, Sahlins mocked the neo-Marxists:

> [S]omething like that possibility was afflicting Marxist theory also, which set it to producing epicycles of reciprocal influences of ideology on the economy or other such Ptolemaic complications of the original. (This was the period of 'the discovery of the true Marx', what Marx really said, or else Marx Himself, 'Himself' being the one whose theory actually covered all the apparent exceptions and was known solely to the author of the commentary demonstrating this, as opposed to the others who were acquainted with Marx, but not Himself).
>
> (p. 17)

If both matter and form are realised as structures through the mediation of cultural schemes, Sahlins then asks, what is left of the 'primacy of infrastructures'? He came away from all this with the idea that 'economic activity or the all-determining praxis was a particular functional expression of a more general *cultural* logic' (p. 19). Praxis was not a direct experience of the world through the immediacy of work; it was always mediated.[6]

The continued relevance of Marxist/neo-Marxist anthropology

In Western academe today, almost no anthropologists use the categories of the 1960s and 1970s, yet it would be deeply mistaken to say that neo-Marxism is irrelevant to contemporary practice. Some of its proponents are still writing (Sahlins, Godelier) and it is clear that the earlier 'Marxist' layers in their thinking were essential springboards for their and others' theoretical development. And should we regard the substantive studies done in neo-Marxist mode as superseded, in the way that a proposition in physics is discarded in favour of one that explains the same matter better? Anthropology is, or should be, cumulative, but in a different way. This is why the brilliant ethnography-based studies done in the past continue to inspire new research despite the fact that they have almost always been subjected to numerous critiques in the meantime (as shown by the success of the journal *HAU*, that is devoted to this idea). This is also true of the theoretical propositions. In fact, denounced as 'myths' (Firth 1972: 192) and trashed a thousand times they may have been, but the stalwart 'modes of production' of the *Grundrisse* are still ideas that can be turned inside out and put to new uses – viz. David Graeber's essay (2006) on why the capitalist mode of production is a transformation of the slave-owning one.

A way of understanding the relevance of Marxism is to perceive its influence in genealogies of thought that branch out and end up in quite different places. One example of this will have to suffice: the trajectory of Terence Turner – though similar accounts could be given of at least one prominent track of feminist anthropology[7] and equally of the trail of Marxist geography from Lefebvre to David Harvey. To work backwards: Turner is well known now as having been a politically engaged anthropologist and civil rights activist, renowned for his writing and films on the resistance of the Kayapo people of the Amazon to encroachment on their way of life. Yet his theorising of the ways in which Kayapo fought their battles was made possible by the insights of feminist and aesthetic anthropology; these in turn were enabled by the underlying understanding that producing food, shelter, fighting techniques or ornamentation are all elements of larger productive processes that fashion social persons and the relations between them. Turner's brilliant essay on the Kayapo 'social skin' (2012 [1980]) discusses body paint, the meanings of colours, symbolic marks, penis-sheaths and so on as a social filter that manifests the conceptual structures by which socially adult men and women are made evident to themselves and others. These marks constitute values for people – a conclusion that goes back to Turner's earliest as well as later research (2008) concerning the question of value as such, and how that can be theorised for a non-capitalist society (see discussion in Graeber 2006: 71–73). Turner's work has been remarkably consistent and yet it demonstrates how a Marxist approach can lead to insights in very diverse fields, from aesthetic matters to political confrontation.

Still, perhaps it is Marxism's revolutionary potential and its ability for self-transformation that make it an ever-lively source for opponents of neoliberalism. Anthropologists' objections to Marxism were outlined long ago by Firth (1972: 206): its 'outworn evolutionary schemes, economic determinism, apocalyptic vision and demand for total intellectual commitment'. But it was exactly these shibboleths that were overturned by neo-Marxist ethnographic investigations and consequent rethinking in the 1960s and 1970s. In later decades, Marxism-influenced thinking in anthropology has been open to ('on the side of' or 'part of') new theory – emerging from philosophy, feminism, psychoanalysis and radical geography – that expands ways of understanding humanity and above all ameliorating injustice in the writer's own society. In this work, Marxism may be only a deep – and critically examined – stratum. But it is there.

Notes

1 http://openanthcoop.ning.com/group/theoryinanthropology/forum/topics/from-marxist-anthropology-to. Consulted August 2016. Hart was correct only in respect of a certain zeitgeist. The important meeting of the American Anthropological Association on Marxist approaches was

held in 1984, and influential books such as Maurice Bloch's *Marxism and Anthropology: The History of a Relationship* (1983) and debates continued into the 1980s.

2 These ideas are found and elaborated in Marx, but they have a long genealogy and many earlier meanings. Marx's thought derived much from Hegel (1770–1832), Morgan (1818–1881) and Engels (1820–1895). A clear summary of Marxist concepts is given in Firth (1972).

3 For a vivid outline of these ideas, phrased as a polemic against Marvin Harris's 'cultural materialism', see Friedman (1974).

4 The term 'peasant' has a heavy political history, that was suffered and understood in the societies that were classified in these terms in the 20th century. In China and Russia, such implicit judgments give an edge to historical analysis and self-descriptions still today. See the extended discussion in Leonard and Kaneff (2002).

5 Sahlins was somewhat unfair to Worsley, whose subtle analysis is directed primarily against Fortes's notion that Tallensi kinship is irreducibly its own moral system. Worsley argues that this is inadequate: a morality that enjoins harmony is an essential part of an encompassing social structure evolved under specific historical and economic conditions (1956: 63).

6 Sahlins' *Culture and Practical Reason* was strongly criticised by Roger Keesing (1980), on the grounds that it reifies 'culture' as an ill-defined monolith, thus creating illusions of which Sahlins seems to be unaware.

7 The trajectory here would start with the impetus given by the peasant mode of production, via the querying of the 'natural' character of the household, to a new research agenda concerned with women and patriarchy (see chapter 12).

References

Anderson, Perry 1974. *Passages from Antiquity to Feudalism*. New York, NY: Verso.

Clammer, John 1978. *The New Economic Anthropology*. London: Macmillan.

Donham, Donald 1981. Beyond the domestic mode of production. *Man* NS 16 (4): 515–541.

Firth, Raymond 1972. The sceptical anthropologist? Social anthropology and Marxist views on society, *Proceedings of the British Academy*, www.britac.ac.uk/pubs/proc/files/58p177.pdf.

Fortes, Meyer 1949. *The Web of Kinship Among the Tallensi*. London: Oxford University Press.

Friedman, Jonathan 1974. Marxism, structuralism and vulgar materialism. *Man* NS 9 (3): 444–469.

——— 1984 [1975]. Tribes, states and transformations. In M. Bloch (ed.), *Marxist Analyses and Social Anthropology*, pp. 161–202. London: Routledge.

Geertz, Clifford 2002. An inconstant profession: The anthropological life in interesting times. *Annual Reviews in Anthropology* 31: 1–19.

Godelier, Maurice 1977 [France 1973]. *Perspectives in Marxist Anthropology*, translated by Robert Brain. Cambridge: Cambridge University Press.

Graeber, David 2006. Turning modes of production inside out; Or why capitalism is a transformation of slavery. *Critique of Anthropology* 26 (1): 61–85.

Hindess, Barry and Hirst, Paul 1975. *Pre-Capitalist Modes of Production*. London: Routledge.

Hobsbawm, Eric 1964. *Pre-Capitalist Economic Formations*. London: Lawrence & Wishart.

Humphrey, Caroline 1998. The domestic mode of production in post-Soviet Siberia? *Anthropology Today* 14 (3): 2–7.

Jal, Murzban 2014. Asiatic mode of production, caste and the Indian left. *Economic and Political Weekly* XLIX (19): 41–49.

Kahn, Joel 1978. Perspectives in Marxist anthropology (review). *Journal of Peasant Studies* 5 (4): 485–496.

Keesing, Roger 1980. Review of Sahlins' *Culture and Practical Reason*. *American Anthropologist* 82 (1): 130–131.

Leonard, Pamela and Kaneff, Deema (eds.) 2002. *Post-Socialist Peasant? Rural and Urban Construction of Identity in Eastern Europe, East Asia and the Former Soviet Union*. New York, NY: Palgrave.

Marx, K 1887 [1867] *Capital*. Volume 1, translated by Samuel Moore and Edward Aveling, edited by Frederick Engels. Progress Publishers, Moscow, online, www.marxists.org/archive/marx/works/download/pdf/Capital-Volume-I.pdf.

Meillassoux, Claude 1981 [1975]. *Maidens, Meal and Money: Capitalism and the Domestic Economy*. Cambridge: Cambridge University Press.
Nugent, Stephen 2007. Some reflections on anthropological structural Marxism. *JRAI* 13 (2): 419–431.
Robinne, Francois and Sadan, Mandy(eds.) 2007. *Social Dynamics in the Highlands of Southeast Asia*. Leiden and Boston, MA: Brill.
Sahlins, Marshall 1972. *Stone Age Economics*. New York, NY: de Gruyter.
——— 1976. *Culture and Practical Reason*. Chicago, IL: University of Chicago Press.
——— 2000 *Culture in Practice: Selected Essays*. New York, NY: Zone Books.
Seddon, David, ed. 1978. *Relations of Production*. London: Frank Cass.
Terray, Emmanuel 1969. *Le Marxisme devant les sociétés 'primitives'*. Paris: Maspéro.
——— 1974. Long-distance trade and the formation of the state: The case of the Abron kingdom of Gyaman. *Economy and Society* 3 (3): 315–345.
Turner, Terence 2008. Marxian value theory, http://ant.sagepub.com/content/8/1/43.short.
——— 2012 [1980]. The social skin. *HAU, Journal of Ethnographic Theory* 2 (2): 486–504.
Wolf, Eric 1966. *Peasants*. New York, NY: Prentice Hall.
Worsley, Peter 1956. The kinship system of the Tallensi: A re-evaluation. *JRAI* 86 (1): 37–75.

Chapter 4

From transactionalism to practice theory

David Sneath

In the late 1970s, Anglophone anthropology saw a growing interest in theoretical stances that emerged in reaction to structuralist social thought; termed 'practice theory' by Ortner (1984: 144). The most influential of these theorists was Pierre Bourdieu, whose *Outline of a Theory of Practice* was published in English translation in 1977. Fredrik Barth's transactionalist anthropology was an earlier alternative to structuralism that focused on actors and their choices. In each case, these 'actor-oriented' perspectives emerged in reaction to, and in conversation with, theoretical paradigms that emphasized notions of 'structure' in explaining social life.

Problems with structure

Structural-functionalism inherited Durkheimian conceptions of social structure as patterns of social relations that somehow reproduce themselves and/or the wider social order (see chapter 1). Sociology was, for Durkheim, the science of 'institutions' – the beliefs and modes of behaviour instituted by society and reproduced by individuals, be that the French language or shaking hands. The proper subject of sociology was not the individual, who was better studied by psychology and other disciplines, but 'social facts', that is, 'manners of acting or thinking and feeling external to the individual, that are invested with a coercive power by virtue of which they exercise control over him' (Durkheim 1982 [1895]: 52). So in his classic study *Suicide*, for example, Durkheim (1897) accepts the uniqueness of individual cases, but seeks to identify broader 'structural' forces that increase the likelihood of someone taking their own life, such as the breakdown of the social and moral bonds he termed *anomie*. Orthodox Marxism also presented a structural account of society, focusing on class and enduring structures of exploitation.

By contrast, Max Weber conceived of sociology as the interpretive understanding of social action. Although concerned with structure, Weber advocated a *verstehen* (interpretive) sociology that included the subjective meanings that individuals attributed to social action. In sociology, the interest in individual ascription of meaning was developed by the School of Symbolic Interactionism, popularized by Herbert Blumer and inspired by the work of George Mead (1863–1931). Rather than studying statistical data to detect abstract 'social facts', or assuming the stability and uniformity of 'collective representations' and their social meanings, the symbolic interactionists were concerned with the meanings that individuals attach to things and happenings. Furthermore, these meanings are not stable or essential, but arise from processes of social interaction and personal interpretation that are fluid and changeable (Blumer 1969: 86–89). This interactionist perspective shared much common ground with the phenomenologically informed school of ethnomethodology founded by Harold Garfinkel (1967) that examined the ways in which people actively made sense of their experiences to produce a shared social order.

In social anthropology, Raymond Firth gently critiqued functionalist notions of social structure that reflected assumptions as to the relative importance of different sets of social relations, what they accomplish and to what ends (1951: 35). Firth distinguished 'social structure' – enduring sets of relations – and 'social organization', which he defined as 'the systematic ordering of social relations by acts of choice and decision' (1951: 40). Where accounts of social structure focused upon formal institutions, rules and norms, descriptions of social organization refer to patterns of action 'in practice'. For example, in a given society virilocal residence after marriage might be a dominant norm or 'custom', but for all sorts of reasons the majority of households might not follow the rule. The actual pattern of residence, then, would count as social organization.

Whereas Malinowski had been critical of universalist economic claims, Firth saw potential in the wider application of economically informed notions of rational choice: 'Most social relations have an economic aspect. The exercise of choice in social situations involves economy of resources in time and in energy' (p. 130). In Firth's work, we see several of the key elements later developed in transactionalism – that individual acts of choice can be seen as a mechanism that creates wider social patterns; that such choices take place within a socially specific set of values and are conditioned by them;[1] that many choices can be seen in 'economic' terms as reflecting an interested rationality. As we shall see, in this last point, and in his critique of structuralism as preoccupied with norms and rules, Firth also anticipated points Bourdieu makes in proposing his theory of practice.

The transactionalism of Fredrik Barth

While Firth was content for his notion of choice-driven 'social organization' to complement structural analysis, Fredrik Barth developed a more axiomatic alternative to structural-functionalism. Invited by Firth to give a series of lectures, Barth published these in 1966 as *Models of Social Organization*, laying out his theoretical perspective in formal terms, and marking what Kapferer (1976: 2) described as a 'paradigm shift' in British social anthropology.

Rather than assuming general structural forms, Barth argued, anthropologists should develop what he called 'generative models' (1966: v), an account of the particular mechanisms of interaction that produce the wider social patterns we observe. This was to be a description of how society is continually produced 'bottom-up' as it were, by individual actions, rather than an account of 'social structure' as an abstract plan.

> The structural-functional model seeks to generalize the overt form of a society and show its coherence with people's own cultural representations. A generative model tries to identify sets of regular events, processes, that lead toward the emergence of such an observed form in a local or regional social system.
>
> (Barth 2007: 8)

For Barth, society was a process, rather than an object, and the anthropologist's task was to describe regularities in social behaviour by attending to the concrete happenings that produced the regularities of social life.

Many of the elements of this approach are found in Barth's earlier work. His 1959 work, *Political Leadership among Swat Pathans*, presented an analysis of political organization in the Swat valley, Pakistan, as an outcome of the choices and strategies of the actors involved, rather than a structure based on rules and norms.

Barth on Swat politics

At the time of Barth's fieldwork in 1954, Swat was a former princely state that had retained considerable autonomy within the newly independent Pakistan. The 'Pathans', or more properly Pakhtuns, were a landowning strata amongst a series of ranked hereditary groups that Barth describes as 'castes' because they were largely endogamous (1959: 10). The Pakhtun 'landed gentry'[2] constituted something between 10 and 20 percent of the population in different regions (pp. 30, 112; Asad 1972: 75), but Barth considered the remaining 'four fifths' of the Swat population to be encompassed in the political system he described, because members of the lower orders were all 'allocated as political clients to some Pakhtun' (1959: 30). The region was divided into districts, villages and sub-village wards that were usually governed by a Pakhtun landowner with noble title (*Nawab*, *khan* or *malak*), all of which Barth terms 'chiefs' (p. 72).

Barth's analysis should be seen in the context of the influential Africanist structural-functionalist ethnography of the time. Evans-Pritchard's celebrated account of the Nuer (see chapter 1) seemed to show a society made up of corporate descent groups and an acephalous political system that was a product of a segmentary kinship structure. In Swat, Barth found neither corporate descent groups nor a political system based on an extended kinship structure (pp. 3, 22), despite a lack of centralized authority in relatively recent memory. Instead, he found an apparently fluid series of political factions that nevertheless displayed regularities. The 'primary groups' in Swat politics were not clans or lineages, he argued, but 'chiefs' or Muslim 'saints' and their followings of Pakhtun landowners with their entourages.

Barth stressed the voluntaristic nature of political relations (p. 3). The local khan or 'chief' was 'confirmed in his position by all landowners of the ward' (p. 11) and the size of his following reflected his relative standing and popularity. Leading landowners maintained 'men's houses' and engaged in lavish hospitality, that Barth (p. 12) explained in terms of a competition for regard analogous to the 'potlatch institutions' that he saw as oriented towards political rather than economic gain. Landowners might lose tenants or attract more, depending on their popularity. Political relations were largely contractual. In the case of disputes, and conflict over land was common, local landowners were expected to mobilize their dependents to, if need be, defend their interests by force. Most of these dependents would have a contractual relationship with their landlord as tenants who farmed his land and/or lived in one of his houses (most of which were owned by the 'gentry'). In the case of the followings of Muslim 'saints', followers had a 'contractual' relationship in which they displayed loyalty in return for 'religious tutelage' (p. 3).

Despite the lack of a formal structure governing relations between 'chiefs', Barth found the numerous local factions appeared to consistently coalesce into 'two large, dispersed, internally co-ordinated alliances or blocs, as a result of the activities and choices of individual leaders' (p. 3). These Swat-wide political camps had never been instituted in any formal way, and yet everyone knew about them. How could this be explained?

Barth turned to the theories of rational choice that he found in Neumann and Morgenstern's 1944 work *Theory of Games*. Their model of a multi-person game of alliance presented a pervasive two-party coalition as a logical outcome of strategizing actors engaged in a rule-bound game. In this model, it makes sense for any given leader to align with more distant allies in disputes with a rival local faction over resources such as land. The larger the alliance, the better, as size increases the chance of success. The logical outcome is the formation of two enormous blocs. These two opposing alliances would be stable features because as one faction increased in size, the spoils available to each member shrank. This limited defections from one side to the other and prevented either faction from entirely overwhelming the other.

This, then, was Barth's alternative to accounts of social structure; rather than the behaviour of political actors being dictated by their structural position in, say, a kinship structure, Barth tried to show how political forms were produced by the choices made by the actors involved.

Modelling social process

From a theory of games, Barth moved to what he termed 'a processual model of interaction' (1966: 5). His aim was not to make a perfect description of social reality but to model it so as to be able to make statements that could be tested against observable reality. To do this, Barth proposed a set of analytical terms and principles. The most basic unit of choice was *not* the individual, Barth was later at pains to point out (1981: 85), but the *status*, the set of rights and obligations attached to a given social position. A person was, in this sense, a combination of statuses and Barth's perspective allowed statuses to be treated as *assets* (pp. 122–125) in social processes.

The next step was to construct rules as to how various statuses interact:

> One basis for such rules may be found in the transactional nature of most interpersonal relations, in the reciprocity we impose on ourselves and others. In any social relationship we are involved in a flow and counterflow of prestations, of appropriate and valued goods and services.
>
> (Barth 1966: 3)

By *transaction*, Barth meant 'those sequences of interaction which are systematically governed by reciprocity' (p. 4). This notion of reciprocity, moreover, is of a particular kind. For Barth, the parties of a transaction 'systematically try to assure that the value gained for them is greater or equal to the value lost' (p. 4). The statuses engaged in interaction could be treated as strategizing maximizers of subjective value. Barth concedes that some social relations are not of this 'transactional' form (p. 8) but nevertheless gives the impression that the important ones are.

Many of Barth's points resonated with the formalist position in the formalist–substantivist debate within economic anthropology. For those such as Herskovits (1952 [1940]), Cook (1966) and Schneider (1974), some form of individual maximizing behaviour could be treated as universal, and such 'economic' behaviour – the rational pursuit of ends given limited means – could be studied using elements of formal microeconomic theory. Of course, people's choices reflected personal preferences that differed in various cultural and historical settings. But the idea that individuals choose between alternative means to maximize desired ends was seen as a cross-culturally applicable heuristic.

Rather than microeconomic theory, however, Barth looked to the sociology of Erving Goffman (1959) and drew out the notion of *impression management*, the process by which people try to control the way others perceive them. For Goffman (p. 17), social interaction is a performance by which actors attempt to manage the impressions others have of them in ways that are consistent with their desired goals. For Barth, this provided rules by which broader social patterns could be seen to emerge. 'We may thus construct a model whereby complex and comprehensive patterns of behaviour (roles) may be generated from simpler specifications of rights (statuses), according to a set of rules (the requirements of impression management)' (1966: 3).

The other key term in Barth's analytical scheme was *prestation*, borrowed from Mauss to describe that which was notionally exchanged in transactions – that is, almost anything said or done designed to have meaning for others, from goods and services to remarks and facial expressions. He uses the term 'token' prestations where these communicative acts or objects are offered as signs of intentionality and assurance. The notion of transaction, Barth (p. 5) argues:

> gives us a logically consistent model of an observable social process. It is a model whereby one may *generate* forms according to the rules of strategy, given the parameters of value; and these forms generated by the model may then be compared to the empirical patterns which one has observed.

Transactionalism applied: Barth's herring boats

Barth (pp. 6–11) illustrates his approach with an analysis of the social dynamics of Norwegian herring boat crews. Herrings arrive off the coast of Norway in the late winter in large shoals and if a fishing boat can find a shoal the crew can net fish worth large sums of money. In the early 1960s, fishing boats searched for shoals using sonar. If herring were found, two smaller boats were launched to manoeuvre nets coordinated by a specialist called a netboss in a motorboat. Barth argues that the behaviour of the herring boat crews generally conformed to certain stereotypes, and sets out to explain their behaviour.

The formal statuses of the crew are defined in custom and contract as follows:

a The *skipper* is the captain of the ship and has, formally, almost total command over the crew. He receives two to three times the share of the profit of the fishermen who work under him.
b The *fishermen* do the skilled manual work on the ship. Each gets one share of the gross catch.
c The *netboss* controls the tricky netting operation, and receives from two to five times the share of the fishermen – sometimes more than the skipper.

The actual behaviour of crews is somewhat surprising. Skippers avoid relations of command and subordination with their crewmembers, and rather than the bridge acting as the skipper's private preserve, the crew tends to congregate there, looking at the sonar, searching the horizon with binoculars and listening to the radio. The skippers tend to be taciturn, exuding an air of deep experience and considered judgement. In contrast, however, the netboss is stereotypically extrovert and spontaneous, cracking jokes and bragging about escapades.

Barth explains the patterns of behaviour in terms of transaction. Each crew member is engaged in managing the impressions of those around them so as to appear to be increasing the chance of success in the joint enterprise. The fishermen crowded watchfully on the bridge are making token prestations indicating their willingness to work. The skipper gives an impression of competence and wisdom because this is what the crew wants if they are to make a good catch. The netboss has to be able to command the respect of the crew when the time comes, but his extrovert manner suggests leadership in a way that does not compete with the skipper's air of understated authority. This demeanour is transformed during the netting process, however, when netbosses are notorious for their furious swearing. All this 'transactional' behaviour, then, is about generating and maintaining relations of trust.

The most important decision, however, is where the ship should go, and in the end, the reputation of the skipper depends upon his knowing how to find fish. Given that the herring shoals might appear almost anywhere over a wide expanse of ocean, one might expect the fishing vessels to spread out to maximize the chances of finding one. But this is not what happens. Instead, most boats bunch together in groups and much effort is spent trying to follow other vessels. Why?

This too, Barth explains, is the result of impression management. Those skippers with the highest reputations can sign up the best crew-members, and will be trusted by them to sail where they choose. Most skippers, however, are not in this position. They need to nurture the trust of their crew, and to sail off alone and then fail to find fish risks a damaging loss of reputation. If, however, their boat joins a group who *all* fail to find herring, then the skipper has not been shown to be any less competent than the other captains in the group. And if the fish are found then the skipper's judgement is vindicated. The resulting pattern is self-confirming. Those skippers with high reputations set off looking for herring, and are then followed by less confident skippers. Being first in the group means the leading skippers have the best chances of a large catch, thus reinforcing their reputations.

Political entrepreneurship and the Sami

Barth's method emphasized the 'economic' interpretation of behaviour as reflecting actors' evaluations of the relative gains and losses in social interaction. This offered the prospect, he argued, of explaining processes of cultural change and integration through the concrete actions of individuals, rather than making claims in terms of generalized abstractions.

In economics, the role of the entrepreneur – the profit-seeking innovator – is well recognized. New products appear on the market as a result of entrepreneurial individuals who devised ways to place them there. The biggest profits are to be found where there are large disparities in values; where the entrepreneur can buy cheap and sell dear. And by establishing a new flow of goods, Barth argues, the entrepreneur initiates a process that eventually reduces this disparity. Processes of cultural integration, Barth argues, can be approached in the same way: entrepreneurs can be seen as agents of change, creating new flows of value and new patterns of preference and production.

This is not limited to economics; we can recognize political entrepreneurs who put together new combinations of value in an attempt to appeal to their constituencies. Barth describes the post-war plight of Sami ('Lappish') speakers in Finnmark, an economically impoverished and politically disorganized minority within Norway as a whole. At first, political mobilization was achieved by a politician promising to bring the welfare state to the region; but this required the representation of Finnmark as an underprivileged periphery. This image, Barth argued, threatened Sami identity, so a second political entrepreneur emerged with an electoral 'product' of ethnic pride, cultural assertion and bilingual schooling. These different 'package deals' of values (p. 19) were presented to voters throughout the

1950s by rival politicians with an interest in manipulating Sami identity for their own ends. This was a mechanism for cultural change.

Critical reflections

Barth's position provoked criticism on a number of different grounds. His Swat analysis was deconstructed by Asad (1972), who reanalyzed his ethnographic material from a loosely Marxist perspective. Barth applies a sort of 'market model' to society, Asad notes, in which maximizing individuals make free choices within certain rules. But this gives an illusion of choice in a situation that is largely governed by inheritance and descent. As only Pakhtuns may own land, the remaining four-fifths of the population are in effect forced to become tenants and/or clients of one or other landowners.

> If Swat political organisation is to be viewed as a consensual rule-structured activity, the question arises: who defines and applies the rules of the game? . . . Bath's answer is quite clear: it is Pakhtun landowners who carry out these functions . . . collectively, in relation to the rest of society, they occupy the position of a ruling authority, a sovereign class, with distinctive interests, privileges and powers . . . This means the system does not regulate itself as it were after the consent of all participants has been obtained. It is regulated by a dominant class of landowners who exploit the landless.
>
> (p. 82)

Barth describes the hereditary Pakhtun landowners as a 'dominant, cosmopolitan 'gentry', 'and notes that they received between three quarters and four fifths of the gross crop grown by their tenants – the 'parochial, subordinate population serving a succession of different lords' (Barth 1959: 10).[3] So when Barth claims that 'Most statuses and rights are usually defined by contractual agreements between persons; that is to say, they are achieved' (p. 23), Asad points out that this does not apply to the most fundamental statuses of caste and lineage. In elaborating a model of the individual choices of landowners while deliberately ignoring class, then, Barth has missed the most telling features of life in Swat.

Barth's analysis also drew fire from Ahmed, who knew Swat well because his wife was related to the ruler, and found Barth's ethnography inaccurate and ethnocentric.[4] The analysis was a sort of make-believe account of Swat politics in an ahistorical equilibrium. The 'generative model', for example, assumed that larger khans did not eliminate rival factions by driving them out, but historically such leaders frequently did so (Ahmed 1976: 28). And rather than the two blocs (*dalla*) predicted by games theory, there seem to have been four at that time (p. 144). Swat underwent continuous historical change in the 20th century, but Barth keeps the temporal setting of his analysis hazy and unspecific (p. 23). But the most serious mistake, Ahmed notes, was that Barth ignored the importance of centralized rulership and the state.

Swat had been subject to the centralizing and state-building efforts of rulers both outside and within the British Raj and Pakistani state structures. For Ahmed (p. 131), Barth had presented a 'Khan's eye-view of politics' that ignored the importance of central power structures and exaggerated the role of the leading landowners. Structural-functionalists had been critiqued for leaving the colonial state out of their accounts of African societies (Asad 1973: 18; see chapter 2), but for Ahmed (1976: 15), 'minimizing the impact of the Wali [ruler] on Swat socio-political organization as an epiphenomenon is a more serious misreading than that which functionalists are accused of in leaving out Political Agents and District Commissioners'. Barth assumed the state was imposed upon a 'tribal structure' that was distinct from central authority. This distinction, Ahmed notes, is unsustainable. 'Centralized authority in Swat is not a compromise with acephalous tribal forms but a surrogate for it with all its social and structural ramifications' (p. 26).

The vision of the acephalous tribe as the ideal anthropological subject, then, had framed Barth's study. To challenge Evans-Pritchard's model of a society without a state, Barth had to create his own, based on a reconstruction of a disappeared era in the history of Swat. But this meant ignoring centralizing power and the state, on the one hand, and minimizing the importance of the underlying caste/class relations on the other.

Scholars concerned with theories of process, transaction and interaction also critiqued Barth's position. Kapferer (1976) pointed out that Barth's modelling method assumes the cross-cultural applicability of certain notions of self-interested rationality, reciprocity and maximization. But what if these are products of particular intellectual traditions rather than universal principles? Second, by setting out formal modelling principles *a priori*, Barth exacerbates the tendency for the model to inform the material the anthropologist collects in the field. And when it comes to analysis, the axiom that social interaction can be treated as value-maximizing transaction looks worryingly circular and self-confirming. Kapferer cites Cancian on this point:

> If [the analyst] cannot see the act as maximization, he immediately assumes that his statement of norms, motives etc., and the conditions are not yet correct and seeks to 'balance the equation' so that it will work. He does not reject the idea that people will maximize, for it is the basis of his scientific strategy.
>
> (Cancian 1968: 231, cited in Kapferer 1976: 4)

Finally, Barth's approach requires the specification of relative value preferences (to show that some courses of action bring more value to an actor than others). But people may not have clearly ranked objectives, and often reinterpret their own motives in the light of changing circumstances. So 'declared motives and value scales, rather than generating their relationships, can be retroactively produced by those relationships' (Kapferer 1976: 7).

Evens (1977) critiqued Barth's methodological focus on the 'individual', that entailed a set of assumptions regarding self-interest and rationality. In this sense, 'the interactionist's "individual" is as grandly abstract as the functionalist's "group"' (Evens 1977: 585). Barth's transactionalist method failed 'to get from an essentially presocial individual to a genuine or objectivated sociocultural phenomenon' (p. 589). This style of modelling could only try and account for 'how and in what form norms get actuated, but not how and in what form they get constructed' (p. 589).[5]

On reflection, Barth's transactionalism had a good deal in common with the structural-functionalist anthropology of his era. Both were concerned with building models of something conceived of as 'tribal' society, and Barth was also engaged in a 'salvage' project – trying to reconstruct social forms thought to be already disappearing in the face of colonial and state interference.[6] Furthermore, both sorts of models treated societies as closed systems in a sort of timeless equilibrium that ignored historical process in effect, if not in theory (Asad 1972: 90; Kapferer 1976: 6; see chapters 1 and 5).

In his vision of two political blocs in perpetual equilibrium within an apparently acephalous political system, Barth almost seems to be trying to match Evans-Pritchard's model of segmentary opposition. Indeed, it is not hard to imagine how one model could be translated into the other. If among the Nuer, for example, the 'value' gained to an actor for supporting another in a dispute was greater for closer agnatic kin, then something like Evans-Pritchard's model of segmentary opposition could be made to appear a result of choice and decision-making, rather than 'rules and norms'. The 'matrix of values and statuses' (1966: 15) devised by the analyst begins to look rather structural.[7]

Although his excursion into transactionalist modelling did not result in an enduring theoretical school, Barth's contribution should not be underestimated. His approach inspired other work such as that of Bailey (1969; 2001) who approached political systems as analogous to games. And by offering a critique of structuralism and proposing an analytical alternative in grand terms, Barth stimulated wider debate. As a scholar, Barth's best-known contribution was his work on ethnicity. The concise introduction to his edited volume *Ethnic Groups and Boundaries* (1969) became a classic in its field, making the point that it is 'the ethnic boundaries that define the group, not the cultural stuff that it encloses' (p. 15). It was, ironically perhaps, one of Barth's least transactionalist texts. However, his rejection of essentialism and his interest in instrumentalism prepared the way for a new generation of critical studies of ethnicity and nationalism (Brubaker 2009: 29).

Pierre Bourdieu's theory of practice

Pierre Bourdieu is widely recognized as one of the leading social theorists of the 20th century. Rather than attempt a comprehensive treatment of his work, this chapter will concentrate upon the theoretical position he set out in his 1972 work *Outline of a Theory of Practice,* published in English translation in 1977. Bourdieu shared some ground with transactionalism. Like Barth, he argued that behaviour could be treated as, in some sense, 'economic' – reflecting evaluations of costs and benefits. He also borrowed concepts from economics for analytical purposes – not the figure of the entrepreneur, as Barth did, but the more institutional concept of 'capital'. However, Bourdieu was not methodologically individualist, and he was critically concerned with class structures and the symbolic practices of power.

In observing the social lives of others, Bourdieu argued, anthropologists are like art historians – writing about art but not actually making it themselves – and therefore 'condemned to see all practice as spectacle' (p. 1). As non-practitioners, they had to adopt modes of representation used by actors who lack a practical mastery of something; they had to make 'rules' about how things were done, deduced from what they observed. Just as an outsider who does not know their way around a place might make a map, so anthropologists ended up describing 'culture' as a sort of social map. What was needed, then, was an understanding of society as *practice* rather than representation.

Modes of theoretical knowledge

Bourdieu grounded his own approach in a dense critique of existing social theory, drawing on an occasionally daunting array of intellectual and philosophical thought. He begins with an examination of the conditions of sociological knowledge, identifying three modes of theoretical knowledge of the social world.

Phenomenological knowledge, that he identified with ethnomethodology in sociology, 'sets out to make explicit the truth of primary experience of the social world ... which, by definition, does not reflect on itself and excludes the question of the conditions of its own possibility' (p. 3). This form of experience-based knowledge cannot, then, account for the conditions that produced it, or the influences on thought that the thinker is unaware of. The personal experience of factory work, for example, does not itself include the historical processes that produced industrial capitalism.

'Objectivist knowledge', on the other hand 'constructs the objective relations (e.g. economic or linguistic) which structure practice and representations of the world ... This construction presupposes a break with primary knowledge' (p. 3) This form of knowledge makes sense of personal experience in terms of 'objective' general structures and processes that may not be experienced directly as such. This 'gap' between representation and experience prevents 'objectivist' treatments from being full accounts of social practice. For Bourdieu, this includes various structuralist schools of social theory, including Levi-Straussian structuralist anthropology and orthodox Marxism.

Having elaborated his own theory of social theory, Bourdieu proposes his theory of practice as the third mode of generating theoretical knowledge of the social world. This is to be a 'science of the dialectical relations between the objective structures to which the objectivist mode of knowledge gives access and the structured dispositions within which those structures are actualized and tend to reproduce them' (p. 3). So the structures of class and capital produce the conditions for factory work as lived experience, and the practices of factory workers in turn produce forms of capital and class.

Examples: Gift exchange, honour, and parallel-cousin marriage

Bourdieu illustrates the deficiency of structuralist analysis (which he terms 'objectivist') by taking on the classic anthropological topic of *gift exchange*. Since Mauss, anthropological analysis has tried to identify principles that explain the 'native' explanation of gift-giving. For Lévi-Strauss (1963), this was the unconscious principle of the obligation to give, to receive and to reciprocate. But, Bourdieu points out, the objectivist or structuralist explanation directly contradicts the phenomenological or native explanation. For the actors involved, if a gift *must* be reciprocated (or is 'reversible' in Bourdieu's terminology), it is not really a gift but some sort of bribe or loan. The whole point of a gift is that it is seen as generous ('irreversible') and not self-interested.

Bourdieu argues that the interval – the time delay – is the crucial factor that separates the gift from other sorts of transactions.

Actors need to have some sense of an obligation to reciprocate, but by making this explicit in the form of a general law or principle, the anthropologist is misrepresenting the practice. It is the sense of *timing* of gift and counter-gift that is essential, just as *tempo* is essential to music. Timing changes the meaning of the acts involved. This is a fine sensibility, knowing not to return a gift too soon, or to leave it too long. Furthermore, there is clearly no law that determines reciprocity; the gift may not have the desired effect, it may not be reciprocated. It is not enough for the analyst to conclude that this is just an instance of a 'failed gift'. We should recognize that actors are engaged in 'the 'art' of the *necessary improvisation*' of social life (1977: 8). Attempts to formulate rules for this improvisation are in effect retrospective, the present outcomes as 'necessary' by some general rule, whereas the outcomes were not certain. The basis of gift exchange, Bourdieu concludes, is '*institutionally organised and guaranteed misrecognition* ... intended to transmute, by the sincere fiction of a disinterested exchange, the inevitable, and inevitably interested relations imposed by kinship, neighbourhood, or work, into elective relations of reciprocity' (p. 171).

Bourdieu examines the 'sense of honour' among the Kabyle in Northern Algeria, where he did ethnographic research in the 1950s. Kabyle male honour, he argues, cannot be described properly in terms of rules and norms. Like Barth, Bourdieu makes use of the notion of 'strategy', and invokes games theory (p. 12) to explain patterns without resorting to rules. But, rather than aiming for a stripped-down model of behaviour based on game-theory logics, Bourdieu moves in the opposite direction, building a rich sense of the sensibilities and subtleties of honour. So:

> the driving force of the whole mechanism is not some abstract principle . . . still less the set of rules that can be derived from it, but the sense of honour, a disposition inculcated in the earliest years of life and constantly reinforced.' (. . . It is this) 'cultivated disposition, inscribed in the body schema and in schemes of thought, which enables each agent to engender all the practices consistent with the logic of challenge and ripost'.
>
> (pp. 14–15)

To press his case, Bourdieu critiques structuralist analysis of parallel-cousin marriage, taken as a preferential form in Arab and Berber society. The assumption that marriage patterns are the results of preferential or prescriptive marriage rule systems, he argues, can be misleading:

> To consider regularity . . . as the product of a consciously laid-down and consciously respected ruling . . . or as the product of an unconscious regulating by a mysterious cerebral and/or social mechanism, is to slip from the model of reality to the reality of the model.
>
> (p. 29)

When it comes to marriage, Bourdieu argues, 'far from obeying a norm which would designate an obligatory spouse from among the whole set of official kin, the arrangement of marriage depends directly on the state of the practical kinship relations' (p. 52).

Like Barth, Bourdieu uses the analogy of the game, and employs the metaphor of the market to explain the processes of choice and decision-making.

> The matrimonial game is similar to a card game, in which the outcome depends partly on the deal, the cards held . . . and partly on the player's skill: that is to say, firstly on the material and symbolic capital possessed by the families concerned . . . and secondly on the competence which enables the strategists to make the best use of this capital, practical mastery of the (in the widest sense) economic axiomatics being the precondition for production of the practices regarded as 'reasonable' within the group and positively sanctioned by the laws of the market in material and symbolic goods.
>
> (p. 58)

This logic may not be made explicit. Agents can employ 'officializing strategies' (p. 40) to describe their actions as conforming to an approved norm.

Rather than setting aside structural accounts of the distribution of power and wealth, as Barth tended to, Bourdieu attempted to explain how these shaped both actors and their choices, and how these practices in turn created and reproduced the wider structures. He saw the 'objective' structures of class, for example, as internalized by actors who then reproduced them in their daily lives. His theory of practice aimed to be 'an experimental science of the dialectic of the internalization of externality and the externalisation of internality, or, more simply, of incorporation and objectification' (p. 72).

Habitus and doxic knowledge

To explore 'incorporation and objectification', Bourdieu introduced a series of analytical terms. Perhaps the most important of these was *habitus*, a term used by Mauss (1935) and Elias (1978 [1939]), among others, to mean ingrained habits of the body, and learned character. For Bourdieu, habitus referred to the embodied dispositions created by processes of socialization, that could be seen to structure the socialization of succeeding generations in their turn. Hence 'the structures constitutive of a particular type of environment (e.g. the material conditions of existence characteristic of a class condition) produce *habitus*, systems of durable, transposable *dispositions*, structured structures predisposed to function as structuring structures' (1977: 72) Through habitus, then, the past is present in current practices and shapes the future. This produces enduring features of social life that are identified by structuralist/objectivist analysis, but that cannot be properly explained by it.

Habitus is also about embodiment, and Bourdieu uses the term *body hexis,* for the way in which our bodies are conditioned to move, rest, gesture and talk. Our upbringing is a 'structural apprenticeship that leads to the em-bodying of the structures of the world, that is, the appropriating by the world of a body thus enabled to appropriate the world' (p. 89). Growing up in a Kabyle house, for example, forms people in particular ways. The organisation of domestic space reflects a social order that becomes internalized. The house is 'the principal locus of the objectification of the generative schemes ... this tangible classifying system continuously inculcates and reinforces the taxonomic principles underlying all the arbitrary provisions of this culture' (p. 89).

The habitus concept does a lot of work for Bourdieu, partly because he defines it in such a way that it is bound to accomplish the processes of internalization and externalization that he proposes. It also, by definition as it were, explains regularity without rules. Habitus entails:

> principles of the generation and structuring of practices and representations that can be objectively 'regulated' and 'regular' without in any way being the product of obedience to rules ... collectively orchestrated without being the product of the orchestrating action of a conductor'.
>
> (p. 72)

As with jazz music, the parts actors play in society are in some sense spontaneous, but they conform to a certain order and a sense of what is appropriate. Habitus provides this sense, it is 'the durably installed generative principle of regulated improvisations' (p. 78).

Habitus does not dictate social practice, but it does orient and constrain it, particularly through expectations and a sense of the limits of the possible. 'Through the habitus, the structure which has produced it governs practice, not by the process of a mechanical determinism, but through the mediation of the orientations and limits it assigns to the habitus's operations of invention' (p. 95). The experience of social class, age or gender, for example, sets limits as to what is considered possible, within which agents improvise their social practices.

In exploring the limits of the thinkable, Bourdieu introduces the concepts of *doxa*, the universe of the undisputed. Within the universe of argument, or discourse, one can identify a range of positions from the most orthodox on the one hand to the most heterodox on the other. Standing outside the spectrum of the debatable, however, stands the universe of the undisputed and undiscussed; that which is taken for granted by all. This he terms *doxa*:

> Every established order tends to produce . . . the naturalization of its own arbitrariness . . . Systems of classification . . . make their specific contribution to the reproduction of the power relations of which they are a product . . . This experience we shall call *doxa* . . . The instruments of knowledge of the social world are in this case (objectively) political instruments which contribute to the reproduction of the social world . . . to the world of tradition experienced as a 'natural world' and taken for granted.
>
> (p. 164)

Bourdieu finds the Marxist notion of ideology too narrow to capture the diverse processes that legitimate the established order. In addition to explicit 'ideological' forms such as law codes, Bourdieu is concerned with the powerful implicit meanings of symbols, and with the institutional forms that stand behind knowledge claims and induce respect (p. 188). But he thinks of class struggle as a major historical process, as Marx did. Even the most radical critique of the status quo will share a certain amount of doxic knowledge with the established order. However, dominated classes have an interest in pushing back the limits of doxa and challenging that which has been taken for granted – the privileges of aristocracy, for example. In the face of critique, defenders of the status quo may have to produce explicit justifications for what had previously been accepted without question. In this way, that which had been unquestioned doxa can enter the field of opinion and argument, and become the subject of orthodox and heterodox views.

> It is only when the dominated have the material and symbolic means of rejecting the definition of the real that is imposed on them . . . i.e. when social classifications become the object and instrument of class struggle, that the arbitrary principles of the prevailing classification can appear as such and it therefore becomes necessary to undertake the work of conscious systematization and express rationalization which marks the passage from doxa to orthodoxy.
>
> (p. 169)

Forms of capital

Bourdieu takes a loosely Marxist notion of capital and broadens it for application to non-industrial and 'archaic' societies, and to immaterial and symbolic arenas more generally. Social domination is maintained through forms of capital – instruments that allow for the lasting appropriation of other agents' labour, services and deference. The form of wealth we think of as economic capital, then, is only one form of enduring social resource among many. Bourdieu uses the term *symbolic capital* to indicate forms of prestige, authority and political legitimacy that allow the 'legitimate exercise of violence' (p. 41). In societies that are not fully monetized, economic wealth may not be the most important resource. So in Kabylia, for example, the patrimony of a prominent family or lineage includes not just the 'economic capital' of its land holdings and material goods, but its clientele, kin and networks of alliances with all the attendant rights and duties of these relationships. Thus:

> [S]ymbolic capital, which in the form of the prestige and renown attached to a family and a name is readily convertible back into economic capital, is perhaps *the most valuable form of accumulation* in a society in which the severity of the climate (the major work – ploughing and harvesting – having to be done in a very short space of time) and the limited technical resources (harvesting is done with the sickle) demand collective labour.
>
> (1977: 179)

Following Polanyi, Bourdieu considers the notion of the disembedded economy, as a distinct and bounded field, to be a historical product of capitalism. This segregation has left us with a restricted conception of 'economic interest' and the assumption that rational means–ends evaluations only apply to economic activity. Like Barth, however, Bourdieu is keen to apply profit–loss logic more widely; 'contrary to naively idyllic representations of "pre-capitalist" societies (or of the "cultural sphere" of capitalist societies), practice never ceases to conform to economic calculation even when it gives every appearance of disinterestedness' (1977: 177). By treating relations of authority and prestige as assets attached to particular agents, Bourdieu extends the calculation of benefit to the sorts of behaviour dismissed by economists as irrational. If we could build a '*comprehensive balance-sheet* of symbolic profits' we could see the 'economic rationality of conduct' (p. 181).

Bourdieu also develops the notion of *cultural capital*, assets within the field of cultural production. This entails a certain type of prestige: 'academic qualifications are to cultural capital what money is to economic capital' (p. 187). But a suitcase of money is of little use without a means of spending it, and such assets require wider systems to operate:

> Just as economic wealth cannot function as capital until it is linked to an economic apparatus, so cultural competence in its various forms cannot be constituted as cultural capital until it is inserted into the objective relations between the system of economic production and the system producing the producers (which is itself constituted by the relation between the school system and the family)'.
>
> (p. 186)

Bourdieu links this to literacy and in particular to educational systems that enable the formalization and transmission of cultural competencies. So prestigious private schooling can convert the economic capital of the wealthy into the cultural capital of their children, but this is only an asset in certain social conditions. Academic qualifications can then be taken as demonstrating that the elite 'deserve' their top jobs. 'The educational system helps to provide the dominant class with what Max Weber terms 'a theodicy of its own privilege' (p. 188).

Critical reflections

Bourdieu's thought has led to countless critical engagements, both within anthropology and beyond it. Edward Said (1989: 223) accused him of ignoring colonialism and the

Algerian state in his treatment of Kabylia.[8] Said's point echoes Michael Herzfeld's critique in which he notes that Bourdieu:

> hermetically separates the values of the local community from those of the encompassing bureaucratic and religious institutions ... His thesis turns on the assumption that local societies lack permanent hierarchy ... [behind this lies] an argument that treats the values of local societies as relatively simple features surviving from a prestate era.
> (1987: 8)

In this respect, Bourdieu's approach resembles that of Barth and Evans-Pritchard. He looked to Kabylia for lessons as to how society operated in the absence of entrenched institutions of industrial capitalism and the nation-state. Since evolutionist social theory fell from favour, this aspect of Bourdieu's work appears increasingly dated.

The theoretical apparatus of *Outline* also provoked criticisms.[9] Brubaker (1985: 749–753) notes Bourdieu's bundling together of a range of different problems under his catch-all terms 'objectivism' and 'subjectivism'. The notion of *habitus* does so much work for Bourdieu that it risks becoming a sort of explanatory do-it-all. 'Doubts inevitably arise about the usefulness of any concept so vague and versatile' (p. 760). For Free the characterization of habitus as 'structured structures predisposed to function as structuring structures' is ultimately opaque; 'such formulae are fudges that obscure the questions not only of how exactly anything is caused ... but also, perhaps more importantly, of who or what specifically causes it' (1996: 400).

Furthermore, Bourdieu's concepts of 'symbolic' and 'cultural' capital' effectively reify status or esteem, so that 'a series of relations and evaluative discourses that are termed "status" by Weber are treated by Bourdieu in terms of "capital", a quasi-thing that is held rather than being immediately dependent on recognition' (1996: 402). Presenting symbolic and cultural capital as 'objective' disguises differences of opinion as to, for example, how much prestige an individual or group may have. Indeed, Bourdieu's notion of symbolic action as 'economic', governed by self-interested calculation and competition, leads to the sort of problem that Bourdieu critiqued in structuralist analysis: the actors' account still contradicts that of the analyst. Gift giving relies upon 'the sincere fiction of a disinterested exchange' (1977: 171) but the notion of 'misrecognition' seems to explain away the actors' subjective accounts as did Marx's notion of 'false consciousness' (see chapter 3). The concept of symbolic capital depends upon a notion of misrecognition that resembles mystification, and Bourdieu's theory sometimes looks like an attempt to correct, or perhaps perfect, Marx's dialectical materialism. In this treatment, agency is so constrained by the dispositions that reproduce structural inequality that it almost appears to be a vehicle for them.

Bourdieu's theory of practice left us with an enduring analytical perspective and a new lexicon for social analysis; however, in many ways, it represented a less radical and ambitious break with structuralist thought than Barth's transactionalism. Where Barth bracketed out social structure, trying to model mechanisms that explained its appearance, Bourdieu internalized it. In doing so, he directed attention to practice, process and embodiment as subjects of analytical thought, and helped usher in a new wave of social science scholarship.

Notes

1 In some passages Firth's position appears to resemble that of the symbolic interactionalists. He stresses, for example, 'sociality is vitally important. The choice, the behaviour, the values of

any one person are all conditioned by other people. [...] the actions of these other persons give sense to the conceptual and symbolic system of the individual' (Firth 1951: 124).

2 Barth (1959: 9).

3 Barth concedes that the level of exploitation was extreme, as 'non-landowners are already living barely above subsistence level, and would be below it were not a part of the taxes they pay returned to them in the form of hospitality in the men's houses' (1959: 68).

4 Ahmed points out that Barth's notion of 'saint' for example, was confused (1976: 17, 56) and that he made mistakes in translating local terms (1976: 148) (see also 1976: 9, 13, 59).

5 Barth responded to some of this criticism in a later work, rejecting the charge that he failed to acknowledge the importance of power or pay attention to extra-individual factors. Rather, he saw these factors as 'brought into play *through* social action' (Barth 1981: 88).

6 When it came to testing his models against observed reality, Barth - disappointingly perhaps – explained discrepancies (such as new patterns of behaviour among khans) not by modifying his model but by invoking changing 'strategic factors' external to it, such as expanding grain markets and the increasing power of the ruler. Barth notes that on a revisit to Swat in 1960, he found 'men's houses serving mainly as places for the demonstration of wealth towards co-members of the wealthy elite, and clientship disappearing as a vital relationship' (1966: 17). This removes a key part the original model, in which men's house hospitality by 'chiefs' was a central mechanism sustaining their clientelistic followings. However, for Barth, the mismatch was relatively unimportant, because he had set out to explain 'traditional' tribal politics, just as many structural-functionalists had, not present-day Swat. His determination to treat Swat as if it were an acephalous segmentary 'tribal' society led to many of the features of his work that were later critiqued.

7 Similarly, the work done by Barth's notion of 'status' in his model of Swat looks a lot like social structure. The statuses of the Pakhtun landowner and the landless tenant inform the resources, aims, values and options of the actors so profoundly that a game theory of their 'transactions' seems to tell us less about their social life than the structures that define status.

8 Go (2013: 49), however, argues that Bourdieu did have a theory of colonialism.

9 Throop and Murphy, for example, are critical of Bourdieu's characterization of phenomenology, and point out that 'Bourdieu does little to acknowledge the fact that consciousness must itself be somewhat structured for there to be the very possibility of the internalization of external structure' (2002: 201).

References

Ahmed, Akbar 1976. *Millennium and Charisma Among Pathans: A Critical Essay in Social Anthropology.* London: Routledge & Kegan Paul.

Ahmed, Akbar and Gustaaf Houtman 2009. Swat in the eye of the storm. *Anthropology Today* 25 (5): 20–22.

Asad, Talal 1972. Market model, class structure and consent: A reconsideration of Swat political organisation. *Man* (N.S.) 7 (1): 74–94.

——— 1973. Introduction. In Asad, T. (ed.), *Anthropology and the Colonial Encounter.* London: Ithaca Press.

Bailey, Frederick G. 1969. *Stratagems and Spoils: A Social Anthropology of Politics.* Oxford: Basil Blackwell.

——— 2001. *Treasons, Stratagems, and Spoils. How Leaders Make Practical Use of Beliefs and Values.* Boulder, CO: Westview Press.

Barth, Fredrik 1959, *Political Leadership among Swat Pathans.* London: Athlone Press.

——— 1966. *Models of Social Organization.* London: Royal Anthropological Institute of Great Britain and Ireland, Occasional Papers no 23.

——— 1969. *Ethnic Groups and Boundaries: The Social Organization of Culture Difference.* Oslo: Universitetsforlaget.

——— 1981. *Process and Form in Social Life: Selected Essays of Fredrik Barth,.* London: Routledge & Kegan Paul.

——— 2007. Overview: Sixty years in anthropology. *Annual Review of Anthropology* 36: 1–16.
Blumer, Herbert 1969. *Symbolic Interactionism: Perspective and Method*, Englewood Cliffs, NJ: Prentice-Hall.
Bourdieu, Pierre 1977. *Outline of a Theory of Practice*. Cambridge: Cambridge University Press.
Brubaker Rogers 1985. Rethinking classical theory: The sociological vision of Pierre Bourdieu. *Theory and Society* 14 (6): 745–775.
——— 2009. Ethnicity, race, and nationalism. *Annual Review of Sociology* 35: 21–42.
Firth, Raymond 1951. *Elements of Social Organization*. London: Watts and Co.
Cancian, Frank 1968. Maximization as norm, strategy, and theory: A comment on programmatic statements in economic anthropology. In Edward LeClair and Harold Schneider (eds.), *Economic Anthropology*, pp. 228–233. New York, NY: Holt, Rinehart & Winston.
Cook, Scott 1966. Maximization, economic theory, and anthropology: A reply to Cancian. *American Anthropologist* 68 (6): 1494–1498.
Durkheim, Émile 1897. *Le suicide: étude de sociologie*, Paris: Les Presses universitaires de France.
——— 1982 [1895]. *The Rules of Sociological Method*. New York, NY: The Free Press.
Elias, Norbert 1978 [1939]. *The Civilizing Process, Vol. 1: The History of Manners*. New York, NY: Urizen Books.
Evens, Tim 1977. The predication of the individual in anthropological interactionism. *American Anthropologist* 79 (3): 579–597.
Free, Anthony 1996. The anthropology of Pierre Bourdieu. *Critique of Anthropology* 16 (4): 395–416.
Garfinkel, H. 1967. *Studies in Ethnomethodology*. Englewood Cliffs, NJ: Prentice-Hall.
Go, Julian 2013. Decolonizing Bourdieu: Colonial and postcolonial theory in Pierre Bourdieu's early work. *Sociological Theory* 31(1): 49–74.
Goffman E. 1959. *The Presentation of Self in Everyday Life*. New York, NY: Anchor.
Herskovits, Melville 1952 [1940]. *Economic Anthropology: The Economic Life of Primitive Peoples*. New York, NY: Norton.
Herzfeld, Michael 1987. *Anthropology through the Looking-Glass: Critical Ethnography in the Margins of Europe*. Cambridge: Cambridge University Press.
Kapferer, Bruce 1976. Introduction: Transactional models reconsidered. In Bruce Kapferer (ed.), *Transaction and Meaning: Directions in the Anthropology of Exchange and Symbolic Behavior*, pp. 1–13. Philadelphia, PA: Institute for the Study of Human Issues.
Mauss, Marcel 1935. Les techniques du corps. *Journal de psychologie normale at pathologique* 32: 271–293.
Lévi-Strauss, Claude 1963. *Structural Anthropology*. New York, NY: Basic Books.
Neumann, John and Morgenstern, Oskar 1944. *Theory of Games and Economic Behaviour*. Princeton, NJ: Princeton University Press.
Ortner, Sherry 1984. Anthropological theory since the sixties. *Comparative Studies in Society and History* 26 (1): 126–166.
Said, Edward 1989. Representing the colonized: Anthropology's interlocutors. *Critical Inquiry* 15 (2): 205–225
Schneider, Harold 1974. *Economic Man: The Anthropology of Economics*. New York, NY: The Free Press.
Throop Jason and Murphy, Keith 2002. Bourdieu and phenomenology: A critical assessment. *Anthropological Theory* 2 (2): 185–207.

Chapter 5

Anthropology and history

Susan Bayly

Introduction

Anthropology and history have never been untroubled allies. But the tensions between them, and the battles about how and whether anthropologists should embrace historical tools and perspectives, have produced far-reaching theoretical innovations. It is now widely accepted that anthropology cannot develop conceptually without making history a key concern, but in terms excitingly unlike those of most historians.

The historian may productively ask: why did this or that war, revolution or other momentous happening take place in the way it did, at the time and in the particular place it did, and with what consequences and effects? When anthropologists address matters of time and change in the light of our fieldwork findings, we tend to probe and question what historians see as universal and self-evident when they test their particular kinds of evidence for what they define as truth or plausibility.

Our historical concerns have therefore led us to see that for those we engage with in the field, history may be embodied in living ancestors, and hence 'continuous in the present' (Lambek 2002: 66) rather than impersonal, remote and finalized. Many of us work in contexts where history may be constituted in the realm of poetic imagination and moral reasoning, though often in ways that interpenetrate with objective and analytical reckonings of the past. This leads us to ask how time itself is conceptualized in the contexts where we work. We use tools and models equipping us to understand how our interlocutors negotiate the continuities and ruptures of a volatile and uncertain world, and what they recognise about the markers and contours of a life lived temporally: in and through the passing of time, with a sense of eventfulness and change in the experience of an uncertain world.

As a result, we ask such questions as:

- What do those living through times of crisis and dislocation understand about the ways their past and future connect and relate?
- What does it mean for an experience to be perceived as new and transformative or familiar and recurrent, and hence for any particular instance in the flow of life to be understood as an event, occurrence or happening?
- What novel ideas about the flow of time and the meaning of shared and individual pasts come into being in situations of dramatic change and uncertainty, such as those of colonial conquest and global capitalist penetration?
- Who holds the power to label selves and others in temporal terms, for example, as modern or traditional; backward or progressive; 'primitive' or forward-looking; and what are the sources and effects of such power?

Related to all of these concerns is the further question of how authority may be constructed and/or contested through command of historical narrative and the regulation of chronologies and time regimes. The term temporality is widely used to convey this notion of the power that comes from control of the meanings and markers of time. It also reflects the recognition that time may be experienced in radically different or interlocking forms and modes, for which anthropologists have drawn extensively on philosophers' models of time-consciousness, especially those of phenomenology's critics and heirs. Thus time may be cosmic, an entwining of vast and numerous scales and levels of duration impacting everyday existence in ways that may be either replicated or reconfigured through the performance of ritual (Kapferer 2013) It can also be experienced as both sequential, progressive and calendrical, and as mythic, episodic and 'revelatory'. This is what Strassler found in Indonesia, where charting the past can mean catching a prophetic glimpse of the future, as well as recalling the occurrences of bygone times (2010).[1] And, like the notion of 'multiple modernities' (Faubion 1988; Thomassen 2010), concern with temporalities has been a hallmark of the attempt to add ideas about the dynamics of agency, fragmentation and power to the ways we conceptualize the discipline's key building-block concepts: society and culture (Lambek 2002).

Anthropologists do interact with historians in the framing of our theoretical approaches, and with other history-conscious fields, including archaeology and historical sociology.[2] We are on especially close terms with those historians who ask themselves 'post-positivist' epistemology questions much like our own about the validity of their knowledge strategies, and what it is that constitutes legitimate ways to define and assess evidence (Hastrup 2004). Yet like our predecessors, today's anthropologists still seek to understand the ongoing and the customary in the ethnographic contexts we explore, but without taking for granted the existence of systems, structures or regularities in our interlocutors' lives. We use a wide range of fast-changing theoretical frameworks to grasp both the short-term and longer lasting flows and fluxes of social and cultural life. And we also study the act of studying the past, and take a keen interest in representations of past lives and epochs as critical sources of moral and political authority in contemporary affairs (e.g. Humphrey 1992).

It has become axiomatic for us that even where clocks, calendars and written records are scarce or unfamiliar, there is no such thing as a 'people without history'. This was a phrase used as an ironic dismissal of outdated views by Eric Wolf in a landmark study showing that the African tribal identities once thought of as timeless ethnographic givens were in fact products of Europe's brutal colonial impact on the continent: particularly the traumatic dislocations set in train by the North Atlantic slave trade (1982). Sidney Mintz was another anthropologist who explored the dark underside of global capitalist modernity, dealing ethnographically with issues of taste and class identification to document the ways Europeans acquired and satisfied their growing taste for slave-produced sugar in the age of empire (1985; see also chapter 3).

Today's ethnographers are well aware that our discipline has been accused of denying 'coevalness' to others, that is, by writing about those we have called informants in the ethnographic present tense. This, according to Fabian, implies self-perpetuating sameness in their lives and therefore places them outside the dynamic 'temporal space' that we ourselves inhabit (1983). As a result, many anthropologists write in depth about the shifting power relations occurring over time in their interactions with interlocutors (e.g. Birth 2008). This has been one of the many productive responses to the claim that the discipline has a tainted history of its own as colonialism's 'handmaiden', inheriting and perpetuating the subjugating knowledge strategies that underpinned imperial rule (Asad 1973; see also chapter 1).

Important new theoretical tools have been forged out of these contestations, especially the use of Marxist and poststructuralist theories of culture and consciousness to situate ethnography in the context of the memory of traumatic 'critical events' (Das 1995) and the effects and legacies of colonialism.[3] Key contributors to the field have used these perspectives to ask how changing environments and awarenesses are engaged and negotiated in a world of action, agency, performance and practice. Anthropologists now seek to understand how meanings and power relations are made or contested, and with what benign or pernicious effects. The idea that practices and identity markers conventionally thought of as age-old traditions are really constructs originating in specific historic circumstances is a key instance of a concept taken over from historians (Hobsbawm and Ranger 1983) and reshaped for anthropological use.[4]

Anthropologists have come to understand virtually all traditions and identities as processual and constructed, even those perceived in our fieldwork sites as timeless and fundamental. The attempt to show how the collectivities of class, gender, ethnicity, caste, tribe, nation and 'indigenous' community are made and unmade, and to whose advantage or disadvantage, has become an area of both convergence and fierce debate between historians and anthropologists.[5] There are other arenas too where we attempt to explain the often surprising ways that the temporal, the processual and the historic are perceived and reckoned with, and what it is that can make matters of time, change and rupture either troubling and destructive, or empowering and inspiring for those we engage with in our fieldwork.

Historicity

Anthropology did not accept these concerns readily or uncontentiously. The rest of this chapter therefore explores the ways in which issues of change and historicity have been defined and fought about by those with very different views about how to add history-mindedness to our theoretical repertoire. It notes the repudiation and subsequent revival of history in the work of key early-20th-century pioneers; the models of time and change underpinning the 'culture and practice' debates of the 1980s; and the howls of disparagement generated by Marshall Sahlins's call for ethnographies to be reframed as event-conscious 'structuralist histories' (1985). It also addresses theorisations of temporality, narrative and memory within and beyond the contexts of colonialism, revolution and postsocialist transformation, and anthropologists' attempts to add theoretical sophistication to our understanding of modernity, post-modernity, and the contentious concept of 'multiple modernities'.

The birth of anthropology in its modern forms took place in an arena of polemic about history-mindedness (see chapter 1). Like Malinowski's functionalism, Radcliffe-Brown's vision of what came to be called structural-functionalism rested on a fierce repudiation of the works of his evolutionist and diffusionist predecessors. His dismissive label for what they did was 'conjectural history' (1941), an enterprise he thought deadening and wrong-headed, reducing those whom anthropologists called their subjects to 'living ancestors' or dehumanized embodiments of pan-historical culture traits. Both functionalism and structural-functionalism were therefore declarations of war against historical concerns, which was taken to mean visions of human types or races engaged in giant movements across time in the form of diffusionist interactions, or neo-Darwinian competition favouring the 'fit' over the evolutionarily 'unfit' (Carneiro 1973).

Yet the functionalist and structural-functionalist traditions accepted much of what anthropologists now problematize about Western modernity, including the idea that modern understandings of time and change are calendrical and historic, thus wholly different from those once categorized as traditional, 'pre-literate' or 'primitive'. This is the view animating Evans-Pritchard's celebrated account of the 'cattle clock' (1940: 103) by which the Nuer people of the Nilotic Sudan were said to regulate their daily round, using markers such as 'milking time', rather than impersonal measured units.[6]

A closely related view held by Malinowski and his successors was that when 'pre-literate' peoples thought about their past, they did so in mythic terms. This too was a notion well suited to what had become the prevailing vision of anglophone anthropology: that it was the discipline uniquely equipped to understand 'the history of non-events' (Fogelson 1989, cited in Hoffman and Lubkemann 2005: 317), meaning the remarkable ways 'structural continuities' were shaped and perpetuated in human life. Being and continuing to be social, to share an existence in communities, to follow a way of life with structure, ordering principles and shared schemes of meaning was a wonder like those of the non-human natural world, and equally demanding of sophisticated study and analysis.

The origin stories of people in 'simple societies' were therefore of interest to early-20th-century anthropologists when it could be shown that they were narratives that served structural or functional ends: for example, as 'charters' legitimating the authority of chiefs and elders (Fortes 1953). 'Tribal' communities were thus assumed to lack a sense of historicity in any Western or modern form: that is, the past understood as a sequence of definable historic moments relating to and also succeeding one another as episodes in a unidirectional flow of calendrical or clock-governed time. When historical concerns returned to the anthropological mainstream in the 1970s, these assumptions were reframed as a means of addressing the challenging new questions being tackled by a new generation of history-minded ethnographers. One was: what is literacy as a social phenomenon, and how transformative is its acquisition for those acquiring new technologies to record their accounts of the ancestral and foundational?[7] Closely related was the rediscovery of Halbwachs' 1920s work on social memory, prompting wide-ranging interest in processes of memorialization and remembrance, as in contexts where forgetting as well as remembering a traumatic past can be understood as powerful and empowering forms of moral practice (Connerton 1989; Antze and Lambek 1996).

Yet not all of anthropology's early 20th-century believers in structure and function were anti-historical. Thirty years before the study of colonialism became an established feature of the discipline, Evans-Pritchard made a bold stab at producing an ethnography of historical change in the context of the epic upheavals that ravaged Libya's Cyrenaica region before and after the Second World War, dealing with the impact and aftermath of British colonial rule as well as the brutal operations of the region's Italian fascist occupation force. His theoretical framing was very much of his time: an adaptation of his model of segmentary political order to account for the transformation of 'tribal' structure into centralized kingship (1949).[8]

For all its novelty in insisting that there were real histories to be written about 'simple' and 'tribal' peoples, the basis of Evans-Pritchard's study was an unproblematized notion of bounded societies as the core units of ethnographic study. Edmund Leach's 1954 ethnography of the political order of Burma's Kachin highlanders was far more radical in conception and has been much more widely read. What Leach found was a way to make historical concerns achieve something novel and compelling: a rethinking of the concept of structure, and a repudiation of his predecessors' understanding of societies as structured equilibrium systems ordered by any single structuring principle such as segmentation, egalitarianism or hierarchy.

Leach found Kachin social life to be neither timelessly self-reproducing, nor an arena of decisive once-and-for-all transformation like Evans-Pritchard's Cyrenaica. The history he recounted was of change but not rupture: a pattern of continual oscillations between the contrasting forms of Kachin political order he identified as hierarchical (*gumsa)* and egalitarian (*gumlao*). This was certainly a world of dynamism rather than stasis. But it was not what subsequent generations of either Marxists or non-Marxists would look for in such a setting. Both Marxist and non-Marxist alternatives to Leach's oscillation version of change could entail a critical break with established modes and norms: mass resistance to colonial rule, or at least the forging of novel ideologies and subsistence strategies in response to the impact of Christianization, or the effects of deforestation, intrusive cash cropping and the ravages of mechanized warfare. Nor was Leach's history-mindedness an anticipation of works such as those of Marshall Sahlins, as discussed below. Sahlins asked provocatively whether there are such things as 'other times – other historicities' (1985), meaning distinctive ways history and change are perceived in local contexts, especially in times of upheaval and transformation when structures in the very different sense in which the term came to be used – as prevailing cultural principles or power differentials – may shift and change in dramatic ways.

Thus, Leach's elegant oscillation model did not appeal to key contenders in the battles about practice and agency that have gripped the field since the 1980s, or to those making colonialism and global capitalism central to their concerns. Yet it is notable that there were some historically minded anthropologists whose concerns and theoretical tools resonated with Leach's. Goody, Macfarlane and Gellner also asked in bold and sweeping terms how change was patterned in the contexts they explored: conceptual and material transformations in both regional and worldwide globalizing contexts.[9]

But instead of embracing the structuralist perspectives that became central to Leach's later work, they built on the great modernist social theorists, particularly Weber, in their attempts to provide anthropological answers to the classic social theory questions about the animating principles of modernity. Even so, one of Gellner's most provocative works finds its answer to the global modernity question in a highly distinctive use of the principle of oscillation, which he derived from a fusion of his reading of Weber and the medieval Islamic historian Ibn Khaldun. This was the basis of his much-challenged account of the history of Islam as a 'pendulum-swing' of long-durational alternations between dispersed and centralized models of political order (1981) until the moment of 'unhinging' when the pendulum swing was broken. This he saw as having been achieved by Muslims themselves through the rise of purist Islam, thus as the Muslim world's moment of decisive rupture when its populace achieved its own self-generated variant of modernity.[10]

These were not the historical questions attracting the far more numerous anthropologists who were gripped by the field's turbulent engagements with structuralism, feminist and Orientalist critique, Gramscian Marxist and Foucauldian theories of power and subjectivity, and the subsequent impact of postmodern and poststructuralist culture theories. The bombshell effect of 1960s French structuralism as pioneered by Claude Lévi-Strauss (see chapter 2) was a declaration of war against existing understandings of both society and culture: that is, social orders as schemes of mappable roles and statuses to be documented through structural-functionalist methodologies; and bounded 'cultures' as decodable webs of meaning (Geertz 1973).

Lévi-Strauss's structuralist analyses were historically framed, but in a highly original sense. His concern was not with cultures in the plural, but with 'culture', the human capacity to make and convey meanings in patterned and dynamic ways. He applied the term structure to the deep-seated structural regularities of the human mind acting culturally: doing so by symbolizing; attaching meanings to what minds register, doing so dynamically, in temporally active ways over very long periods of time, through patterns of endless binary oppositional pairings (life–death, nature–culture, raw–cooked); perpetually mediating and resolving those oppositions so as to form further patterns of structured conceptual pairs. That was the history Lévi-Strauss revealed through semiotic analysis of myth and other products of creative, thinking human life: the continual long-term structural change that he referred to as transformations, transformation in such forms as the symbolic conceptual fragments that formed the structures of mythic narratives.

Lévi-Strauss's (1966) widely cited contrast, between 'hot' or change-oriented societies, and stasis-based contexts where thought is mythic and 'cold', may appear very close to the structural-functionalist vision of the mythic as 'primitive', and the historical as a modern or Western mode of thought. This has often been said about his account of *bricolage*: the mental operations of the 'tinkerer' or 'potterer', in contrast to the break-the-mould thinking of the modern scientist or engineer.[11] In fact, he intended this as a picture of the mental operation present in every human mind grappling to order the flow of life and experience with all its painful dilemmas and complexities, though more accessible to the anthropologist in 'objectified' cultural forms, as in myth.

Lévi-Strauss also built on the concept of the dialectic, a key conceptual tool of Marxism, though for him a means to understand the dynamics of human thought engaged in the ever-shifting patterning and recombining of semiotic material. So this was emphatically not a means of asking the kinds of history questions a Marxist might ask: in what changing material and social conditions did the myth arise? Who created and controlled it? Whose interests did it serve? Who gained or lost from its use in particular sites and settings? What changed or did not change as a result?

Historicizing in action

Yet, anthropologists have found ways to merge structuralist methods with historically framed political economy, as in Gow's account of an Amazonian myth incorporating the historic experiences of colonialism into its symbolic framework (2001). Marxist understandings of ideology and class domination have also been updated by means of historical ethnography. Bloch (1989) identified the symbolic language of Madagascar royal ritual as a source of ruling class ideological power disguising the historic origins of domination, thus making kingship appear timeless and unchallengeable because it was temporally distant from the realm of dateable historic events.[12]

The study of symbolism in structuralist and other forms has thus been a foil and a resource for ethnographies of colonialism, gender inequality and the dislocations of global capitalism. Its use has also raised provocative questions about the ways anthropologists should understand and problematize the notion of culture. Comaroff's study of the spirituality of

South Africa's Tshidi people under colonialism and apartheid rule was an early effort to deal anthropologically with the possibility of resistance in conditions of oppression and subjugation, combining semiotic analysis with Gramscian Marxist conceptions of cultural power and subaltern consciousness (1985).[13] Comaroff saw Tshidi believers' distinctive reworking of Christianity as 'syncretic bricolage', their colour symbolism and commodity taboos expressions of an unarticulated 'cultural logic' of resistance to disempowerment and proletarianization.

Other historically framed ethnographies have also sought to show that those forcibly integrated into systems of domination may use language, ritual or the mimetic possibilities of self–other imaginings to subvert hegemonic authority. There remains much debate about whether such challenges may be discerned even for those who may appear trapped and voiceless in the face of power in its insidious modern forms: in particular, those forms of power that constitute modern subjects as instruments of their own subjugation (Scott 1990; Brown 1991; Kaplan 1994; Sivaramakrishnan 2005).

Marshall Sahlins made bold but very different use of structuralist semiotics through his deployment of the concepts of 'mythopraxis' and 'structure of the conjuncture' to explore epic moments in the history of empire. These include his celebrated account of the killing of the British explorer-navigator Captain James Cook by Hawaiian islanders in 1778 (1985). There are also his provocative interpretations of the opposing 'cosmologies of capitalism' structuring Chinese and British diplomatic encounters at the court of the Qing (Manchu) emperors (Sahlins 1994), and the entanglement of colonial missionaries in Fiji's bloody 19th-century chiefdom wars (Sahlins 2004).

Sahlins was thus another history-minded anthropologist who insisted that the world of violent colonial interactions should be central to the discipline, but as the raw material for something he meant to be radically new: the enterprise he called 'structuralist history'. He intended this as a means to ask far-reaching questions about history and culture themselves, though his work has also nourished debate about colonial power relations and the means by which empire and its legacy may be resisted or expunged.

The wider framing for Sahlins's project was the long-running 'culture and practice' debate (see chapter 4) about whether those taking action in environments of unpredictability and change should be seen as world-making agents, unconstrained by symbolic codes or meaning systems. Key contributors included Renato Rosaldo, whose account of the intermittent pursuit of headhunting by the Ilongot people of the Philippines found a 400-year history of colonial invasion and warfare relived through oral narratives recited as the group traversed their forest landscape (1980). Rosaldo presents this as conclusive evidence for a view of culture as continual episodic invention, rather than 'webs of meaning' or durable collective scriptings (1980).

Sahlins took a radically different tack. In contrast to his accounts of the domestic mode of production and the 'original affluent society', the new work proclaimed the birth of a history-based anthropology of culture, built on a melding of structuralism with resources only minimally overlapping with Comaroff's use of Gramscian Marxism. This was the work of Fernand Braudel, leader of the group of French historians known as the *Annales* school. These scholars' 'total histories' of mentalities and civilizations deployed a notion of divergent registers or rhythms of change interacting in complex temporal relations that the *Annalistes* called conjunctures (Golub et al. 2016: 18) But their concern was with the slowest of these converging rhythms. *Annales* thinkers paid little attention to what happened at the point of such conjunctures, thus taking minimal interest in the historian's conventional concern with dateable events. Their focus was the deep, slow, eventless changes of environment and geography occurring on a vast scale of 'long duration' (*longue durée*).

What this meant for Sahlins, however, was the possibility of investigating episodes of entangled global history in order to turn the *Annales* enterprise on its head: reframing the notion of conjuncture, and posing the 'what happened to those people at that moment' question that *Annalistes* thought superficial and unimportant. Sahlins's misleadingly simple concern was with precisely those one-off moments in time involving named actors and dramatic individual episodes, that in his view would answer an even more fundamental question: what exactly is an event? Sahlins's answer was everything Rosaldo repudiated: an event is what culture shapes and defines, 'a culturally interpreted happening' (Strathern 1990: 160); hence what is perceived from the perspective and scripting that participants bring to their understanding of a given occurrence. This underpinned his telling of all the stories of bloodshed and cosmological collision that he excavated from the archival record. The killing of Cook was thus carried out on Hawaiian terms and for distinctively Hawaiian reasons. It was this apparently determinist 'culturology' that made his account so provocative to his opponents and critics, even though his notion of structure of the conjuncture was intended as a rethinking of the notion of culture to allow for action and change: the structural forces and energies of culture to be understood as forged rather than inhabited, imbued with dialectical energy and subject to perpetual patterned change.

What he claimed was thus that anthropology's existing notions of culture could be transformed and reconfigured by asking what it is that marks out a body of experiences and perceptions as a distinct moment in time. It was for this critical aim that he coined his key term 'structure of the conjucture'. By this he meant the point of contact or collision between a pre-existing script or framework by which experiences are ordered and actions taken, and the novelty of any given moment in time when such framings are put to the test.

Sahlins's analysis of the Hawaii episode rested on the claim that the islanders brought a distinctively structured cultural logic to the fatal encounter with Cook, seeing him not in mundane 'trade and empire' terms as an enemy invader, but as the incarnation of their god-king Lono. This cosmic being was Hawaiians' guarantor of the perpetual renewal of life and fertility, existing in eternal structural opposition to Ku, god of war and death, living and dying forever in a never-ending sequence of sacrificial endings and returns. Hawaiian time sense was therefore mythic, a cyclical recapitulation of recurrent archetypes rather than unique calendrical instances. Hence Sahlins's account of the killing as the recapitulation of a mythic foundational event: the generative sacrifice through which the world is eternally made and remade.

His key claim, however, was that this was a highly dynamic form of cultural scripting, dialectically active as a moment of conjuncture between existing terms of reference, and the novelty of what is experienced at the moment when those perceptions are deployed. His neologism mythopraxis echoed the Marxist notion of praxis: the critical point at which thought and action may converge to bring about the transformation of an existing form or structure. Hawaiians experienced the great events they celebrated in ritual as recurrences of mythic archetypes, in an endless recapitulation of eternally repeated foundational events relived in the present. No happening on the scale of Lono's coming to the world was a novelty: their sense of time was cyclical rather than progressive and directional.

None of this was acceptable to Sahlins's attackers. For Obeyesekere, the only real myth in the Cook story was that of the age-old self-glorifying White man's view of himself as a figure of godlike wonder to the 'natives' encountered in the brutal world of empire building (1992).

Strathern raised an even more compelling possibility about the nature and perception of events: that even something as highly charged as a first encounter with pale-skinned strangers may not be an occasion for explanatory framing of any kind. She proposes for the Hagen context that what is brought to such events is visual rather than analytical perception. The relevant experience is that of a spectacular performance of masked dancing, meaning that such an event is registered the way an image impacts a viewer. This means its capacity to force the viewer to look, see and register all at once: as a one-off, that is, through an instantaneous impact on eye and sight, rather than tapping into one's capacity to interpret or explain as one does with experiences that are inserted into a verbal narrative, relating them to other such instances in a flow of durational or on-going time (Strathern 1990; compare Hirsch 1999; also Peel 1984).

Ongoing developments

Historicity remains an abiding concern in our field. The embrace of deconstructive techniques drawn from poststructuralist identity and language theory has generated wide-ranging debate about historically minded anthropologists' methodologies and knowledge practices. Thus, when anthropologists employ the typologies used by historians for such purposes as distinguishing between the effects of different forms of colonial rule, their efforts may be denounced: for example, as a defining of difference reproducing the coloniser's silencing of colonized subjects through the structural violence of 'naming power'.[14] There have been equally fierce exchanges about the extent to which historical perspectives can equip anthropology to defend or challenge the assault on outdated theories of modernity and modernization through engagement with the idea of multiple, cross-cutting or vernacularized 'modernities'.[15]

For those seeking to problematize anthropologists' concern with on-going structured norms, contexts such as those of Christianization have been provocatively explored through a focus on visions of rupture and radical discontinuity (e.g. Robbins 2014). Another key account is Humphrey's exploration of what she calls 'decision events' in the context of Mongolia's turbulent revolutionary upheavals and historic ruptures. This is a richly nuanced reflection on the much-debated 'death of the subject' achieved through engagement with the philosopher Alain Badiou's treatment of decisive transformation events such as Pauline conversion (Humphrey 2008; see also Humphrey and Hurelbaatar 2013).

New possibilities are still emerging in our engagements with history, and the issues addressed through our deployment of historical perspectives are likely to carry on bearing rich theoretical fruit. There are certainly powerful insights to be gained by making historicity a key concern of our burgeoning new work on the anthropology of morality and ethics. A notable instance of this is in the study of how national histories are perceived and challenged in forms identifying them as moral narratives calling for complex understandings of the transhistorical nation and the self as citizen and ethical agent of a shared or contested nationhood. Thus, the past and its imaginings are now and will certainly remain a critical terrain for anthropological theorizing: it is clear that our discipline will continually rethink its own past, present and future through engagement with historical perspectives in ever-changing and productive forms.

Notes

1 With the possibility of productive 'traffic' between them. For other examples, see Gell (1992), Munn (1992), Ssorin-Chaikov (2006) and Guyer (2007).

2 For example, engagements with Polanyi's concept of 'great transformation' (1944), Wallerstein's 'world systems' theory (1976) and Anderson's historical sociology of nationalism (1983) by anthropologists including Eriksen (2002), Herzfeld (2005) and Ingold (1993). There has also been fertile interaction with the work of Hayden White and other historians inspired by postructuralist critique to challenge positivistic modes of historical analysis, for example, using Barthes and Ricoeur to direct their concerns towards processes of representation and the dynamic operations of narrative, text and discourse (Ulin 1994).

3 For example, Appadurai (1981), Werbner (2002), Comaroff and Comaroff (1991), Kaplan (1994), Luhrmann (1996), Stoler and Strassler (2000) and De L'Estoile (2008).

4 Including recognition of fractures and contestations as well as dynamism and creativity involved in the handling of painful 'culture contact': for example, intrusive foreigners' claims about the supposed 'culture traits' of peoples and societies in such forms as cannibalism and the ritual killing of widows (Thomas 1992; Hawley 1994).

5 As in the battles about whether caste is an elemental fact of Indian civilization, an invention of Orientalist knowledge practices or a dynamic construction arising from enduring colonial and global 'entanglements' (Thomas 2009; see also Bayly 1999; Gupta 2005. Compare Friedman 1992; Li 2000). Historians and anthropologists interact with varying degrees of enthusiasm. Historical anthropologists and anthropologically minded historians (e.g. Burke 2005) have been both praised and attacked by those who think historians treat anthropology as a licence to reify 'culture', and that anthropologists decontextualize and overtheorize when they discuss historical trends or episodes.

6 'Events follow a logical order, but . . . are not controlled by an abstract system' (Evans-Pritchard 1949: 103). It is now widely held that this account overstates the distinction between 'our time' that 'passes and progresses', and that of people who see time in unmeasured, non-sequential or mythic terms (Adam 2013). Radcliffe-Brown and his contemporaries saw modernity as a force irresistably destructive of the 'pre-literate' peoples and ways of life. What this meant, he believed, was that what the ethnographer observed – structure, equilibrium, all the elements (institutions) of a social order serving a 'socially integrating' function – would lose their integration and collapse into structural breakdown, leading to a form of traumatic disorientation for which he coined the term 'social dysnomia' as a consequence of 'contact with White civilization'. Intriguingly, he found a way to turn his notions of socially functional thought and practice into a vision of the rise of new religious movements among such people that differed greatly from that of his contemporaries who thought them only a sign of 'primitive' irrationality and mental or social collapse in the face of disorienting 'culture contact': for Radcliffe-Brown, they were functional and positive, 'spontaneous movements towards reintegration' (1935: 399).

7 Goody (1977), Ong (1982) and Bloch (1989); see also Messick (1996); on the ethnographic study of numeracy, see Crump (1990), and the extensive literature on the deployment of statistics as a source of modern rulers' subjugating power-knowledge: Scott (1998).

8 Instead of seeing the Sanusi people's transformations as a story of native fragility, that is, their indigenous social forms obliterated by an all-powerful Western modernity, Evans-Pritchard identified the source of change in the Sanusi's own political order. Segmentation, he said, was an ordering principle animated by a dynamic of replication, and what he claimed was that the upheavals of their colonial and wartime experiences led to this dynamic of replication being directed to a newly available model. This was the centralizing logic of their fascist Italian oppressors, which he believed they had successfully appropriated and remade for their own use.

9 Goody (1983; 2012), Macfarlane (1978; 2000) and Gellner (1981).

10 He followed this with an equally bold overview of the rise of modern nationalism (1983), and a representation of the whole of humanity's history as a giant tripartite sequence of transformations taking shape as contrasting civilizational orders of the 'plough, sword and book' (1988).

11 Thinking like a *bricoleur* means putting existing symbolic elements together in different patterns and configurations: like the 'jack-of-all-trades' or 'handyman' making whatever comes to hand

serve his needs; thus never thinking in truly novel terms, never inventing a new tool or method to serve a new and different need. So in *bricolage*, existing signs and symbols serve their needs as a means of ordering and apprehending the world, using and reusing them in ever-changing patterns and configurations. This is what he believed he had found in his studies of tribal myth and art. For a challenge to oversimplifications of the myth–history distinction, see Herzfeld (1985).

12 Studies of ritual continue to make historicity a central concern (e.g. Lambek 2002).

13 This is the usual English translation for Gramsci's key term *subalterno*, the subordinated. It is an important conceptual tool within and beyond anthropology, central to debates both using and contesting the perspectives of both Foucault and Gramsci on the cultural foundations of power in colonial and other contexts (see e.g. Gledhill 2001).

14 Taken from Bayly (2016: 6). See Krautwurst (2003).

15 Miller (1994), Englund and Leach (2000), Osella and Osella (2006) and Thomassen (2010). And see Donham's exploration of divergent narrativizations of revolutionary modernity in the context of Ethiopia's turbulent 20th-century history (1999).

References

Adam, B. 2013. *Timewatch. The Social Analysis of Time*. Cambridge: Polity Press.

Anderson, B. 1983. *Imagined Communities. Reflections on the Origin and Spread of Nationalism*. London: Verso.

Antze, P. and M. Lambek 1996. *Tense Past: Cultural Essays in Trauma and Memory*. London: Routledge.

Appadurai, A. 1981. The past as a scarce resource. *Man* (N.S.) 16: 201–219.

Asad, T. (ed.) 1973. *Two European Images of Non-European Rule, Anthropology and the Colonial Encounter*. New York, NY: Humanity Books.

Bayly, S. 1999. *Caste, Society and Politics in India from the 18th Century to the Modern Age*. Cambridge: Cambridge University Press.

——— 2016. Colonialism/Postcolonialism. In F. Stein, S. Lazar, M. Candea, H. Diemberger, C. Kaplonski, J. Robbins and R. Stasch (eds.), *The Cambridge Encyclopaedia of Anthropology*, online: http://preview12.admin.cam.ac.uk/entry/colonialism-postcolonialism.

Birth, K. 2008. The creation of coevalness and the danger of homochronism. *The Journal of the Royal Anthropological Institute* 14 (1): 3–20.

Bloch, M. 1989. The disconnection between power and rank as a process. An outline of the development of kingdoms in central Madagascar. In M. Bloch (ed.), *Ritual, History and Power. Selected Papers in Anthropology*, pp. 46-88. London & Atlantic Highlands, NJ: Athlone Press.

Brown, M. 1991. Beyond resistance: A comparative study of utopian renewal in Amazonia. *Ethnohistory* 38 (4): 388–413.

Burke, P. 2005. *The Historical Anthropology Of Early Modern Italy*. Cambridge: Cambridge University Press.

Carneiro, R. 1973. Structure, function, and equilibrium in the evolutionism of Herbert Spencer. *Journal of Anthropological Research* 29 (2): 77–95.

Comaroff, J. 1985. *Body of Power, Spirit of Resistance: The Culture and History of a South African People*. Chicago, IL: University of Chicago Press.

Comaroff, J. L. and J. Comaroff. 1991. *Of Revelation and Revolution: Christianity, Colonialism, and Consciousness in South Africa*. Chicago, IL: University of Chicago Press.

Connerton, P. 1989. *How Societies Remember*. Cambridge: Cambridge University Press.

Crump, T. 1990. *The Anthropology of Numbers*. Cambridge: Cambridge University Press.

Das, V. 1995. *Critical events: An anthropological perspective on contemporary India*. Delhi: Oxford University Press.

De L'Estoile, B. 2008. The past as it lives now: An anthropology of colonial legacies. *Social Anthropology* 16 (3): 267–279.

Donham, D. 1999. *Marxist Modern*. Berkeley, CA: University of California Press.

Englund, H. and Leach, J. 2000. Ethnography and the metanarratives of modernity. *Current Anthropology*. 41(2): 225–248.

Eriksen, T.H. 2002. *Ethnicity and Nationalism. Anthropological Perspectives*. London: Pluto Press.

Evans-Pritchard, E.E. 1949. *The Sanusi of Cyrenaica*. Oxford: Oxford University Press.

Faubion, J. 1988. Possible modernities. *Cultural Anthropology* 3(4): 365–378.

Fortes, M. 1953. The structure of unilineal descent groups. *American Anthropologist* (N.S.) 55 (1): 17–41.

Friedman, J. 1992. The past in the future: History and the politics of identity. *American Anthropologist* 94 (4): 837–859.

Geertz, C. 1973. *The Interpretation of Cultures. Selected Essays*. New York, NY: Basic Books.

Gell, A. 1992. *The Anthropology of Time: Cultural Constructions of Temporal Maps and Images*. Oxford: Berg.

Gellner, E. 1981. *Muslim Society*. Cambridge: Cambridge University Press.

Gellner, E. 1983. *Nations and nationalism*. Oxford: Blackwell.

Gledhill, J. 2001. Deromanticizing subalterns or recolonializing anthropology? *Identities* 8 (1): 135–161.

Golub, A., D. D. Rosenblatt and J.D. Kelly 2016. Introduction. *A Practice of Anthropology. The Thought and Influence of Marshall Sahlins*, pp. 3–39. Quebec: McGill-Queen's University Press.

Goody, J. 1977. *The Domestication of the Savage Mind*. Cambridge: Cambridge University Press.

——— 1983. *The Development of the Family and Marriage in Europe*. Cambridge: Cambridge University Press.

——— 2012. *The Theft of History*. Cambridge: Cambridge University Press.

Gow, P. 2001. *An Amazonian Myth and its History*. Oxford: Oxford University Press.

Gupta, D. 2005. Caste and politics: Identity over system. *ARA* 43: 409–427.

Guyer, J. 2007. Prophecy and the near future: Thoughts on macroeconomic, evangelical, and punctuated time. *American Ethnologist* 34 (3): 409–421.

Hastrup, K. 2004. Getting it right. Knowledge and evidence in anthropology. *Anthropological Theory*. 4 (4): 455–472.

Hawley, J. (ed.) 1994. *Sati, the Blessing and the Curse: The Burning of Wives in India*. Oxford: Oxford University Press.

Herzfeld, M. 1985. Lévi-Strauss in the nation-state. *The Journal of American Folklore* 98 (388): 191–208.

——— 2005. *Cultural Intimacy. Social Poetics in the Nation State*. Oxford: Routledge.

Hirsch, E. 1999. Colonial units & ritual units. Historical transformations of persons and horizons in Highland Papua. *Comparative Studies in Society and History* 41 (4): 805–828.

Hobsbawm, E. and T. Ranger (eds.) 1983. *The Invention of Tradition*. Cambridge: Cambridge University Press.

Hoffman, D. and S.C. Lubkemann. 2005. Warscape ethnography in West Africa and the anthropology of 'events'. *Anthropological Quarterly* 78 (2): 315–327.

Humphrey, C. 1992. The moral authority of the past in post-socialist Mongolia. *Religion, State and Society* 20 (3–4): 375–389.

——— 2008. Reassembling individual subjects. Events and decisions in troubled times. *Anthropological Theory* 8 (4): 357–380.

Humphrey, C. and U. Hurelbaatar. 2013. *Monastery in Time. The Making of Mongolian Buddhism*. Chicago, IL: University of Chicago Press.

Ingold, T. 1993. The temporality of the landscape. *World Archaeology* 25 (2): 152–174.

Kapferer, B. 2013. Montage and time. In Suhr, R. and R. Willerslev (eds.), *Transcultural Montage*, pp. 20–39. New York, NY and Oxford: Berghahn.

Kaplan, M. 1990. Meaning, agency and colonial history: Navosavakadua and the Tuka movement in Fiji. *American Ethnologist* 17: 3–22.

——— 1994. Rethinking resistance. Dialogics of disaffection in colonial Fiji. *American Ethnologist* 21 (1): 123–151.

Krautwurst, U. 2003. What is settler colonialism? An anthropological meditation on Frantz Fanon's 'concerning violence'. *History and Anthropology* 14: 55–72.

Lambek, M. 2002. *The Weight of the Past. Living With History in Mahajanga, Madagascar*. New York, NY: Palgrave.

Lévi-Strauss, C. 1966. *The Savage Mind*. Chicago, IL: University of Chicago Press.

Li, T.M. 2000. Articulating indigenous identity in Indonesia: Resource politics and the tribal slot. *Comparative Studies in Society and History* 42(1): 149–179.

Luhrmann, T.M. 1996. *The Good Parsi: The Fate of a Colonial Elite in a Postcolonial Society*. Cambridge, MA: Harvard University Press.
Macfarlane, A. 1978. *The Origins of English Individualism: The Family, Property and Social Transition*. London: Wiley-Blackwell.
——— 2000. *The Riddle of the Modern World*. Basingstoke: Palgrave Macmillan.
Messick, B. 1996. *The Calligraphic State. Textual Domination and History in a Muslim Society*. Berkeley, CA: University of California.
Miller, D. 1994. *Modernity. An Ethnographic Approach. Dualism and Mass Consumption in Trinidad*. Oxford: Berg.
Munn, N. 1992. The cultural anthropology of time: A critical essay. *Annual Reviews of Anthropology* 21: 93–123.
Obeyesekere, G. 1992. *The Apotheosis of Captain Cook*. Princeton, NJ: Princeton University Press.
Ong, W. 1982. *Literacy and Orality: The Technologizing of the Word*. New York, NY: Methuen.
Osella, C. and F. Osella. 2006. Once upon a time in the West? Stories of migration and modernity from Kerala, South India. *JRAI* 12 (3): 569–588.
Peel, J.D.Y. 1984. Making history: The past in the Ijesha present. *Man* (N.S.) 19: 111–132.
Radcliffe-Brown, A.R. 1935. On the concept of function in social science. *American Anthropologist* 37 (3): 392–402.
——— 1941. The study of kinship systems. *JRAI* 71 (1–2): 1–18.
Robbins, J. 2014. The anthropology of Christianity. Unity, diversity, new directions. *Current Anthropology* 55 (S10): S157–S171.
Rosaldo, R. 1980. *Ilongot Headhunting*. Palo Alto, CA: Stanford University Press.
Sahlins, M. 1985. *Islands of Hstory*. Chicago, IL: University of Chicago Press.
——— 1994. Cosmologies of capitalism. In Dirks, N., G. Eley and S. Ortner (eds.), *Culture/Power/History. A Reader in Contemporary Social Theory*, pp. 412–456. Princeton, NJ: Princeton University Press.
——— 2004. *Apologies to Thucydides. Understanding History as Culture and Vice Versa*. Chicago, IL: University of Chicago Press.
Sivaramakrishnan, K. 2005. Some intellectual genealogies for the concept of everyday resistance. *American Anthropologist* 107 (3): 346–355.
Scott, J. 1990. *Domination and the Arts of Resistance. Hidden Transcripts*. New Haven, CT and London: Yale University Press.
——— 1998. *Seeing like a State*. New Haven, CT and London: Yale University Press.
Ssorin-Chaikov, N. 2006. On heterochrony: Birthday gifts to Stalin, 1949. *Journal of the Royal Anthropological Institute* 12 (2): 355–375.
Stoler, A.L. and K. Strassler. 2000. Castings for the colonial: Memory work in 'New Order' Java. *Comparative Studies in Society and History* 42 (1): 4–48.
Strassler, K. 2010. *Refracted Visions. Popular Photography and National Modernity in Java*. Durham, NC and London: Duke University Press.
Strathern M. 1990. Artifacts of history. Events and the interpretation of images. In J. Siikala (ed.), *Culture and History in the Pacific, The Finnish Anthropological Society, Transactions*, No. 27, pp. 25–43. Helsinki: Finnish Anthropological Society.
Thomas, N. 1992. The inversion of tradition. *American Ethnologist* 19 (2): 213–232.
——— 2009. *Entangled Objects. Exchange, Material Culture and Colonialism in the Pacific*. Cambridge, MA: Harvard University Press.
Thomassen, B. 2010. Anthropology, multiple modernities and the axial age debate. *Anthropological Theory* 10 (4): 321–342.
Trouillot, M.-R. 1995. *Silencing the Past*. Boston, MA: Beacon.
Ulin, R.C. 1994. The anthropologist and the historian as storytellers. *Dialectical Anthropology* 19 (4): 389–400.
Werbner, R. 2002. Introduction: *Postcolonial Subjectivities: The Personal, the Political and the Moral*. In R. Werbner (ed.) *Postcolonial Subjectivities in Africa*. London and New York: Zed Books, pp. 1–21.

Chapter 6

From the extended-case method to multi-sited ethnography (and back)

Harri Englund

Method as theory

Is there anything that holds together a discipline whose practitioners seem as prepared to study on-line worlds in California as dream worlds in Aboriginal Australia? For many anthropologists, the answer lies not in any of the theories that the practitioners might bring to bear on their diverse topics but in the *method* with which anthropological research proceeds. On closer inspection, the theory–method divide appears somewhat suspect, because if the disciplinary core lies in the protocols of how it is practised, then surely those protocols exert greater influence on knowledge production than methods as mere techniques or skills would do. The point has been well recognized in the history of social science. Over half a century ago, C. Wright Mills (1959: 216) likened academic disciplines to the practice of a craft. In sociocultural anthropology, the craft has come to be known as participant observation.

The topic of this chapter – the extended-case method – presents an influential approach to method as theory in sociocultural anthropology. Developed by mid-century anthropologists associated with the Manchester School, the extended-case method serves to remind us of the recursive nature of anthropology's preoccupations. Among other things, it prefigured contemporary debates about how to demarcate the field in which the anthropologist would carry out his or her research. Its emphasis on the emergent properties of social life, on following events and developments as they unfold, issued an early challenge to the programmatic inclination in structural-functionalism to take societies as stable wholes (see chapter 1).

The extended-case method was integral to the conceptual ferment for which the Manchester School is renowned. Concepts such as networks, situations, social dramas and social fields came to augment the work of the extended-case method in introducing anthropology to new possibilities in what it could and should study. Urbanization, industrialization, race relations, colonialism and conflict were some of the issues that made methodological and conceptual innovation run parallel in the work of these anthropologists. The issues both extended the remit of anthropology and made its practitioners acutely aware of the need to recognize certain limits to how far such extension should reach. The lessons to be drawn from this ferment include sobering reflections on the apparently epoch-defining quality of multi-sited ethnography in more recent anthropology.

1940

Far more profound issues were at stake in the world in 1940 than the theory and method of sociocultural anthropology. Yet the year holds some significance to disciplinary history too because of two separate developments. One was the presidential address A.R. Radcliffe-Brown gave to the Royal Anthropological Institute. In what has come to be known as one of structural-functionalism's founding charters, he asserted the distinction between theory and method in no uncertain terms:

> Science (as distinguished from history or biography) is not concerned with the particular, the unique, but only with the general, with kinds, with events that recur. The actual relations of Tom, Dick and Harry or the behaviour of Jack and Jill may go down in our field note-books and may provide illustrations for a general description. But what we need for scientific purposes is an account of the form of the structure.
> (1952: 192)

Contained in this statement are several precepts with which structural-functionalism has been associated (see also chapter 1). The emphasis on 'events that recur', for example, came to be restated by Meyer Fortes in 1953. The structural frame of reference, he wrote, 'gives us the procedures for investigation and analysis by which a social system can be apprehended as a unity made up of parts and processes that are linked to one another by a limited number of principles of wide validity in homogeneous and relatively stable societies' (p. 39). From Radcliffe-Brown's view of science as being concerned 'only with the general' to Fortes's emphasis on 'a limited number of principles of wide validity', the consequence for field-based knowledge production was clear. 'The actual relations of Tom, Dick and Harry' could only be ephemera from which illustrations for general principles could be extracted but which would not become the very means of knowledge production. No wonder that the result would be an account of a 'homogeneous and relatively stable society' when the method by which information was obtained already assumed as much![1]

A very different design for anthropological research was put forward in a study whose first edition was published in 1940. Max Gluckman's *Analysis of a Social Situation in Modern Zululand*, republished in 1958, had as its focus precisely an event that would *not* recur – the opening of a bridge. The event was not a customary celebration but the first one of its kind and something that no participant anticipated to be held again in the future. For Gluckman, the official opening of the bridge provided a way of examining complex social and political relations in South Africa that was in the throes of racial segregation. In a nutshell, the Zulu and white participants in the ceremony were as interlinked as they were antagonistic. Zulus were divided, among other things, into Christians and non-Christians, nobles and commoners, while tensions were apparent between the white South African officials and Swedish missionaries. Gluckman wrote in detail about how these divisions could be observed in bodily comportment, modes of dress and speech and in the spatial organization of the ceremony. At the same time, his close engagement with the ceremony and its participants did not preclude attention to social and economic forces beyond the ceremony itself. He wrote about Zulu history, South African racial politics and the world economy to demonstrate how the multiple divisions and tensions he could observe had the dominant black–white antagonism at their apex. The objective was not to reduce the rich detail into a single structural principle but, on the contrary, to show how the opening of the bridge afforded a particular perspective on the South African social field. Another set of events could have yielded another perspective on the alliances and cleavages that constituted that social field.

Before considering further how the extended-case method came to surpass this early attempt at situational analysis, it is worth remarking on the novelty of Gluckman's genre itself. He characterized his subject matter as 'a series of events as I recorded them on a single day' (1958: 8), but it was precisely the 'I' in his writing that set the work apart from anything else that British social anthropology in particular had to offer at the time. Where Bronislaw Malinowski had in *Argonauts of the Western Pacific* (1922) published a photograph of his tent to establish his own presence in the field, Gluckman made no secret of the extent to which his own person afforded a particular perspective. He described his journey with a number of Zulus, who had to eat in the kitchen of the establishment they visited while he ate with other whites in the hall. A sense of racial discord became apparent when he visited the kitchen to chat with people there. This was not the 'I-witnessing' that Clifford Geertz used to characterize the 'highly "author-saturated", supersaturated even, anthropological texts' (1988: 97) that had begun to appear in the aftermath of anthropologists' critiques of their own textual practices in the 1980s (Clifford and Marcus 1986; see chapter 8). Gluckman's own person was, rather, one instrument among others to explore the social situation, not a narcissistic ploy to make the anthropologist the focus of attention. This mode of writing, possibly more successful from the outset than the subsequent texts critiqued by Geertz, pioneered a more personal engagement in anthropological writing than was possible in both Malinowski's and Radcliffe-Brown's visions of science. This was not the only approach that Gluckman's 1940 study pioneered. It was also one of the first anthropological studies to show in detail how ethnicity, race and religion were contingent upon historically and politically informed situations. But what its situational analysis did not pioneer was the extended-case method.

From situations to extended cases

In 1949, Gluckman became the first professor of social anthropology at the University of Manchester. His appointment heralded decades of intense methodological and conceptual innovation on the basis of fieldwork mostly in Southern and Central Africa but later also in Israel and Europe. He brought with him a number of anthropologists who had been associated with the Rhodes-Livingstone Institute in Zambia (then Northern Rhodesia). Gluckman had been its second director, but the institute continued to serve as a focal point for Manchester anthropologists' fieldwork in the region. The kind of research they wished to pursue was becoming increasingly difficult to do in South Africa after the imposition of apartheid. Zambia and Malawi (then Nyasaland) provided conditions that were somewhat more conducive for anthropological research on emerging topics such as urbanization and trade unionism as well as on what were then the discipline's core interests such as ritual, kinship and law. Their practice advanced sociocultural anthropology as a field science in which, as Lyn Schumaker (2001) has emphasized, the role and contributions of Africans were crucial. I return below to the interface between politics and knowledge production that the Manchester anthropologists were acutely aware of.

The progress from the situational analysis introduced in Gluckman's 1940 study to the extended-case method was the result of this collaborative effort. Clyde Mitchell, another key figure in the development of the extended-case method, remarked how the Manchester School may have looked as such from the outside, but internally intellectual and personal differences were often felt to be more important (Werbner 1984: 158). Yet the sense of a common project was evident in the forewords that Mitchell and Gluckman wrote for their colleagues' and pupils' monographs published by Manchester University Press, in the notoriously acerbic seminar discourse that awaited visitors and in Gluckman's insistence that all anthropologists congregated at Old Trafford to watch Manchester United Football Club. As mentioned, from this intellectual and social ferment arose not only the extended-case method but also other innovations in anthropological research, notably network analysis that Mitchell and J.A. Barnes (1954) initially developed in response to the perceived complexity of new field sites but that also enabled them and other Manchester anthropologists to combine participant observation with statistical methods. A traffic in ideas and approaches between anthropology, sociology, case law and social psychology was another aspect of this intellectual ferment, although the precise influence of cognate approaches, such as sociology's Chicago School, in its uses of the case method, remains a moot point. Elizabeth Colson, for example, brings her contemporary witness to bear when she notes, *pace* David Mills (2006: 171–172), that the Chicago School was 'less important than [Mills] thinks' (2008: 336). She mentions another pioneer in the sociological uses of the case method – George Homans, whose book *The Human Group* (1951) she recommended to Gluckman, who subsequently invited Homans to Manchester for a term.[2]

These brief remarks serve to indicate that the use of case studies was neither invented by, nor unique to, the Manchester School. Certainly, Gluckman's own trajectory had included encountering the case method in law and psychoanalysis. From these diverse influences, and not least from the practice of fieldwork, arose his and his colleagues' realization that the 1940 analysis of a social situation fell short of an extended case. In his subsequent reflections on the evolution of the case method, Gluckman paid tribute to Malinowski for raising fieldwork to 'a professional art' (2006: 14). Yet his work revolved around, in Gluckman's view, 'the method of apt illustration' rather than the case method. Apt illustrations were used to highlight specific customs or social relationships in an argument based on some general principles. Note how Gluckman describes the crafting of anthropological arguments in this vein: 'Each case was selected for its appropriateness at a particular point in the argument; and cases coming close together in the argument might be derived from the actions or words of quite different groups or individuals' (2006: 15). In the method of apt illustration, there was no suggestion that by paying close attention to specific instances of social life, and to specific persons or groups, the anthropologist might do more than illustrate something already established without a case. In the evolution of the case method, Mitchell (2006) saw Gluckman's situational analysis in 1940 as an important step away from apt illustration (see also van Velsen 1967). Yet his reason for regarding the extended-case method as a yet further stage in the evolution is illuminating. Here the anthropologist is interested in 'a sequence of events sometimes over quite a long period, where the same actors are involved in a series of situations in which their structural positions must continually be re-specified and the flow of actors through different positions specified' (2006: 28–29). Gluckman's analysis of a social situation in Zululand did not meet these criteria, because it made no attempt to follow the protagonists beyond the events of a single day.

In Mitchell's (1956) study of *The Yao Village*, for example, an eight-page account of witchcraft accusations over a six-year period provided an extended case of personal and sectional relations in a single village. Indeed, the method proved indispensable to studying accusations and conflicts of various kinds, from witchcraft (Marwick 1965) to disputing (Gluckman 1955; 1965) and strike action (Epstein 1958; Kapferer 1972). Gluckman himself singled out Victor Turner's *Schism and Continuity in an African Society* (1957) for showing the method 'at its best' (2006: 19). It demonstrated the affinity between the life-history approach and the extended-case method by following the trials and tribulations of Sandombu, an ambitious but ultimately doomed figure whose fate Turner compared to Greek tragedies. It is a study that also indicated the close link these anthropologists drew between the method – whether it revolved around a single personality or a set of actors – and the analysis of social structures. As will be discussed further below when considering the limitations of extended cases, Manchester anthropologists saw no need to dispense with the concept of social structure. For Sandombu, the fates in his tragedy were dealt by the structural contradiction between virilocal residence and matrilineal descent.

The same method could lead to different emphases, with Jaap van Velsen constructing his *The Politics of Kinship* (1964) around a greater sense of manipulation and contingency than the other Manchester anthropologists seemed to advocate.[3] Turner (1968; 1974) himself was to develop his approach towards the study of performance in his ground-breaking work on ritual. Already in *Schism and Continuity*, he had coined a concept with theatrical resonances: social dramas. It captured both the interest in conflict and the methodological principle of following events through time. In Turner's famous definition, 'the social drama is a limited area of transparency on the otherwise opaque surface of regular, uneventful social life' (1957: 93). Social dramas did not furnish the anthropologist with apt illustrations. They were moments invested with transformative potential in the flow of social life.

The demands that the extended-case method placed on the fieldworker were considerable. Intrinsic to the method was the desire to produce observations and other information in excess of what the argument might require. For van Velsen, one objective was to engage the reader in assessing and interpreting fieldwork findings:

> By this method the ethnographer not only presents the reader with abstractions and inferences from his [sic] field material but he also provides some of the material itself. This should put the reader in a better position to evaluate the ethnographer's analysis not only on the basis of internal consistency of the argument, but also by comparing the ethnographic data and the inferences drawn from them. Particularly when several or most of the actors appear again and again in different situations, the inclusion of such data should reduce the chances that cases become merely apt illustrations.
>
> (1964: xxv–xxvi)

Analytical argument, no less than the features of social life themselves, was an emergent property of fieldwork-based knowledge production, not a matter of high theory spatially and temporally removed from it.

As the above remarks on the Manchester School's intellectual influences have already indicated, it was not hostility to theory *per se* that drove this methodological and conceptual innovation. Just as the genres of writing up extended cases could diverge, so too did the theoretical debates these anthropologists wanted to address. At one end lurked transactionalism with the apparent compatibility between the extended-case method and a focus on individuals (Kapferer 1972). At another was phenomenology with its attention to embodiment, the senses and experience, an approach for which Turner again appeared as a trailblazer (Werbner 1984: 177). More recently, the processual view entailed in all applications of the method has been recovered as one early response to anthropology's neglect of historical analysis (see chapter 5). High theory has its attractions for certain reconsiderations of the Manchester School, such as when continental philosophers are paraded as intellectual resources to augment this or that aspect of its legacy. A 'Hegelian/Marxist logic' is found to be 'inherent in Gluckman's concepts of structural cleavage, systemic contradiction, etc.', while Turner's work stands to be 're-expressed in the post-structuralist terms of Deleuze and Guattari' (Kapferer 2006: 136). The unexpected heights to which this anthropology can be taken also include the way in which Martin Heidegger gets deployed to reveal the radical ontological implications of the extended-case method (Evens 2006). Such enthusiasm to push the frontiers of anthropological theory by revisiting the discipline's classics can only be commended. Whether it does anything to remind us of the demands the extended-case method put on the anthropologist as a field scientist is another matter.

The modernism of multi-sited ethnography

The method undoubtedly continues to leave open critical methodological and theoretical issues, such as the question of whose definition of the situation or event it is that drives fieldwork – the anthropologist's or his or her interlocutors'? Rather than seeking to provide answers to this and other outstanding questions, it is worth considering, within the confines of this chapter, how the intellectual legacy has been recognized in other attempts at innovation in social sciences. From the 1970s onwards, social theory came to be saturated with a concern for praxis, variously conceptualized in terms of structure and agency or practice theory (Ortner 1984; see chapter 4). Some of these theorists seemed aware of the Manchester School's foundational interest in related matters. Pierre Bourdieu, for example, attributed to van Velsen a 'regress to the pre-structuralist stage of the individual and his choices' (1977: 26), while consigning to a footnote the more generous comment that 'van Velsen's analyses are essentially consistent with my own analysis of the strategic uses made of kin relations' (1977: 202). As far as the method is concerned, the most relevant development came to be known as 'multi-sited ethnography' (Falzon 2009).

Fresh from the controversy he helped to generate regarding anthropology's construction of its own texts (Clifford and Marcus 1986), George Marcus embarked on reimagining the practice of fieldwork in the 1990s. 'Multi-sited ethnography' became the phrase to define the new practice in an article published in *Annual Review of Anthropology* in 1995 and then in Marcus's 1998 book on ethnography. The anthropology he sought to address was notable for its American bias, with additional flavour derived from French post-structuralism and certain developments in anti-disciplinary fields such as cultural studies. Yet its flirtation with postmodernism could not conceal a veritable modernist thrust in Marcus's announcement of an epochal shift in the way anthropology should conceive its research practices. Modernism inheres in his evocation of the term 'traditional' as an epithet for the practices that require profound rethinking, if not rejection. Frequent mention is made of 'traditional ethnographic concerns' (Marcus 1998: 82), 'traditional fieldwork' (p. 85), 'anthropology's traditional subjects' (1998: 87) and, of course, 'traditional single-site fieldwork' (p. 99). The words 'brave new worlds' with which Marcus (p. 99) ends his essay indicate ironical distance to the new topics and subjects that the contemporary world appears to throw at anthropology, but there can be little doubt about the earnestness with which Marcus wants to reform anthropology's 'traditional' practices. Such reform is motivated by a range of intellectual and political imperatives, perhaps the most pressing of which is inspired by the idea that a 'world system' and 'the global' are to be made a part of ethnographic research. By tracing connections and associations between spatially dispersed subjects, the anthropologist does not assume those abstractions as contexts of study but precisely seeks to render them ethnographically knowable and, as such, emergent rather than pre-given (Marcus p. 81).

The extended-case method gets bracketed in this campaign for methodological reform, quite literally so. Among Marcus's methodological guidelines for practising multi-sited fieldwork is the injunction to 'follow' subjects (people, things, metaphors) across multiple sites. When he suggests conflict as another subject to be followed by the well-travelled anthropologist, the extended-case method makes its shadowy appearance: 'In small-scale societies, this has been an established technique ("the extended case [sic] method") in the anthropology of law' (p. 95). Note the three restrictive moves that accompany this brief mention. The extended-case method appears to be restricted to the study of *conflict* in *small-scale societies* within the sub-field of the *anthropology of law*. I have already adumbrated some of the challenges from which the extended-case method arose and how it has to be situated in conceptual as well as methodological innovation. Conflict and disputes were indeed of particular interest to the Manchester School, but their study served more than narrow sub-disciplinary objectives. At issue, as mentioned, was an alternative to the abstraction of 'homogeneous and relatively stable societies' in structural-functionalism. The small scale applied to participant observation that revolved around the same sets of people across time. There was no suggestion that the resultant study would necessarily be *about* a 'small-scale society'. Even in Gluckman's (1958) original use of situational analysis, the events observed in a single day provided a perspective on South Africa where black and white were integrated into a single, antagonistic and large-scale social field.

Modernist reformism is notorious for its zeal to expunge traces of the past,[4] but little is gained if anthropology is also caught up in such fervour. Lost, among other things, would be early insights into the world society that mid-century Southern and Central Africa generated among anthropologists affiliated with the Rhodes-Livingstone Institute (see Wilson 1941). Consider, for example, how A.L. Epstein (1958) was drawn to develop the extended-case method during his fieldwork on the Zambian Copperbelt in the early 1950s. In a racially charged situation, strikes, boycotts and public meetings were the stuff of everyday life for many African men in a mining town. While crafting his extended cases around 'personal rivalries' and 'frequent changes in allegiance' (p. 179), Epstein did not lose sight of the larger point that 'urban Africans were staking out their claim for recognition as full members of an industrial society' (p. 157). As if to dispel any lingering assumption among his Euro-American readers that he was dealing with 'tribesmen' in a small-scale society, Epstein added: 'By their acceptance of trade unions, Africans showed that they were fully committed to an industrial system, and were engaged in a struggle to improve their position in that system'. Africans in cities had to be studied as full-blown, if often exploited, members of the urban society, not as 'natives' displaced from their rural habitat. Those who see mid-century anthropology practised in Africa only through the lens of colonial subjugation need to be reminded of Gluckman's criticism that 'the tradition of anthropology is still "tribalistic", and with it goes a tendency to make the tribe and the tribesman the starting-point of analysis' (1961: 69). Hence his famous maxim: 'An African townsman is a townsman, an African miner is a miner'.

Marcus's (1998) survey of new and emerging topics that call for multi-sited ethnography include development and humanitarian projects as well as media products. By confining the extended-case method to the study of conflict in small-scale societies, he misses the opportunity to consider how such topics might benefit from it. To draw upon my own research trajectory, my early work on war, displacement and repatriation on the Mozambique–Malawi borderland had the extended-case method at its core. I used the method to draw 'attention to refugees and their local hosts as persons whose relationships had histories beyond the immediate challenges of an unprecedented crisis' (Englund 2002: 164). The 'unprecedented crisis' had emerged from regional, if not global, geopolitics in which the apartheid regime in South Africa, with right-wing sympathizers in the US, had encouraged the formation of a violent insurgency movement against the Marxist–Leninist government in a newly independent Mozambique. By the time of my fieldwork in the early 1990s, humanitarian relief aid had become integral to the complex ways in which kinship and other personal ties shaped the processes of war, displacement and repatriation. These processes, in other words, had profoundly trans-local dimensions, but the analysis of their actual unfolding required extended cases of relationships both during and beyond the crisis.

In a more recent project, I deployed the extended-case method to study the production and reception of a popular radio programme by Malawi's public broadcaster (Englund 2011). *Nkhani Zam'maboma* (News from Districts) broadcast every evening listeners' stories of misconduct and scandal in their localities. The extended-case method proved productive not only as a way of investigating editorial dilemmas but also how the media product got woven into, and shaped further, relationships among listeners. For example, I followed how a story broadcast about an errant village headman influenced a similar predicament in a rural area not mentioned on the programme. Here the village headman had engaged in sexual transgressions and arrogant behaviour over a long period of time, and had eventually been stripped of his public roles. The broadcast story coincided with his return to public life, rather against the villagers' will, and served to remind him and his subjects how precarious his authority was. *Nkhani Zam'maboma*, I wrote, 'had an effective local presence that was likely to have defused the specter of a more confrontational approach to [the village headman's] misdemeanors' (Englund 2011: 196). The extended-case method brought into one integrated framework localized and personal conflicts, nation-wide radio broadcasts and the effects on moral imagination by stories from elsewhere. The method was key to the anthropologist's version of a 'reception study'. By extending the focus beyond a single media product, the various scales of analysis acquired a processual character, so often absent from reception studies by media researchers.

The limits to extension

Implicit in this processual approach is not only the expansion of research in spatial terms but also attention to temporality in social life. Marcus has recently sought to shift the emphasis from multi-sitedness to multi-temporality in fieldwork, but his sources of inspiration remain curiously presentist. 'The study of globalization', he writes, 'is associated with the temporality of emergence or the contemporary as the study of change in a dynamic present that exists at the intersection of the recent past and the near future' (2010: 32). As a question of temporality, emergence, as has been seen, hardly needed to wait for the study of globalization in order to enter anthropology. No less than when he consigned the extended-case method to the study of small-scale societies, Marcus misses here the opportunity to consider the precise challenge that studying globalization might pose to anthropologists, who already had a multi-temporal, as well as a multi-sited, methodology at their disposal. The issue is not to be cavalier about the new questions that an abstraction such as 'globalization' may suggest. The issue is that the modernist and presentist thrust of Marcus's anthropology makes it impossible to assess with any precision what might strike anthropologists as 'new' in those questions.

Underlying this interest in spatial and temporal extension are other questions. What are the limits to such extension? When does the anthropologist know that enough is enough? Marcus (2010: 34) approvingly cites Marilyn Strathern, who in her reflections on crises and the unforeseen states that 'social anthropology has one trick up its sleeve: the deliberate attempt to generate more data than the investigator is aware of at the time of collection' (2004: 5). She further describes the disciplinary practice thus: 'Rather than devising research protocols that will purify the data in advance of analysis, the anthropologist embarks on a participatory exercise which yields materials for which analytical protocols are often devised after the fact' (pp. 5–6). These are 'open-ended modes of study', that 'allow one to recover the antecedents of future crises from material not collected for the purpose' (p. 7). Warning against methods by which anticipation becomes 'routinized', Strathern extols the virtues of social anthropology in anticipating 'a future need to know something that cannot be defined in the present' (p. 7). Recall van Velsen's (1964: xxv–xxvi) call for 'data' in excess of what an argument may require. While his primary objective may have been to draw the reader into assessing the validity of the argument – and perhaps to extend it further – the extended-case method has provided anthropologists with means 'to recover the antecedents of future crises' by attending to events and relationships with no necessary expectation that any of them will become relevant to analysis. For analysis as for fieldwork, the question is where to stop.

Inevitably, the fieldworker will concentrate on a limited number of persons, or sets of persons, whose relationships he or she traces across time. Because the extended-case method does not separate data collection and analysis into temporally distinct tasks, review and reflection are important moments during fieldwork even as its objectives remain open ended. The method's early proponents were well aware of the need to delimit the scope of what they studied. Commenting on the concept of network, for example, Mitchell recognized that 'some limit must be put on the number of links to be taken as definitive for any specific network; otherwise it would be co-extensive with the total network' (1969: 40).[5] The limits to one's knowledge received a particularly sophisticated discussion in the volume Gluckman (1964) edited on what would now be called 'inter-disciplinarity'. While welcoming collaboration across the disciplines, he and his colleagues emphasized the need to limit the field of study any one of those disciplines would call their own and to make what they called 'naïve' assumptions about other specialist fields in order to produce their analyses of particular subjects. The contributors to the volume explained 'how they [had] decided that they had reached the appropriate limits of their own field and how they could, or could not, fruitfully stop their analyses at these limits' (Devons and Gluckman 1964: 18).

It is a paradox of the reforms advocated by Marcus and his cohorts (Clifford and Marcus 1986; Marcus 1998) that while their critiques of ethnographic writing may have dented anthropologists' sense of authority, their modernist impulse to embrace 'new' topics recognized no limits to what the anthropologist might study. In the wake of these reforms, anthropology could, as Matei Candea has noted, 'talk about anything, anywhere, and in any way' (2007: 170). Against this 'new holism', Candea asserted the value of the bounded site, 'whose strength is not in *fulfilling*, but rather in perpetually *deferring* closure' (p. 179; emphases original). In new holism as in old, 'the imagination of wholes is a rhetorical imperative for ethnography since it is this image of wholeness that gives ethnography a sense of fulfilling "closure" that other genres accomplish by different rhetorical means' (Thornton 1988: 286).[6] The limits to extension in the extended-case method do not effect a closure in order to produce a view on a whole society. Rather, by foregrounding process and emergence, the method imposes limits upon itself in order to 'perpetually defer closure'.[7]

This is where some Manchester anthropologists' personal background becomes relevant. Racial segregation in South Africa loomed large in much of what Gluckman in particular saw as anthropology's proper remit. His early critique of Malinowski's (1945) work on 'cultural change' rejected the view that the South African predicament was about multiple, distinct cultures in uneasy contact with each other (Gluckman 1949; see also Kapferer 2006: 129–130). This line of critique reappeared the year he died, when he attacked Edmund Leach, a mid-century anthropologist based in Cambridge, for seeing no middle ground between an emphasis on radical difference in human cultures and the sentimental insistence that all humans are alike (Gluckman 1975). Drawing a troubling parallel between Leach's preference, at that point in his work, for a structuralist study of internally coherent cultural grammars and the seemingly benevolent attitude of some apartheid protagonists towards distinct cultures, Gluckman signed off his attack with these words:

> It is possible in the cloistered seclusion of King's College, Cambridge, to put the main emphasis on the obstinate differences; it was not possible for 'liberal' South Africans confronted with the policy of segregation within a nation into which 'the others' had been brought, and treated as different – and inferior.
>
> (p. 29)

Concepts such as social fields and networks were corollaries of this scholarly and political urge to make the extended-case method break the confines of conventional categories such as 'society' and 'culture'. Just as there were limits to how far cases could and should be extended, so too did Gluckman bemoan those orientations that broke the confines to the point that any structural analysis became redundant. In another intervention at the end of his life, Gluckman noted the way in which the case method had come to focus the anthropology of law on disputes instead of rules: 'If the study of disputes is erected as a slogan, it can be as stultifying as the reporting of rules on their own' (1973: 636). He remarked that those who saw only social dramas in Turner's (1957) famous book missed out on something important: 'Almost all anthropologists who have thus used this book seem to have overlooked the first 90 pages of the book which contain a detailed, and often quantified, analysis of the external environment and the structure of Ndembu society' (Gluckman 1973: 636). On one hand, the limits to the extended-case method were set by other, equally important methods that every fieldworker had to consider using. On the other hand, at issue was the insight, incipient in *Analysis of a Social Situation in Modern Zululand*, that even the most fleeting, small-scale social interactions had to be understood within a field of relationships of another order, structured and yet not determined by political, economic and ideological forces not immediately apparent to the participant observer.

Conclusion

To discuss the extended-case method in the twenty-first century is to confront some of the questions that have defined sociocultural anthropology since its inception as a field science. What is the status of the ephemera of everyday life as well as its crises when it comes to producing analytical and comparative insights? How to situate the fieldworker in the writing that ensues from fieldwork? Indeed: How separate are fieldwork and writing (or producing a film or a podcast) as moments in the anthropological research process?

If anthropological knowledge is to reach beyond the spatial and temporal boundaries of a single field site, how does it recognize limits to what it can study? These and other questions are not the prerogative of anthropologists who deploy the extended-case method, nor does the method provide exhaustive answers to them. As always in discussing the development of anthropological theory, excavating the history of theory and method can yield fresh insights. Exploring the development and legacies of the extended-case method should certainly temper the modernist impulse to consider our era of globalization as one that poses altogether novel challenges to anthropological knowledge.

Notes

1 It should be noted, however, that Fortes's 1953 overview of 'unilineal descent groups' recognized, on the basis of work by anthropologists associated with the Manchester School, that social life in Central Africa did not conform to the model of corporate descent groups (p. 37). Indeed, few actual studies were as slavishly tied to the programmatic statements as subsequent dismissals of structural-functionalism would lead us to believe. Fortes's (1945; 1949) own corpus on the Tallensi of Ghana, for example, is a towering example of what participant observation can yield and remains an inspiration for anthropological studies about moral sentiment, kinship, parenting, bereavement and regional cults.
2 Others have similar stories of exposing Gluckman to influences that were new to him. Ronald Frankenberg, a 'second-generation' Manchester anthropologist, has said that he gave him a copy of Erving Goffman's *The Presentation of Self in Everyday Life* (1956) five years before Goffman came to give a seminar in Manchester (Mills 2006: 175).
3 Rather idiosyncratically, van Velsen also called his approach 'situational analysis', claiming that the word 'case had been used with so many different meanings' (1964: xxv). A further indication of the presence of strong personalities in the Manchester School beyond Gluckman himself can be found in the 1990 obituary for van Velsen. A former student remembered him as 'a giant of a man; towering above people who were usually thought tall, booming his irreverent criticisms in seminars. Diffident undergraduates were liable to think that he might eat them for breakfast' (Shurmer-Smith 1990: 26). An amusing insight into anthropology as a field science also arises from the recollection that 'the Tonga and Jaap were, in that curious way that so often occurs in ethnography, perfectly fitted for one another – "agin [sic] the government" (any government!), lacking in ritual, argumentative, adventurous'.
4 For an excellent study of such zeal in West Africa, see McGovern (2013).
5 A parallel realization came more recently to accompany interest in actor-network theory, which Strathern (1996) queried for its apparently limitless tracing of networks. For discussions about 'cutting and connecting' relations and networks in the anthropologies of Africa and Melanesia, see Myhre (2016).
6 For an argument that replaces the idea of a singular holism in anthropology with a variety of holisms, see Bubandt and Otto (2010).
7 As an illustration of how related ideas circulate among anthropologists with or without attention to the extended-case method, consider Liisa Malkki's (2007: 179–186) argument about a 'tradition of improvisation' in anthropological knowledge production. True to its American provenance, the argument draws on Geertz rather than Gluckman, but the point about improvising within limits appears there too.

References

Barnes, John A. 1954. Class and committees in the Norwegian island parish. *Human Relations* 7 (1): 39–58.
Bourdieu, Pierre 1977. *Outline of a Theory of Practice*, translated by Richard Nice. Cambridge: Cambridge University Press.
Bubandt, Nils and Ton Otto 2010. Anthropology and the predicaments of holism. In Ton Otto and Nils Bubandt (eds.), *Experiments in Holism: Theory and Practice in Contemporary Anthropology*, pp. 1–15. Oxford: Wiley-Blackwell.

Candea, Matei 2007. Arbitrary locations: In defence of the bounded field-site. *Journal of the Royal Anthropological Institute* 13(1): 167–184.

Clifford, James and George E. Marcus 1986. *Writing Culture: The Politics and Poetics of Ethnography*. Berkeley, CA: University of California Press.

Colson, Elizabeth 2008. Defining 'the Manchester School of Anthropology'. *Current Anthropology* 49(2): 335–337.

Devons, Ely and Max Gluckman 1964. Introduction. In Max Gluckman (ed.), *Closed Systems and Open Minds: The Limits of Naïvety in Social Anthropology*, pp. 15–19. Chicago, IL: Aldine.

Englund, Harri 2002. *From War to Peace on the Mozambique–Malawi Borderland*. Edinburgh: Edinburgh University Press for the International African Institute.

——— 2011. *Human Rights and African Airwaves: Mediating Equality on the Chichewa Radio*. Bloomington, IN: Indiana University Press.

Epstein, Arnold L. 1958. *Politics in an Urban African Community*. Manchester: Manchester University Press.

Evens, Terence M.S. 2006. Some ontological implications of situational analysis. In Terence M.S. Evens and Don Handelman (eds.), *The Manchester School: Practice and Ethnographic Praxis in Anthropology*, pp. 49–63. Oxford: Berghahn.

Falzon, Mark-Anthony (ed.) 2009. *Multi-Sited Ethnography: Theory, Praxis and Locality in Contemporary Research*. Farnham: Ashgate.

Fortes, Meyer 1945. *The Dynamics of Clanship among the Tallensi*. Oxford: Oxford University Press for the International African Institute.

——— 1949. *The Web of Kinship among the Tallensi*. Oxford: Oxford University Press for the International African Institute.

——— 1953. The structure of unilineal descent groups. *American Anthropologist* 55(1): 17–41.

Geertz, Clifford 1988. *Works and Lives: Anthropologist as Author*. Cambridge: Polity.

Gluckman, Max 1949. *An Analysis of the Sociological Theories of Bronislaw Malinowski*. Oxford: Oxford University Press.

——— 1955. *The Judicial Process among the Barotse of Northern Rhodesia*. Manchester: Manchester University Press.

——— 1958 [1940]. *Analysis of a Social Situation in Modern Zululand*. Manchester: Manchester University Press for the Rhodes-Livingstone Institute.

——— 1961. Anthropological problems arising from the African industrial revolution. In Aidan Southall (ed.), *Social Change in Modern Africa*, pp. 67–82. Oxford: Oxford University Press for the International African Institute.

——— (ed.) 1964. *Closed Systems and Open Minds: The Limits of Naïvety in Social Anthropology*. Chicago, IL: Aldine.

——— 1965. *The Ideas in Barotse Jurisprudence*. New Haven, CT: Yale University Press.

——— 1973. Limitations of the case-method in the study of tribal law. *Law and Society Review* 7(4): 611–642.

——— 1975. Anthropology and apartheid: The work of South African anthropologists. In Meyer Fortes and Sheila Patterson (eds.), *Studies in African Social Anthropology*, pp. 21–39. London: Academic Press.

——— 2006 [1961]. Ethnographic data in British social anthropology. In Terence M.S. Evens and Don Handelman (eds.), pp. 13–22. *The Manchester School: Practice and Ethnographic Praxis in Anthropology*. Oxford: Berghahn.

Goffman, Erving 1956. *The Presentation of Self in Everyday Life*. Edinburgh: Edinburgh Social Sciences Centre.

Homans, George C. 1951. *The Human Group*. London: Routledge & Kegan Paul.

Kapferer, Bruce 1972. *Strategy and Transaction in an African Factory*. Manchester: Manchester University Press.

——— 2006. Situations, crisis, and the anthropology of the concrete: The contribution of Max Gluckman. In Terence M.S. Evens and Don Handelman (eds.), *The Manchester School: Practice and Ethnographic Praxis in Anthropology*, pp. 118–155. Oxford: Berghahn.

Malinowski, Bronislaw 1922. *Argonauts of the Western Pacific*. New York, NY: Dutton.

——— 1945. *The Dynamics of Culture Change: An Inquiry into Race Relations in Africa*. New Haven, CT: Yale University Press.

Malkki, Liisa H 2007. Tradition and improvisation in ethnographic field research. In Allaine Cerwonka and Liisa H. Malkki (eds.), *Improvising Theory: Process and Temporality in Ethnographic Fieldwork*, pp. 162–187. Chicago, IL: University of Chicago Press.

Marcus, George E. 1998. *Ethnography Through Thick and Thin*. Princeton, NJ: Princeton University Press.

——— 2010. Holism and the expectations of critique in post-1980s anthropology: Notes and queries in three acts and an epilogue. In Ton Otto and Nils Bubandt (eds.), *Experiments in Holism: Theory and Practice in Contemporary Anthropology*, pp. 28–46. Oxford: Wiley-Blackwell.

Marwick, M.G. 1965. *Sorcery in Its Social Setting: A Study of the Northern Rhodesian Ceŵa*. Manchester: Manchester University Press.

McGovern, Mike 2013. *Unmasking the State: Making Guinea Modern*. Chicago, IL: University of Chicago Press.

Mills, C. Wright 1959. *The Sociological Imagination*. Oxford: Oxford University Press.

Mills, David 2006. Made in Manchester?: Methods and myths in disciplinary history. In Terence M.S. Evens and Don Handelman (eds.), *The Manchester School: Practice and Ethnographic Praxis in Anthropology*, pp. 165–179. Oxford: Berghahn.

Mitchell, J. Clyde 1956. *The Yao Village: A Study in the Social Structure of a Nyasaland Tribe*. Manchester: Manchester University Press.

——— 1969. The concept and use of social networks. In J. Clyde Mitchell (ed.), *Social Networks in Urban Situations*, pp. 1–50. Manchester: Manchester University Press.

——— 2006 [1983]. Case study and situation analysis. In Terence M.S. Evens and Don Handelman (eds.), *The Manchester School: Practice and Ethnographic Praxis in Anthropology*, pp. 23–42. Oxford: Berghahn.

Myhre, Knut Christian (ed.) 2016. *Cutting and Connecting: 'Afrinesian' Perspectives on Networks, Relationality, and Exchange*. Oxford: Berghahn.

Ortner, Sherry B. 1984. Theory in anthropology since the sixties. *Comparative Studies in Society and History* 26(1): 126–166.

Radcliffe-Brown, A.R. 1952. *Structure and Function in Primitive Society*. London: Routledge and Kegan Paul.

Schumaker, Lyn 2001. *Africanizing Anthropology: Fieldwork, Networks, and the Making of Cultural Knowledge in Central Africa*. Durham, NC: Duke University Press.

Shurmer-Smith, Pamela 1990. Obituary: Jaap van Velsen. *Anthropology Today* 6 (4): 26.

Strathern, Marilyn 1996. Cutting the network. *Journal of the Royal Anthropological Institute* 2 (3): 517–535.

——— 2004. *Commons and Borderlands: Working Papers on Interdisciplinarity, Accountability and the Flow of Knowledge*. Oxford: Sean Kingston Publishing.

Thornton, Robert 1988. The rhetoric of ethnographic holism. *Cultural Anthropology* 3 (3): 285–303.

Turner, Victor 1957. *Schism and Continuity in an African Society: A Study of Ndembu Social Life*. Manchester: Manchester University Press.

——— 1968. *The Drums of Affliction: A Study of Religious Processes among the Ndembu of Zambia*. Oxford: Clarendon Press.

——— 1974. *Dramas, Fields, and Metaphors: Symbolic Action in Human Society*. Ithaca, NY: Cornell University Press.

van Velsen, J. 1964. *The Politics of Kinship: A Study in Social Manipulation among the Lakeside Tonga of Nyasaland*. Manchester: Manchester University Press.

——— 1967. The extended-case method and situational analysis. In A.L. Epstein (ed.), *The Craft of Social Anthropology*, pp. 129–149. London: Tavistock.

Werbner, Richard P. 1984. The Manchester School in South-Central Africa and beyond. *Annual Review of Anthropology* 13: 157–185.

Wilson, Godfrey 1941. *An Essay on the Economics of Detribalization in Northern Rhodesia, Part II*. Manchester: Manchester University Press for the Rhodes-Livingstone Institute.

itive anthropology as mological critique

Richard D.G. Irvine

Losing our mind

In 1898, when Haddon set off on his ground-breaking ethnographic expedition to the Torres Straits with a team of Cambridge scientists, three of the participants (W.H.R. Rivers, C.S. Myers and W. McDougall) were psychologists, and they took along with them experiments to try out in the field (see Richards [1998] for an account of this work). Rivers, for example, brought with him instruments to measure visual acuity and colour perception, with the aim of determining whether the perception of colour varied among populations.[1]

It is interesting to reflect on the approach taken by the team at this iconic moment when British anthropology is thought to have climbed out of its armchair. Methodologically, there was an eagerness to combine ethnological and psychological enquiry, and a desire to incorporate experimental methods into field research. Theoretically, there was a desire to understand the role of the mind in social life, and to investigate what was universal and what was particular: What are the general universal characteristics of human mental processes and experiences, and what evidence is there of variation between populations? Equally, it is interesting to reflect on what happened to this line of approach. Rivers's genealogical method became a mainstream preoccupation in social anthropology, but, in spite of his major contributions to the field of psychology, his experimental concerns and his desire to understand mental processes as part of anthropological enquiry did not have anything like the same impact on our field. The reason I start with this reflection is to ask: What happened? How can we account for this apparent loss of interest in mental phenomena? Where did the mind go?

One explanation, of course, lies in the shift away from evolutionary models of explanation and the associated loss of interest in general accounts of humanity as 20th-century anthropology moved towards studies of people in their particularity. As Bloch (2005: 7) argued, for anthropologists rejecting evolutionism, it was 'not fundamental essential characteristics of human beings that explain history but the accidents of whom we are with and have been with' – and in this way, the particular social histories of the people studied become disarticulated from the natural history of humans as a species (see chapter 1).

Alongside ethnographic particularism, an additional factor shaping anthropology's sense of its zone of operation (especially in the context of British social anthropology) was the influence of Emile Durkheim and his claim that society is a *sui generis* reality. In his *Rules of Sociological Method*, Durkheim states that social facts 'consist of manners of acting, thinking and feeling external to the individual' (1982 [1895]: 52), and 'thus when the sociologist undertakes to investigate any order of social facts he must strive to consider them from a viewpoint where they present themselves in isolation from their individual manifestations' (pp. 82–83). This leads, in effect, to a declaration of disciplinary independence. Durkheim insists that explanations of social phenomena by their very nature must stand apart from psychological explanations: an attempt at a psychological explanation of a social fact:

> is not applicable... unless one distorts their nature... Since their essential characteristic is the power they possess to exert outside pressure on individual consciousnesses, this shows that they do not derive from these consciousnesses and that consequently sociology is not a corollary of psychology.
>
> (p. 127)

He formulates his proposition that the study of society stands apart from psychology as a rule: 'The determining cause of a social fact must be sought among antecedent social facts and not among the states of the individual consciousness' (p. 134).

That Durkheim did not always follow his own rule has been pointed out, among others, by Evans-Pritchard (1965: 68). In *Elementary Forms of Religious Life*, Durkheim (1915 [1912]: 216) describes the 'effervescence' experienced during religious ritual as the force that brings the social group together, and here it can be objected that his explanation of the unifying force of religious rites relies on the psychology of the emotional excitement of the individual in the midst of a crowd. Yet whatever may be said about Durkheim's consistency on this point, the important thing to observe here is the effect of the rule on the development of anthropological theory. Radcliffe-Brown, whose structural-functionalist approach had a shaping influence on British social anthropology in the first half of the twentieth century (see chapter 1), reiterates Durkheim when he claims that 'Social phenomena constitute a distinct class of natural phenomena' (Radcliffe-Brown 1940: 3). 'Human beings as individuals are objects of study for physiologists and psychologists'; but for anthropology, the object of study is not the human being as individual, but as a person, 'a complex of social relationships' (p. 5).

So a division of labour between psychology and social science might have led to a blindness to the mind in British social anthropology in the early 20th century; but as the discipline moved to embrace greater theoretical heterodoxy, surely other influences might have been thought to bring it back to the fore? For example, the emergence of Geertzian symbolic anthropology (see chapter 8) heralded a move away from Durkheimean thinking towards a Weber-flavoured search for meaning in human 'webs of significance' (Geertz 1973: 5); and indeed, Geertz engaged at length with the evolution of cognition as a condition for symbolic thought in his essay 'The growth of culture and the evolution of mind'. Yet his approach came with its own sense of the limitations of focusing on the mind when attempting to understand human culture. In response to those who might suggest 'that culture is composed of psychological structures by means of which individuals or groups of individuals guide their behavior' (p. 11), Geertz insists that it is necessary to distinguish culture from mental and physical processes – just as it is necessary to distinguish a musical performance from the actions of the performers and instruments on which a performance takes place (pp. 11–12). For Geertz, the study of humans is the study of the 'imaginative universe within which their acts are signs' – our job is to attend to symbols rather than the brains that process those symbols. Yet the problem here is that by uncoupling human culture from human cognition, we risk greatly overestimating the plasticity of the human mind in accommodating cultural variation; and, furthermore, we risk assuming that representations are straightforward expressions of how things are experienced by the human mind. We can all too easily enter into a world where anything goes.

Consummately social?

The cognitive critique calls on us to examine our epistemological assumptions. First, we have to ask whether providing an account of social life independent of an account of individual mental processes leads to an ignorance of the very processes that make that social life possible. Second, we have to ask to what extent our anthropological analyses assume that the human mind is a *tabula rasa* – a blank slate – marked only by the processes of social interaction over the human life-course.

This line of attack is well illustrated by Bloch (1977) in his analysis of Geertz's essay 'Person, time and conduct in Bali'. Here, Geertz starts from the proposition that 'Human thought is consummately social: social in its origins, social in its functions, social in its forms, social in its applications' (1973: 360). One way in which he demonstrates this is by looking at the way in which time is framed as a concept in Balinese life. He describes a permutational calendar consisting of wheel within wheel, running concurrently: a ten-day cycle of names; a nine-day cycle of day-names; an eight-day cycle; and so on down to a one-day cycle, 'the ultimate of a "contemporized" view of time' (p. 392). The interaction of the cycles produce particular conjunctions that have a qualitative significance, determining festivals, temple celebrations, indicating good or bad days to build a house or hold a puppet show and so on. In such a system of reckoning, time is not durational, but punctual. The calendars, Geertz tells us, are not about marking the quantity of time that passes, but telling you what the time means. 'The cycles and supercycles are endless, un-anchored, uncountable, and, as their internal order has no significance, without climax. They do not accumulate, they do not build, and they are not consumed' (1974: 393). Through this he discerns in the Balinese what he terms a 'detemporalized' conception of time (p. 398), 'pulverizing the flow of time into disconnected, dimensionless, motionless particles' (p. 399), linking this to the 'steady-state' of Balinese social life more widely, in which people are not conceived of as living historically unique lives. In short, what Geertz sets out here is a sense of time that is radically different to that of his own American culture, for whom time is durational and cumulative.

Bloch (1977) is highly critical of this approach. He sees it as an impoverished view of social life precisely *because* it relies on an impoverished view of human thought. He takes the source of the problem to be the idea that knowledge is socially determined. In assuming that 'human thought is consummately social', Geertz works from a social representation of time (the permutational calendar) to a model of how time is processed and experienced in the mind. In doing so, he appears to give the impression that time-perception is highly malleable and variable from culture to culture. In opposition to this, Bloch points to the various areas of social life in which the Balinese do indeed draw on temporal concepts that are durational and cumulative; where they rely on the 'arrow of time', measure out lengths of time, and anticipate cause leading to effect with predictable timing; in particular, practical activities such as agriculture rely on this far more familiar sense of time. For Bloch, then, the task is not simply to set out Balinese time as somehow 'different', but rather to note that in practical contexts time concepts 'are based on cognitive universals' (p. 285), while ritual contexts involve a 'special type of communication' (p. 284). There are therefore limits to the social determination of knowledge: ritual forms may obscure 'the systems by which we know the world' (p. 290), but they should not be mistaken *for* those systems.[2]

The extent to which Bloch is justified in seizing on Geertz's statements about time as an apologia for cognitive relativism is debatable. Gell is more cautious:

> Geertz is providing an interpretation of certain prominent themes in Balinese culture, not a positive account of Balinese psychology and cognition . . . Geertz's relativism is not a matter of dogma, but a by-product of literary artifice.
>
> (1992: 82)

Nevertheless, by calling into question what he sees as unexamined assumptions about the 'social determination of thought', Bloch raises a key issue that remains important. To what extent has anthropology decoupled social life from the mental processes that make that social life possible? And to what extent do our analyses presume that the human mind is a blank slate?

Such an idea was dramatically called into question by the Chomskyian revolution in linguistics. Noam Chomsky took as his core issue the problem of the acquisition of language, and the relationship between the data at the child's disposal (that is, all of the language that is encountered in infancy) and the knowledge of language that the child develops:

> The competence of an adult, or even a young child, is such that we must attribute to him a knowledge of language that extends far beyond anything that he has learned . . . Hence the data available as input is only a minute sample of the linguistic material that has been thoroughly mastered.
>
> (Chomsky 1967: 3–4)

Chomsky's argument was that the rapid and uniform acquisition of language could only be explained if some aspect of our capacity to assemble language is innate – part of the architecture of the brain. He therefore postulates a 'Language Acquisition Device' (1965) as a specific innate capacity upon which we rely to assemble learned knowledge. Such an approach has major implications beyond linguistics, leading to a wider examination of the role of innate characteristics. As Bloch (2012: 60) argues, 'The evolutionary implications of the difference between viewing the mind as a blank slate or as already predisposed for certain types of information could not be more fundamental'. Indeed, the psychologist and linguist Stephen Pinker (2002: 37) sets out the implications in a manner that might sound quite provocative to anthropologists: 'Universal mental mechanisms can underlie superficial variations across cultures'.

A cognitive anthropology, then, is one that is grounded in the specific capacities of the human brain as it has evolved, recognising the universal mental bases that make social life possible. As the psychologist Leda Cosmides wrote in collaboration with the anthropologists John Tooby and Jerome Barkow, 'Culture is not causeless and disembodied. It is generated in rich and intricate ways by information-processing mechanisms situated in human minds. These mechanisms are, in turn, the elaborately sculpted product of the evolutionary process' (Cosmides et al. 1992: 3). Again, they take aim at Geertz in order to make their point, seeing him as the very epitome of an implicit assumption within the social sciences that 'human nature is an empty vessel waiting to be filled by social processes' (Tooby and

Cosmides 1992: 29). In Geertz's work, they see a clear causal arrow in which the individual mind is acted upon and wholly shaped by the sociocultural world. To take the view that social processes are reliant on universal mechanisms of the human brain – that social life is generated by specific capacities of the mind, rather than seeking to show how the mind is shaped by social life – is therefore to turn this on its head. A useful way of imagining the view of the mind they put forward is the analogy of the 'Swiss army knife', 'competent in so many situations because it has a large number of components ... for solving a different problem' (Cosmides and Tooby 1994: 60). This is an architecture of the mind that consists of distinct innate modules that have evolved to perform particular functions, such as the Language Acquisition Device discussed earlier. This 'modularity of mind' approach is not without its critics; see especially Karmiloff-Smith (1992) who, while recognising innate dispositions of the human mind, nevertheless rejects the idea of pre-specified domain-specific 'modules', rather suggesting that modularisation occurs as a product of development. An important line of argument here, significant for anthropological concerns with cultural variation, is that such an approach undermines our recognition of the brain's flexibility: How do we reconcile a literal belief in mental modules with fixed roles with our recognition of human neurological plasticity (Buller 2005)? As Hamilton (2008) has argued, 'from an evolutionary perspective, it can be argued that such flexibility is itself the distinctively human adaptation'.

One important point worth making here is that cognitivists such as Cosmides and Tooby point out that the way of life to which this toolkit is an adaptation is not necessarily a present way of life. Rather, it evolved to meet the challenges of the Pleistocene epoch in which we emerged as anatomically modern humans (Tooby and Cosmides 1992: 109), and 'it is important to recognise that behavior generated by mechanisms that are adaptations to an ancient way of life will not necessarily be adaptive in the modern world' (Cosmides et al. 1992: 5–6). This is highly significant, as it recognises that the process of understanding contemporary human life cannot be reduced to simplistic deduction of what our evolved psychology 'predisposes' us to do. Some anthropological critiques of this evolutionary approach, such as that of Susan McKinnon (2005), argue that it reads into our natural history a contemporary economic presumption of rational-choice theory – taking the idea that the individual is 'self-maximising' as a basis of selection – and naturalises certain cultural assumptions about sexual roles, in particular, male dominance. Yet, while such a line of argument would certainly urge greater caution in the interpretation of evidence about our species' history, it would surely be a category error to take an argument about the scientific construction of particular views of our evolutionary past and turn that into a rejection of the role of the evolutionary past itself in shaping who we are today. From this point of view, the broader call to pay attention to cognition reminds us that 'society' and 'culture' are not free-floating realities, but grounded in specific mental capacities, and that a proper understanding of our adapted brain remains essential if we want to make sense of so much of the fabric of human social life.

Religion explained?

In order to understand this claim that 'culture is not causeless and disembodied' and what it might mean for anthropology, it is perhaps helpful to turn to a specific domain of social life where such an argument has been deployed. Justin Barrett, in a book synthesising cognitive approaches to the study of religion, sets out by stating 'Culture cannot explain culture, and religion cannot explain religion' (Barrett 2004: ix). His argument is that an account of religion as a social form needs to be grounded in an understanding of the mental propensities through which the social form takes shape. Work in the 'cognitive science of religion' – much of which is built on the insights of anthropologists – is wide ranging[3] (see Barrett 2004; 2007 for reviews of this work), but here we will focus on a couple of key concepts in order to illustrate how cognitive approaches to the study of religion might proceed.

Pascal Boyer describes religion as 'parasitic' upon our cognitive framework. That is, it makes use of capacities that people have anyway and use in other domains of activity (Boyer 2001: 231). What kind of capacities are being referred to here? One suggestion in this direction comes from the work of Stuart Guthrie (1980). Noting the recurrent observation that the gods in which humans believe are given human characteristics, Guthrie seeks to make this not simply a feature of religion, but its grounding: 'religions are instances of generalized and systematized anthropomorphism' (p. 192).[4] His argument (further developed in his book *Faces in the Clouds* [1993]) is that our tendency to find human characteristics in the world stems from a deep-seated perceptual strategy; in uncertainty it makes sense to assume agency. As other agents are our greatest threats, and also our greatest opportunities for survival, he argues that a perceptual bias in favour of agency detection would bestow survival advantages, and thus could be considered evolutionarily advantageous.

Barrett (2000) builds on this theory and, noting classic studies of human eagerness to attribute agency even to dots and shapes (Heider and Simmel 1944), and experimental evidence that suggests that infants at 5 months old appear to recognise self-propelled and purposeful-looking movement as agentive (Rochat et al. 1997), refers to an 'Agency Detection Device' in human cognition, that is prone to find agents around us given modest evidence (hence 'Hyperactive Agency Detection Device'). The tendency to attribute superhuman agency – here taken to be characteristic of religion – is thus grounded in properties of human cognition.

Boyer (2001: 147–148) recognises that the anthropological record is full of such attributions of agentive behaviour. For example, drawing on his own ethnographic work with the Fang of Cameroon, he notes the way in which occurrences are attributed to the actions of the spirits of the dead (*bekong*). Yet here it is important to note that the prior concept makes sense of the experience: the sense that *bekong* exist does not emerge from the attribution of agency in a state uncertainty, but precedes it – and indeed can exist in people's mind without people ever having had an experience of having encountered them. Boyer therefore agrees with Guthrie and Barrett that an intuitive understanding of agency is important; yet he insists if we are to explain a *social* phenomenon, it is necessary is to explain how the concepts people draw upon take a stable form that can be transmitted from person to person.

For Boyer, key to this process of transmission is the violation of intuitions about objects and beings in our environment. He suggests that the most memorable signals in communication are those that are 'minimally counterintuitive'[5] (MCI): fulfilling intuitive assumptions about a given class of objects, but also violating some of those assumptions in small ways that make the resulting concepts particularly attention grabbing. Consider the following examples: The mountain towers over us ... *and hears us*. A woman gives birth ... *without having had sex*. Both of these match intuitive assumptions in one domain but violate them in another, and for Boyer, such signals are ripe for successful transmission because, in many respects, they activate our expectations (thus allowing us to understand without too much explanation) while at the same time containing an element that is memorable *because* it violates our expectations. In this way, anthropomorphism itself can be seen as an MCI concept (Boyer 1996).[6] For example, to treat a tree as a listening and communicating thing[7] is to draw on intuitive concepts about biology (the tree remains a tree, with all the botanical assumptions that go along with that), and to draw on intuitive concepts about psychology (our sense of what human minds can do) – while at the same time violating intuition by applying this psychology to an organism that would not usually be seen as having a 'mind' in this way. 'Projections of intuitive intentional psychology make use of the richest domain of inferences available to human minds, whilst violating central aspects of intuitive ontology' (p. 96). Experimental work with American College students (Barrett and Nyhof 2001) has supported Boyer's claims about the memorability of MCI concepts, and the same was found cross-culturally in studies with the Fang of Gabon and Buddhist monks in Tibet (Boyer and Ramble 2001), lending weight to the claim that the memory effects may be universal.

So a cognitive approach seeks to account for the material causes that underlie a social form (in this case, religion), grounding it in universal 'mechanisms' in the brain. The question remains: How far does this get us? Laidlaw (2007), while agreeing that cognitive science provides us with an account of some mental operations that are involved in religious thought and action, nevertheless urges caution. He suggests that cognitive science's contribution is roughly equivalent to that which materials science can make to aesthetics and the history of art. By all means, it is useful to know the material basis of a particular form – the recognition that some things are by-products of the evolved architecture of our brains – but this does not account for the historically and culturally contingent way in which such a material basis is put to work. On the basis of anthropological work on Jainism and Theravada Buddhism, Laidlaw suggests that many of the fundamental features of such religious traditions elude a cognitive science characterisation. So, for example, while agency detection and the salience of minimally counter-intuitive signals might be ways of accounting for the role of deities, ghosts and spirits that figure in these traditions, it would be a mistake to see such features as central to 'religion' as practitioners understand it. What *is* central, according to Laidlaw, is how and why people aim at certain values and virtues – and here, it is suggested that a cognitive approach will struggle to provide an adequate account. The cognitive approach may seek to *explain* mechanisms, but that does not necessarily lead to an *understanding* of 'contingent historical creations of reason, imagination, and will' (2007: 223).

A similar critique is offered by Jenkins (2014), who explicitly endorses Durkheim's declaration of the independence of social science from psychology, discussed above. Cognitive explanation, observes Jenkins, proceeds by identifying mental devices, but does not seek to account for how thought is translated into meaningful action. It looks to the mechanisms involved in a belief, but not the purposes to which such a belief is put.

These critiques raise important concerns about the adequacy of any approach that would seek to explain social phenomena with reference to cognitive mechanisms alone. But let us not be in a hurry to reinforce the disciplinary firewall. By saying, as Laidlaw (2007: 242) does, that 'cognitive science will never be more than ancillary to ... what anthropological and/or historical students of religion have to do', there is a danger that we might severely overestimate the extent to which *explicit* self-understandings can be uncoupled from the *implicit* workings of our evolutionarily adapted brain. In order to explore the ways in which an understanding of the relationship between such explicit and implicit levels of thought might be both necessary and productive for anthropology, let us turn to anthropological discussions surrounding theory of mind.

Mindreading

'Theory of mind' is the term used for the ability to attribute mental states – thoughts, beliefs and intentions – to oneself and to others, with the recognition that other people's mental states are distinct from your own.[8] As Rita Astuti (2012) puts it:

> [H]aving a Theory of Mind amounts to having a non-behaviourist way of looking at the world. When you see someone running, you don't just see a physical body in acceleration – you see the intention or the desire to catch the bus or win a medal; when you see a hand reaching for an object, you don't just see a trajectory through space – you see the goal of getting that object; and so on.

Children are thought to demonstrate theory of mind from the age of around 3 to 4. Within developmental psychology, the emergence of theory of mind in young children has been robustly tested by 'false belief' tests, an experimental approach pioneered by Wimmer and Perner (1983) designed to find out whether a child is able to predict the behaviour of another person on the basis of what that other person believes, rather than what the child knows. A classic form of the test (Baron-Cohen et al. 1985) involves two puppets 'Sally' and 'Anne', acting out the following play: Sally takes a marble and hides it in her basket. She then leaves the room. While she is away, Anne takes the marble out of Sally's basket and puts it in her own box. Sally is then re-introduced and the child is asked: 'Where will Sally look for her marble?' The answer to this question is meant to reveal whether the child recognises that Sally holds a mental state distinct to the child's own. If the child recognises that Sally still believes that the marble is in her own basket (even though the child knows differently), this indicates a capacity to recognise mental states other than one's own. While many variations of the test have been developed, they continue to support the claim that in neurotypical children theory of mind is apparent from the age of around 3 to 4 years. (I will discuss this usage of 'neurotypical' shortly.)

In keeping with the modular approach to the architecture of the mind, as described above, Baron-Cohen (1995) sees this as the end-point of the maturation of an innate theory of mind module (ToMM), emerging as part of the developmental trajectory of the child building on earlier capacities. It is suggested, on the basis of experimental evidence, that very young infants have both an 'intentionality detector' and an 'eye direction detector'. These converge in the 'shared attention mechanism' (following someone's attention towards a particular object) that builds triadic representations – that is, representations involving:

1. an agent
2. the self
3. an object

by using available information about someone else's perceptual state. Theory of mind emerges from the shared attention mechanism, considering and acting on dispositions that are made apparent through triadic representations.

Now, the usage of 'neurotypical' that we saw above – the idea that theory of mind emerges in childhood as part of a standard developmental trajectory – might be thought to highlight exactly the kinds of concerns anthropology should have with a cognitive approach that seems to define 'normality' in terms of a universal cognitive developmental trajectory. By whose standards are we defining typicality? This is a particular concern given the tendency of psychological literature to focus only on samples from what Henrich et al. (2010) call 'WEIRD' populations – that is, Western, Educated, Industrialized, Rich and Democratic. For this reason, as we shall see, an important task of anthropological enquiry has been to explore how theory of mind manifests itself in diverse cultural contexts. And indeed, in considering the development of theory of mind, it is of course important to consider the forms of social interaction outside of this apparent 'typicality', as, for example, in the ethnographic work of Ochs and Solomon (2010) on autistic sociality. From such a perspective, assumptions of normality can be turned on their head (although significantly, this is being done through the understanding of a specific neurological basis of cognitive variation), as in the account Bagatell (2010) gives of a spoof of a Diagnostic and Statistical Manual of Mental Disorders entry for 'Neurotypical syndrome' written by someone on the autism spectrum.

Nevertheless, the hypothesised 'theory of mind module' might be thought to occupy a central role in many of the most fundamental social behaviours that concern anthropologists. The part it plays is, after all, a crucial one in our evolution as a species that relies upon social relationships. For example, theory of mind is key to symbolic thought (Hobson 2000; Zunshine 2006), that involves the recognition of other people's representations; and might be thought to play a central role in reciprocity (Sally and Hill 2006; Takagishi et al. 2010), allowing one to attribute mental states to an exchange partner.

But does an account of the mental mechanisms involved in interaction tell us anything about the character of those interactions, and the meaning and form of the social lives that emerge from them? In other words, returning to Laidlaw's art history metaphor, how relevant is knowledge of the material for a proper understanding of that which is made from the material? Two points are worth making here. The first is that when we talk about social life (and the mental processes that make it possible), we are talking about something that is part of the biological story of our survival as a species. Our sociality is not independent of a biological account of our species; on the contrary, it is integral to it. The meaningful coordination of social interaction, and the capacity to build social bonds, enhanced the adaptive fitness of *Homo sapiens* (Dunbar 1998), and the increase in hominid brain size over the past 3 million years is widely associated with the need for greater social intelligence: powerful mental devices were required to deal with the complexities of behaviour in large groups (Baron Cohen 1995: 13–20). As Laidlaw (2007) rightly states, social phenomena are historical phenomena, not simply the by-products of cognitive mechanisms; yet such history is set within, not apart from, our evolutionary history and the crucial issue is therefore 'the relation of history to nature, and therefore the scale of evolutionary time' (Jenkins 2014: 188). By this logic, it would be a mistake to isolate accounts of the varied goals and meanings of human life from the biological account of our species' evolution, as they *constitute* the fundamental sociality that is key to our natural history.

Second, an understanding of the cognitive grounding of social interaction in theory of mind can play a crucial role in the task of anthropological comparison. How does this capacity for the attribution of mental states manifest itself cross-culturally? One fruitful line of enquiry has been to explore those contexts in which 'mindreading' is apparently confounded by an understanding that 'you cannot know what is in another's mind, even if they claim to tell you it in speech' (Robbins 2008: 422). On the basis of ethnographic work with the Urapmin of Papua New Guinea, Joel Robbins speaks of the 'opacity of mind': the absence of speech acts that assume knowledge of other people's mental states is part of communication (such as apologising and lying), and a broad conception that 'people's minds are private places' (2008: 426), with a reaction of 'almost fear or disgust' that one might know what other people are thinking.[9] Similar observations about the 'opacity of mind' have been made elsewhere in Melanesia; see, for example, Schieffelin (2008) on the Bosavi of Papua New Guinea. So does theory of mind manifest itself in different ways in different cultural contexts?

In order to explore this question, Tanya Luhrmann brought together around 30 scholars – mostly anthropologists – to share evidence of ethnographically distinctive models of the mind, considering differences across a number of dimensions, such as: How important is interiority? How 'bounded' or 'porous' is the mind? What senses are seen as important? The models presented by participants were grouped into six 'different theories of mind' (Luhrmann 2011):

1 *The Euro-American modern secular theory of mind*, in which people consider there to be a clear boundary between what is in the mind and what is in the world.
2 *The Euro-American modern supernaturalist theory of mind* (manifest in contexts such as Charismatic Christianity and new age beliefs), in which the mind–world boundary is permeable for entities such as God, or the dead or for specific agentive 'energies'.
3 *The opacity of mind theory* (as discussed above), particularly evident in Melanesia.
4 *The transparency of language theory*, examples of which are drawn from Central America (see Danziger 2006), in which language is understood in terms of its truth-relationship to the world rather than its relationship to the intention of the speaker.
5 *Mind control theory*, an example of which is drawn from Thai Buddhism (see Cassaniti 2006), in which other people's intentionality is thought to have a direct effect on one's own mind, and the most important concern of mind is therefore how well it is controlled by way of a strong awareness of the influence of others.
6 *Perspectivism* (most closely associated with Amazonian ethnography) within which interiority is determined by relationality (see chapter 14).

A central argument here is that these different models should not be subsumed into one; that is, the Euro-American modern secular theory of mind models the world in a particular way that may carry assumptions that run contrary to other models of mind (for example, the mind is understood to be bounded rather than porous, and the inner states of the mind are thought to be independent of representation of those states). If we ignore variations in how the mind is understood, and what it is believed to be capable of, we risk distorting the very grounds upon which people interact. But does this not simply bring us right back to Geertz's assertion that 'Human thought is consummately social: social in its origins, social in its functions, social in its forms, social in its applications' (1973: 360)?

Here is where it is useful to return to the relationship between *explicit* models and *implicit* workings of our evolutionarily adapted brain. Astuti (2012) warns her fellow anthropologists not to be naive about the reach of this kind of evidence. People in the course of reflection may express different folk theories as explicit reflections on how their minds operate in the world, but this should not be mistaken for what psychologists are referring to as theory of mind, that operates implicitly, 'as individuals, automatically and unconsciously, follow somebody's gaze, interpret her utterances, or predict her next move on the basis of her desires, emotions or beliefs' (Astuti 2015).

The most extreme form of this position, as set out by Scholl and Leslie (1999) – that Astuti does not wholeheartedly adopt, but nevertheless says is important to bear in mind when considering ethnographic evidence – is that whatever explicit folk theory about the mind people have, it is going to make no difference to the implicit mind-reading abilities made possible by the (universal) theory of mind. So, for example, when the Urapmin claim that it is impossible to know what is in the mind of another person, this may tell us a great deal about their ideas of propriety and the value they place upon individual autonomy. However, unwillingness to speculate about the mental state of another is not the same as being 'mindblind' and having no knowledge of other minds.[10] Following this logic, an understanding of theory of mind as an innate potential and a product of our species' evolution allows us to recognise cultural variations for what they are: differences in representation. Yet, as Astuti (2015) argues, it remains essential that we follow people out of the lab and into the worlds where they live, interact and put mental mechanisms to work. In this way, an explanation of theory of mind as an evolved adaptation at the level of the species should not foreclose an analysis of the mind in particular ethnographic settings: rather, it provides a context within which we can see how theory of mind plays out in particular developmental trajectories, and a background against which we can understand the exceptional nature of claims such as opacity of mind.

Conclusion

Cognitive anthropology is, fundamentally, a form of materialism.[11] It is characterised by the recognition that social life is made possible by the activity of the brain, and an understanding that this activity relies upon a cognitive architecture that is a product of our evolution as a social species. It refuses to treat culture as disembodied, but grounds it in our mental capacities.

A serious engagement with cognition requires that we scrutinise the epistemological foundations of our discipline. Has the 'declaration of independence' that establishes social processes as an object of analysis in their own right led to a mindblindness within the discipline – an ignorance of the mental processes that lie at the heart of human interaction? And in so much as we do take an interest in the mind, do we imagine it to be culturally determined, the product of socialisation? A cognitively informed approach to anthropology can help us to confront our wilful dependence on the 'blank slate' model of the mind, and in doing so, overcome a tendency to over-exaggerate the extent of cultural variability.

Of course, there is no need to be chauvinistic about the extent to which cognitive approaches will render other forms of anthropological enquiry outmoded. An understanding of our cognitive architecture does not provide an exhaustive account of purpose and meaning, and it cannot foreclose the task of exploring how mental mechanisms are put to work. By all means, we should attend to the historically contingent trajectory of people's lives – but we need to recognise that this history cannot be isolated from our natural history.

Notes

1 Rivers (1901), proceeding cautiously, largely restricts claims of difference in colour perception to the suggestion that the Torres Straits islanders show an insensitivity to the colour blue;

subsequent analysis of the results of the experiments found little to support any claim of particularities in the islanders' colour perception (Titchener 1916), and, in general, the results of the expedition may be seen as a significant weakening of evolutionist claims about major differences between the psychological aptitudes of different populations (Richards 1998).

2 It is worth noting that Bloch's approach here is characteristically Marxist (see chapter 3) as well as characteristically cognitivist. He is suggesting the economic process (agriculture) is understood through the universal reality of time cognition, but mystified by way of the ritual calendar – a distinction between knowledge and ideology.

3 And, in spite of the unifying effect of the banner 'Cognitive science of religion', is characterised by 'a piecemeal approach' (Barrett 2007: 768) – so one should not assume that all work carried out under this name is integrated, or even pulls in the same direction.

4 Here, Guthrie (1980: 193) explicitly notes that his approach is intellectualist, grounding religion in mental processes, contrary to the Durkheimean definition that treats it as a fundamentally social phenomenon grounded in association (Durkheim [1912] 1915).

5 The term 'Minimally Counterintuitive' (MCI) itself was coined by Barrett (2000) in a discussion of Boyer's theory.

6 Note here that Boyer's explanation runs in a somewhat different direction to Guthrie, who treats anthropomorphism as intuitive.

7 By way of illustration, Boyer (2001: 62–63, 69–70) discusses at length 'ebony divination' among the Uduk-speaking peoples of Sudan, as described by James (1988).

8 This usage of 'Theory of Mind' was introduced by Premack and Woodruff (1978). It is interesting to note that it originates in primatology and the task of discerning whether or not non-human primates are able to conceptualise the mental states of self and others. The question they asked – 'Does the chimpanzee have a theory of mind?' – continues to stimulate debate (Call and Tomasello 2008).

9 Thus, in the context of conversation to Christianity, the arrival of new expectations associated with the 'Protestant demand to become subjects who speak sincerely about what they know in their minds and who can also read the minds of other' (Robbins 2008: 425) was particularly disruptive.

10 Such a position is in keeping with Stasch (2008: 448), in his account of 'opacity of mind' among the Korowai of West Papua: 'denials of the ability to know someone else's intentions . . . do not accurately summarize the total sweep of Korowai understandings of mind, so much as they indicate where major cultural sensitivities lie'.

11 See also Sperber (1996) on 'how to be a true materialist'.

References

Astuti, Rita 2012. Some after dinner thoughts on theory of mind. *Anthropology of this Century* 3, online: http://aotcpress.com/articles/dinner-thoughts-theory-mind/.

——— 2015. Implicit and explicit theory of mind. *Anthropology of this Century* 13, online: http://aotcpress.com/articles/implicit-explicit-theory-mind/.

Bagatell, Nancy 2010. From cure to community: Transforming notions of autism. *Ethos* 38 (1): 33–55.

Baron-Cohen, Simon 1995. *Mindblindness: An Essay on Autism and Theory of Mind*. Cambridge, MA: MIT Press.

Baron-Cohen, Simon, Alan M. Leslie, and Uta Frith 1985. Does the autistic child have a 'theory of mind'? *Cognition* 21 (1): 37–46.

Barrett, Justin L. 2000. Exploring the natural foundations of religion. *Trends in Cognitive Sciences* 4 (1): 29–34.

——— 2004. *Why Would Anyone Believe in God?* Lanham, MD: AltaMira.

——— 2007. Cognitive science of religion: What is it and why is it? *Religion Compass* 1(6):768–786.

Barrett, Justin L., and Melanie A. Nyhof 2001. Spreading non-natural concepts: The role of intuitive conceptual structures in memory and transmission of cultural materials. *Journal of Cognition and Culture* 1 (1): 69–100.

Bloch, Maurice 1977. The past and the present in the present. *Man* 12 (2): 278–292.
——— 2005. *Essays on Cultural Transmission*. Oxford: Berg.
——— 2012. *Anthropology and the Cognitive Challenge*. Cambridge: Cambridge University Press.
Boyer, Pascal 1996. What makes anthropomorphism natural: intuitive ontology and cultural representations. *Journal of the Royal Anthropological Institute* 2 (1): 83–97.
——— 2001. *Religion Explained: The Evolutionary Origins of Religious Thoughts*. New York, NY: Basic Books.
Boyer, Pascal, and Charles Ramble 2001. Cognitive templates for religious concepts: Cross-cultural evidence for recall of counter-intuitive representations. *Cognitive Science* 25 (4): 535–564.
Buller, David J. 2005. *Adapting Minds: Evolutionary Psychology and the Persistent Quest for Human Nature*. Cambridge, MA: MIT Press.
Call, Josep, and Michael Tomasello 2008. Does the chimpanzee have a theory of mind? 30 years later. *Trends in Cognitive Sciences* 12 (5): 187–192.
Cassaniti, Julia 2006. Toward a cultural psychology of Buddhism in Thailand. *Ethos* 34 (1): 58–88.
Chomsky, Noam 1965. *Aspects of a Theory of Syntax*. Cambridge, MA: MIT Press.
——— 1967. Recent contributions to the theory of innate ideas. *Synthese* 17 (1):2–11.
Cosmides, Leda, John Tooby, and Jerome H. Barkow 1992. Evolutionary psychology and conceptual integration. In Jerome Barkow, Leda Cosmides and John Tooby (eds.), *The Adapted Mind: Evolutionary Psychology and the Generation of Culture*, pp. 3–15. New York, NY: Oxford University Press.
Cosmides, Leda, and John Tooby 1994. Beyond intuition and instinct blindness: Toward an evolutionarily rigorous cognitive science. *Cognition* 50: 41–77.
Danziger, Eve 2006. The thought that counts: Understanding variation in cultural theories of interaction. In N. Enfield and S. Levinson (eds.), *Roots of Human Sociality: Culture, Cognition and Interaction*, pp. 259–278. Oxford: Berg.
Dunbar, Robin I.M. 1998. The social brain hypothesis. *Evolutionary Anthropology* 6 (5): 178–190.
Durkheim, Emile. [1895]. 1982. *The Rules of Sociological Method*, edited by Steven Lukes, translated by W.D. Halls. New York, NY: The Free Press.
——— 1915 [1912]. *The Elementary Forms of the Religious Life*, translated by Joseph Ward Swain. London: George Allen and Unwin.
Evans-Pritchard, E.E. 1965. *Theories of Primitive Religion*. Oxford: Clarendon Press.
Geertz, Clifford 1973. *The Interpretation of Cultures*. New York, NY: Basic Books.
Gell, Alfred 1992. *The Anthropology of Time: Cultural Constructions of Temporal Maps and Images*. Oxford: Berg.
Guthrie, Stuart 1980. A cognitive theory of religion. *Current Anthropology* 21 (2): 181–203.
——— 1993. *Faces in the Clouds: A New Theory of Religion*. New York, NY: Oxford University Press.
Hamilton, Richard 2008. The Darwinian cage: Evolutionary psychology as moral science. *Theory, Culture and Society* 25 (2): 105–125.
Heider Fritz, and Marianne Simmel 1944. An experimental study of apparent behavior. *The American Journal of Psychology* 57 (2):243–259.
Henrich, Joseph, Stephen J. Heine, and Ara Norenzayan 2010. The weirdest people in the world? *Behavioral and Brain Sciences* 33 (2–3): 61–83.
Hobson, Peter R. 2000. The grounding of symbols: A social-developmental account. In Peter Mitchell and Kevin Riggs (eds.), *Children's Reasoning and the Mind*, pp. 11–36. Hove: Psychology Press.
James, Wendy 1988. *The Listening Ebony: Moral Knowledge, Religion, and Power among the Uduk of Sudan*. Oxford: Clarendon Press.
Jenkins, Timothy 2014. The cognitive science of religion from anthropological perspective. In Fraser Watts and Leon P. Turner (eds.), *Evolution, Religion, and Cognitive Science: Critical and Constructive Essays*, pp. 173–190. Oxford: Oxford University Press.
Karmiloff-Smith, A. 1992. *Beyond Modularity: A Developmental Perspective on Cognitive Science*. Cambridge, MA: MIT Press.

Laidlaw, James 2007. A well-disposed anthropologist's problems with the 'cognitive science of religion'. In Harvey Whitehouse and James Laidlaw (eds.), *Religion, Anthropology, and Cognitive Science*, pp. 211–246. Durham, NC: Carolina Academic Press.

Luhrmann, Tanya 2011. Toward an anthropological theory of mind. *Suomen Antropologi* 36 (4): 5–69.

McKinnon, Susan 2005. *Neo-Liberal Genetics: The Myths and Moral Tales of Evolutionary Psychology*. Chicago, IL: Prickly Paradigm.

Ochs, Elinor, and Olga Solomon 2010. Autistic Sociality. *Ethos* 38 (1): 69–92.

Pinker, Stephen 2002. *The Blank Slate: The Modern Denial of Human Nature*. New York, NY: Viking.

Radcliffe-Brown, A.R. 1940. On social structure. *The Journal of the Royal Anthropological Institute of Great Britain and Ireland* 70 (1): 1–12.

Premack, David, and Guy Woodruff 1978. Does the chimpanzee have a theory of mind? *Behavioral and Brain Sciences* 1 (4):515–526.

Richards, Graham 1998. Getting a result: The Expedition's psychological research 1989–1913. In Anita Herle and Sandra Rouse (eds.), *Cambridge and Torres Strait: Centenary Essays on the 1878 Anthropological Expedition*, pp. 136–157. Cambridge: Cambridge University Press.

Rivers, W.H.R. 1901. Colour vision. In W.H.R. Rivers (ed.), *Reports of the Cambridge Anthropological Expedition to Torres Straits, Volume II: Physiology and Psychology*, pp. 48–96. Cambridge: Cambridge University Press.

Robbins, Joel 2008. On not knowing other minds: Confession, intention, and linguistic exchange in a Papua New Guinea community. *Anthropological Quarterly* 81 (2): 421–429.

Rochat, P., R. Morgan, and M. Carpenter 1997. Young infants' sensitivity to movement information specifying social causality. *Cognitive Development* 12 (4): 537–561.

Sally, David, and Elisabeth Hill 2006. The development of interpersonal strategy: Autism, theory-of-mind, cooperation and fairness. *Journal of Economic Psychology* 27 (1): 73–97.

Schieffelin, Bambi B. 2008. Speaking only your own mind: Reflections on talk, gossip and intentionality in Bosavi (PNG). *Anthropological Quarterly* 81 (2): 431–441.

Scholl, B.J., and A.M. Leslie 1999. Modularity, development and 'theory of mind'. *Mind and Language* 14 (1): 131–153.

Sperber, Dan 1996. Explaining culture: A naturalistic approach. Oxford: Blackwell.

Stasch, Rupert 2008. Knowing minds is a matter of authority: political dimensions of opacity statements in Korowai moral psychology. *Anthropological Quarterly* 81 (2): 443–453.

Takagishi, Haruto, Shinya Kameshima, Joanna Schug, Michiko Koizumi, and Toshio Yamagishi 2010. Theory of mind enhances preference for fairness. *Journal of Experimental Child Psychology* 105 (1–2): 130–137.

Titchener, E.B. 1916. On ethnological tests of sensation and perception with special reference to tests of color vision and tactile discrimination described in the reports of the Cambridge Anthropological Expedition to Torres Straits. *Proceedings of the American Philosophical Society* 55 (3): 204–236.

Tooby, John, and Leda Cosmides 1992. The psychological foundations of culture. In Jerome Barkow, Leda Cosmides, and John Tooby (ed.), *The Adapted Mind: Evolutionary Psychology and the Generation Of Culture*, pp. 19–136. New York, NY: Oxford University Press.

Wimmer, Heinz, and Josef Perner 1983. Beliefs about beliefs: Representation and constraining function of wrong beliefs in young children's understanding of deception. *Cognition* 13 (1): 103–128.

Zunshine, Lisa 2006. *Why We Read Fiction: Theory of Mind and the Novel*. Columbus, OH: Ohio State University Press.

Chapter 8

Interpretive cultural anthropology

Geertz and his 'writing-culture' critics

James Laidlaw

The way of doing anthropology we now know as 'interpretive' was the expression within anthropology of a much wider trend, one that affected a whole range of arts and humanities disciplines. A common expression used at the time for this broader phenomenon was 'the linguistic turn', and it referred in particular to the influence, first within philosophy and then beyond that, of 'ordinary language' or 'linguistic' analytic philosophy. Although the phrase 'ordinary language philosophy' derives from G.E. Moore, the single most influential source of ideas was the later thought of another Cambridge philosopher, Ludwig Wittgenstein. The most important single source was Wittgenstein's *Philosophical Investigations* (1953), along with others of his mostly posthumous publications, and in particular as interpreted by followers such as Gilbert Ryle (1949) and J.L. Austin (1962). A range of ideas from other philosophers, both Anglophone and Continental, were also part of the picture, as, from time to time, were some ideas and concepts from technical linguistics. There was never a single coherent philosophical doctrine that was then 'applied' in other human sciences; apart from anything else, the exegesis of Wittgenstein's thought was too arcane and disputatious for that. But, in addition, ideas did not go all one way. This was a time when scholars in other disciplines also looked to and learned from anthropology. And the single most influential anthropologist in this regard was Clifford Geertz (1926–2006). There were others whose influence within anthropology was comparable – notably Marshall Sahlins and David Schneider (and in different ways and in a different register, Edward Evans-Pritchard) – but no other figure from this era articulated so clearly a vision of anthropology that placed it so centrally in the linguistic turn, or made it seem so instructive, as such, for scholars in other disciplines. And when opposition began to be voiced to interpretive anthropology, it was often articulated as criticism of Geertz. So, it is not unreasonable to tell the story of interpretive anthropology and of what became of it, substantially with reference to Geertz and his ideas.

Geertz received his anthropological training with Clyde Kluckhohn at Harvard, as part of the interdisciplinary Social Relations Department directed by the social theorist Talcott Parsons, whose synthesis of Weberian sociology, Durkheimian functionalism (see chapter 1), and Shutzian phenomenology provided a rather expansive framework. The fieldwork for Geertz's doctorate was conducted in Indonesia, mostly in a small Javanese town, and he subsequently did fieldwork elsewhere in Indonesia, notably in Bali, and in Morocco. His early work, following his thesis and first monograph on Javanese religion (1960), was greatly concerned with questions of political and economic development, as posed by the Social Relations programme, in relation to 'the new nations' (1963a; 1963b; 1965), and at this time also he wrote a historical comparison of Indonesian and Moroccan Islam (1968).

But during the 1960s, he underwent a considerable intellectual reorientation, progressively drawing more on the philosophical than the economic aspects of Max Weber's corpus, and building on Weber's already rich account of what distinguishes human action from other events in the world; namely, its symbolic dimension and the fact that it is inherently meaningful and inter-subjective. In this, and in the range of philosophical and other thinkers he drew upon in doing so, Geertz was increasingly involved in a conversation that crossed disciplinary boundaries as well as the boundary between Anglo-Saxon and Continental thought. The fruits of this reorientation were collected in a volume of essays that must have a claim to be the single most influential such collection by an anthropologist during the 20th century: *The Interpretation of Cultures* (1973). In these essays, together with those published a decade later in *Local Knowledge* (1983), Geertz synthesised an understanding of human action as constitutively meaningful and inherently social that characterised the linguistic turn, with the concept of culture developed in the Boasian tradition of American anthropology. This was a vision to which he remained consistently committed, and that was reiterated in his later essays, lectures and memoirs (1995; 2000; 2012). Thus, the core of Geertz's anthropology may be elucidated by examining just what he meant, respectively, by 'interpretation' and by 'cultures'.

In saying that what anthropology must do is to *interpret*, Geertz was asserting that the kind of explanation it should seek to offer is qualitatively different from that pursued in the natural sciences, which had provided models and inspiration for other schools of sociological and philosophical thought, most proximately in anthropology at the time, in the diverging approaches of A.R. Radcliffe-Brown (1952; see chapter 1), Leslie White (1949) and Claude Lévi-Strauss (1962 [1953]; see chapter 2). The distinction between interpretation and causal explanation has roots in German historical thought, in the idea that the possible forms of explanatory success are fundamentally different in the sciences respectively of nature and spirit (*Naturwissenschaften* and *Geisteswissenschaften*). The foundational claim is that human beings are, in Charles Taylor's resonant phrase, 'self-interpreting animals' (1985). On this view, the ideas and values people have inform their self-descriptions, and those self-descriptions stand not merely in an external causal relation to what they do, but are internally *constitutive* of who they are and what they are doing. If this is so, then explanation of human conduct in terms of causal laws, on the model of the natural sciences, must be a flawed ambition. It is not merely that such ambitions are impossible to achieve in practice; it is a mistake in principle even to aim at them. An entirely different set of criteria is required for success in an interpretive enterprise. This much was common currency in the linguistic turn, and Geertz expressed the general position forcefully. In addition, he argued that it is a fact about human evolution that we have developed in such a way that being shaped by culture is now part of human 'nature', so that without culture, humans would be radically incomplete and unviable. So, for Geertz, there was a firm scientific account of *why* human conduct, being inherently meaningful, could not be subject to scientific explanation.

Of more practical influence on anthropological practice, he developed a comprehensive account of what should be the aims specifically of *anthropological* interpretation: what forms of excellence it could aspire to, and what it required of its practitioners. The argument aligned anthropology with other humane sciences, such as literary or textual analysis and historical study, and distanced it from 'social sciences' such as economics and sociology. What anthropological interpretation should aim at, argued Geertz, borrowing from the Oxford philosopher Gilbert Ryle, was 'thick description': surrounding description of any specific (say, initially puzzling or apparently meaningless) acts or practices, with sufficient contextual information on local beliefs, customs, values and related practices, so as to enable the reader to grasp the *meaning* those practices have, as part of the form of life or 'culture' to which they belong.

In saying that the objects of anthropological interpretation are *cultures*, Geertz was aligning himself with the Boasian tradition of American cultural anthropology (see chapter 1). Unlike in Tylorian or Lévi-Straussian or cognitivist approaches, the concern was not general features of human thought – not then 'culture' as a universal feature of human life – but the distinctive, relatively internally integrated and relatively distinct and autonomous ways of life in specific places and times – hence 'cultures' in the plural. Where Geertz differed from and developed the Boasian programme, and integrated it intellectually into the linguistic turn more generally, was in displacing the psychological idioms of the most prominent exponents in the previous generation (the concerns of Mead, Benedict and others with the modal *personality* of a culture, and processes of 'enculturation') in favour of a Wittgensteinian view of the meaning of symbols not as representations in the mind but as *use* in social practice. The arguments Wittgenstein and others had made against the possibility of a private language (see Kripke 1982) were used to inform a view of culture as located 'in the public square' in actions and practices that, conveniently, are observable by the anthropologist.

Another argument further removed the object of anthropological interpretation from the psychological. If meaning is the use of language and other symbols in on-going social practice and interaction, as Wittgenstein and others had argued, how then might the anthropologist, observing that on-going interaction, distinguish what pertains to the particular individuals taking part or the specific occasion observed, from what is institutionalised in social and cultural life? What about this particular meal, seminar, haircut, tennis match, sacrifice or shopping trip, is owing to the fact that it is that kind of event or practice in that particular culture, and would be more or less true of every other of its meals, seminars, haircut, etc.? What, on the other hand, is happenstance? How do you sort the cultural from the idiosyncratic? Geertz adapted his answer to this from a suggestion by the French philosopher, Paul Ricoeur (1973). The social or cultural, Ricoeur suggested, may be understood on the model of written text. Ricoeur's idea was that in becoming social or culturally institutionalised – insofar as this act of eating is 'a meal', or this discussion 'a seminar' – the meaning of interaction is *inscribed*; it is transformed in roughly the way that spoken discourse changes when it becomes written text. And among the features of this transformation, the most important is that meaning becomes *less* anchored than in spoken discourse by the intentions of the speaker. What a written text means can move farther from the speaker's intentions than is the case in spoken discourse. Established or institutionalised social practices, the 'cultural' in this sense, are therefore in important ways like texts.

Geertz did not endorse Ricoeur's argument unreservedly, and he also emphasised the usefulness of other models for the cultural or social, such as the game (following Wittgenstein and Irving Goffman) and the drama (following Kenneth Burke and Victor Turner). But the model of the text was crucial for him because his conception of what distinguished a good from a less good anthropological interpretation was essentially that of excellence in the reading of a text. His account of how good anthropological interpretation typically proceeds – tacking between observations of details and characterisations of the whole – explicitly parallelled that of giving a reading of a poem. And his idea of how theoretical language should be used – a balance between 'experience-near' and 'experience-distant' concepts – was likewise based on a view of literary-interpretive practice. And as in a literary reading, content and style are inseparable. It matters very much how an anthropologist writes. It followed for Geertz that the skills and capacities required generically for anthropological interpretation – whatever else might have been required for fieldwork in particular places – were the same as for literary commentary or the interpretation of historical texts, and crucially not, as some others had supposed, any very extraordinary powers of interpersonal empathy or psychological insight. As cultural meanings are out there in public practice and not 'in the heads' of individuals, anthropologists need not be telepathists, or psychiatrists or spies.

So, while Geertz echoed Malinowski's celebrated injunction to anthropologists to grasp 'the native's point of view', he insisted that this did *not* involve 'getting inside' the latter's head. Instead, in probably his single most influential essay ('Deep play', in 1973), Geertz asserted that it requires reading the cultural texts that are important in the lives of our informants – the customs, practices and institutions that are important in their way of life – and that this involves looking 'over their shoulders'. A major concern in Balinese culture is the distinction between humanity and animality. Great trouble is gone to, in almost all aspects of everyday life, to contain and repress expressions of the latter; all disruptive and reprehensible conduct being attributed to it. Yet one of the most powerful intermittent events in the life of a Balinese village, held in defiance of a vigorously enforced police ban, is cock-fighting: a dramatically staged display of brute animal ferocity, that brings together the whole village in avid spectatorship, everyone having bet on the outcome. Geertz's reading is that this event compels the Balinese so totally precisely because it represents what it would be to surrender to animality. Such a vividly experienced representation of the violation of their most cardinal virtues and overturning of the moral order is, argues Geertz, an institutionalised part of Balinese moral education. He compares it to *Macbeth*, which shows inhabitants of an individualist culture what happens when its driving competitiveness is not kept within bounds; what it is to gain the whole world but lose one's own soul. Thus Balinese life consists, in part, of repeated events in which ordinary Balinese engage in an emotionally charged, collectively experienced 'reading' of the text of this key cultural practice. And anthropological interpretation involves interpreting that process of interpretation. The comparison with *Macbeth*, even if it means little or nothing to most Balinese, is part of what Geertz sees as he looks over their shoulder, and what enables his readers, if the comparison is just, to grasp the meaning of what he sees there in the texts of their way of life.

Obviously enough, and as Geertz himself loudly insisted, no accumulation of interpretations of this kind is ever going to result in testable hypotheses or predictive laws about either Balinese culture or human life in general. So what, beyond the pleasing and edifying familiarity they might provide with exotic ways of life, are such interpretations for? What is the aim of a discipline dedicated to their production? Geertz's answer was that anthropology aims at 'the enlargement of the universe of human discourse'. It brings other and new cultures 'into the conversation'. He was not ever very explicit about exactly who is having this conversation or what it is about, but it seems clear enough that he meant, roughly speaking, that 'our' anthropologists, whoever exactly 'we' are, enable us well enough to understand the values and ideas, the ethos and worldview, of other ways of life, that they become values and ideas we can think with and imaginatively participate in and appreciate. In the same way that for centuries familiarity with classical Greece and Rome was part of the mental furniture of educated Europeans, anthropological interpretations of other cultures, in this Geertzian mode, would make them part of the 'universe of discourse' of the educated, cosmopolitan members of modern societies. This is, quite self-evidently, an asymmetrical state of affairs. It may indeed include the Balinese in the scope of 'our' conversation, but it does not, of itself, include them as participants (though equally, it does not in principle exclude them either, or preclude Balinese anthropologists writing about the USA, for example). Geertz generally took it for granted that the people anthropologists write about (unlike anthropologists such as himself perhaps?) inhabit their cultural beliefs and practices as taken-for-granted everyday life. The self-conscious and cosmopolitan condition of doing anthropology is assumed to be unusual. If anthropological interpretation overcomes a misunderstanding or incomprehension of cultural difference, it does so only for those to whom the interpretations are addressed; it implicitly leaves those described as they were. As we shall see, this is one of a cluster of features on account of which a generation of anthropologists came to find interpretive anthropology wanting, but at the time, many felt it to be a liberal and civilised alternative to the 'command and control' ambitions of development-oriented social sciences that engaged with distant parts of the world on the premise that learning and adaptation needed to go all the other way: modern experts teaching 'traditional' societies how to 'develop'. Geertzian anthropology was a charter for 'us' to learn from, rather than developing, reforming or improving, 'them'.

This helps to explain Geertz's influence, both within and beyond anthropology, because the 'conversation' to which he sought to contribute was cross-disciplinary and indeed extended beyond the academy to a broader educated reading public. After several years teaching at the University of Chicago, Geertz became one of the few permanent professors at the Center for Advanced Study at Princeton, which hosts visiting academics for limited periods, often while they finish a book or research project. He clearly flourished in this role, and for decades his influence was acknowledged by a succession of scholars across a range of humanities disciplines, figures such as Robert Darnton (1984), Stephen Greenblatt (1980), William H. Sewell (1980) and Natalie Zemon Davis (1983). For many, 'thick description' became a favoured term for what they set out to accomplish in their imaginative construction of specific historical and literary worlds. And as a prolific reviewer in the *New York Review of Books*, Geertz also became a guide, for a wide educated readership, to the works of anthropology that ought to command their attention. Full-scale ethnographic monographs are not always as gripping for others as they can be for professional anthropologists, and Geertz's skill as an essayist no doubt accounts partly for his influence. After his early Java monographs, he published a joint book with his first wife, Hildred Geertz, on Balinese kinship (Geertz and Geertz 1975), and a large collaborative ethnography with her and Laurence Rosen on their fieldwork in Morocco (Geertz, Geertz and Rosen 1979), but these were nowhere near as widely read as *The Interpretation of Cultures* and *Local Knowledge*. The essays in these collections were very widely cited, and many of the more influential ethnographies of the late 1970s and 1980s more or less explicitly presented themselves as being in the interpretive mode, as Geertz had defined it (e.g. Ortner 1978; Meeker 1979; Rosaldo 1980; Lambek 1981; Rosen 1984).

Geertz's final ethnographic monograph, *Negara* (1980), was a work of historical ethnography, an imaginative reconstruction of the indigenous pre-colonial state in Bali. And it too was really a long essay, with the technical scholarship confined to exceptionally lengthy endnotes. The argument was that the gorgeous theatricality of those Balinese polities was not a superficial or illusory aspect, but the very essence of power in those systems. The political scientist's or sociologist's hard-headed 'mechanical' understanding of power was not the universally applicable underlying material reality they took it to be; it simply missed the reality. There is no such thing as a non-cultural political fact of the matter, and no difference between that and a 'cultural' gloss upon it. Grasping how the highly ritualised indigenous states in pre-colonial Bali worked would help us see that the exercise of power in 'our own' societies also takes place through symbolic meaning. It would teach us both that there are profoundly different ways of 'doing the state' in different cultures, but also that politics as we practise it is more 'cultural' than we know. So Geertz's claim was that the Balinese case enables us to gain sufficient distance that we can see the symbolic language in which our own 'texts' of power are written: 'the real is as imagined as the imaginary'.

As is often the Oedipal way with changes in theoretical orientation within a discipline, those who launched the most damaging attacks, when they came, on the kind of interpretive anthropology Geertz had championed, included some among his own former students and followers. The pivotal text was the collection, *Writing Culture* (Clifford and Marcus 1986), a book that had an extraordinarily disorienting effect on many anthropologists in the decade or so after its publication, provoking widespread talk of intellectual crisis, together with fears (and for some, hopes) that anthropology would be overpowered and eclipsed by cultural studies, and some extraordinarily bad writing. There was much talk of a general 'crisis of representation' in which two sets of ideas were combined, and sometimes confused, to call into question both the possibility and the politics of cultural interpretation as Geertz and others conceived it. First, arguments from the newly self-conscious enterprise of 'literary theory' – an amalgam (Eagleton [1983] was an early, influential synthesis) of Marxism and post-structuralist approaches such as Derrida's deconstructionism and Lacanian psychoanalytic theory – were deployed to stress the impossibility of stable reference amid the anarchic play of *différance* and dissemination, and so to undermine Geertzian confidence in shared standards and reasonably consensual judgement on the content and the quality of interpretations. And second, an early rendition of the ideas of Michel Foucault, as tendentiously simplified by Edward Said (1978), led to questioning not so much of the possibility as the political innocence of any act of cultural representation. Europeans' knowledge of non-European societies had been systematically distorted and falsified in the interests of imperial domination, caricaturing and libelling indigenous peoples as an excuse for their oppression, and forcing schematised classifications upon them as an aid to coercion and control. Unconsciously perhaps in some instances but nonetheless ineluctably, anthropological writings about 'exotic' societies were complicit in the political domination of colonial knowledge/power. Now, there is obviously a considerable tension between these two sets of ideas: if stable representation is as impossible as the deconstructionists claimed, it is hard to see how it could be as effective as Said and his followers accused it of having been, and vice versa. But this did not stop protagonists from asserting both at once, and their combined effect was a very uneasy sense that Geertzian ambitions to introduce other cultures into 'our' conversation were both bound to fail and nefariously damaging.

More than one of the substantive chapters in *Writing Culture* took direct aim at Geertz and the programmatic introductory essay by James Clifford (1986), entitled 'Partial truths', presented his style of anthropology as outmoded and in the process of being superseded by more sophisticated and politically correct forms. Largely ignoring the substantive contents of the ethnographic texts examined, Clifford presented a selective history of the discipline as a succession of dominant genres of ethnographic writing, with emphasis on how the ethnographer him or herself was represented in the text. This narrative was structured by a classification of ethnographic genres that was also, despite disclaimers to the contrary, a story of progress and a moral ranking. It began with 'participant–observer ethnography': works such as those by Bronislaw Malinowski and Margaret Mead that presented themselves as first-hand accounts by people who had participated in what they described, expressing a prescribed attitude of loose cultural relativism, showing a preoccupation with problems of language and translation, and with a representational strategy, whatever the thematic focus or theoretical orientation, of holism. Next in Clifford's sequence comes 'interpretive ethnography', that on his account struggled, through a series of evasions and dishonesties remorselessly exposed in Vincent Crapanzano's chapter (1986) on Geertz's famous 'Deep play' essay, to escape the consequences of its political complicity. But salvation is at hand, or at least being worked out, in the form of 'dialogic ethnography', in which the 'positionality' of the ethnographer is directly represented in the text, along with the circumstances and processes through which 'data' are generated in dialogue with informants, and crucially in which the voice of those informants is heard. This typically meant that such ethnographies were the record of dialogues between the anthropologist and either one or a very few others (e.g. Rabinow 1977; Crapanzano 1980; Shostak 1981; Dwyer 1982). While welcoming this new emerging mode of ethnographic writing, for Clifford, its limitation lay in the fact that the anthropologist retained editorial control and therefore the (necessarily colonialist, oppressive, distorting) authority of representing the voice of the other. Thus it remained complicit in the unequal power dynamics of the post-colonial order. But he proposed (and cited a few examples) ways of relinquishing or sharing authorial control in texts that would be internally fragmented mosaics of different voices. Clifford called this emergent new mode of ethnographic writing 'heteroglossic', a term he borrowed, along with 'dialogic', from Russian philosopher and literary theorist Mikhail Bakhtin (1981). He predicted that, as it developed, the limitations of the academic discipline of anthropology would be transcended in a more open, participatory, activist form of cultural engagement (see also Clifford 1988). There was rather more to the *Writing Culture* volume, it must be said, than was encapsulated in Clifford's introduction – it is hard not to read Paul Rabinow's contribution (1986) as a wittily deflationary raspberry blown at the rest of the volume – but it fairly soon became widely accepted as providing its take-home points.

It is uncontroversial, I think, to say that things have not gone quite as Clifford and his admirers predicted or recommended. The exemplars of heteroglossic ethnography he held up for approbation remain in merited obscurity, and largely unimitated. The fashion for dialogic texts ran its course: few anthropologists had the skill or disposition to write so extensively about themselves, in a way they could reasonably expect others to find interesting; most indeed found that what they wanted to say about the people and places they had studied required textual strategies – notably handling of evidence and argument, and engagement with existing literature – that could not be accommodated in the proposed new genre. And a few high-profile disappointments (e.g. Menchu, with Burgos-Debray 1984) took the shine off the assumed moral sanctity of deferred or dispersed authorial responsibility. More broadly, post-modernist literary theory proved of limited utility in addressing the conspicuous problems that the world threw up in the following decade (the collapse of socialism, new ethnic and nationalist conflicts, globalization) and it became hard to sustain interest or a claim to urgency and inclusivity, through what was all too frequently rebarbative and pretentious, jargon-ridden writing. Today, most anthropologists do describe in their writings more about the processes of fieldwork and their research practice than was conventional before the 1980s, but once again less than the extended breast-beating that was common in the years of *Writing Culture*'s immediate influence. The sub-genre of the anthropology of suffering (e.g. Biehl 2005; see discussion by Robbins 2013) is probably where that influence, in the sense of still-live engagement with its principal concerns through the means it brought to prominence, is still most directly visible.

There were, of course, critics of Geertz of a wholly different stamp from the *Writing Culture* group, who thought that with the opprobrium he was subjected to in that volume and elsewhere, he was reaping what he himself had sown. Ernest Gellner, for instance, had launched his career with a broadside attack on the whole movement of Wittgensteinian ordinary-language philosophy from which Geertz drew so much inspiration (Gellner 1959), and had in the meantime maintained that Geertz's account of Moroccan culture was touched as much by romantic idealisation as by philosophical idealism (1981). He charged also that it was Geertz who had opened the flood-gates to the excesses of postmodernism and relativism, as represented by *Writing Culture* (1992). Gellner argued that in exaggerating cultural difference (both the internal homogeneity of 'cultures' and the depth of differences between them) and in ignoring the explanatory importance of hard material political and economic realities, Geertz had abandoned the legacy of the Enlightenment and with it the big historical and comparative questions that it is anthropology's mission to address (for, in some ways, similar critiques of Geertz, though from first a Marxist and then a cognitivist perspective, see Bloch 1989; 2012). While Gellner was not at his most subtle or perceptive in his reading of Geertz, it is true that it is hard to see the commanding authority Geertz exercised in anthropology at the beginning of the 1980s giving rise to a sustained intellectual agenda. Few would-be imitators could replicate the virtuosity of his essays. And his fidelity to the Boasian notion of relatively homogeneous and relatively bounded cultures closed off virtually any possibilities of systematic comparison. He did espouse the notion of 'cultural systems' – the idea that religion, politics, common sense, among others were areas of life with certain common features, even in very different cultures – but that idea never received more than impressionistic expression and it was unclear what Geertz thought followed from it. Insofar as there was a Geertzian research programme, it consisted exclusively of the ambition of producing another interpretation of another culture, as fine as Geertz's own. No sense was developed of sequential or cumulative enquiry. And in one of his most high-profile performances at the height of his influence – the 'Distinguished Lecture' he delivered to the 1983 meeting of the American Anthropological Association (Geertz 1984) – he mounted a defence of the doctrine of cultural relativism that he elsewhere conceded to be 'witless' (2000), on the shallow grounds that he did not want to be associated with the most vocal, philosophically unsophisticated exponents of universalism, even though he knew perfectly well the choice between these positions is a false one. He remained in the end unwilling to abandon his explanatory dependence on the Boasian concept of culture, and there is no cogent way from there to block a relativist conclusion. On this occasion, a preference for indirection and evasion left the subject he was addressing (uncharacteristically) more confused than he had found it.

Interestingly, however, that same year, Geertz delivered a set of lectures – this time at Stanford – that although this was before *Writing Culture* had been published, formed the basis of a rather effective reply to it, when later revised and published as *Works and*

Lives (1988). That reply works on two levels. At the level of the whole book, there is an argument about the relation between forms of ethnographic writing and anthropological truth, that builds on a much more perceptive and sure-footed reading of Foucault than that which informed *Writing Culture*. Geertz summarily dismissed the idea that there ever could be a way of conveying ethnographic truth from which the singularity of the anthropologist as author could be excised. The idea of relinquishing authorship and letting the ethnographic subject and data speak directly is a mixture of a naive political fantasy (you cannot change the world that easily) with an unacknowledged lingering attachment to a positivist conception of truth (the Marxist idea of 'ideology' as motivated 'partial' distortion). The originality and force of the greatest and most influential anthropologists, Geertz sought to show, consists precisely in the fact that they have created new ways of telling, and therefore *new forms* of anthropological truth. This argument is illustrated in the substantive chapters of the book with a discussion of the literary styles, respectively, of Lévi-Strauss, Evans-Pritchard, Malinowski and Benedict. In each case, Geertz takes as his text a piece of writing that is not a formal exercise in academic anthropology. The implicit reply to *Writing Culture*'s claim, in its subtitle, to unveil 'the poetics and politics of ethnography' is that its poetics was simplistic and its politics naive and unrealistic.

But there is a second level to Geertz's reply because his specific remarks about *Writing Culture* occur in his chapter on Malinowski. The general point Geertz makes in that chapter, which concerns the diary Malinowski kept while conducting his Trobriand field-work, and that caused something of a scandal when published posthumously (Malinowski 1967), is that the anxieties expressed so tortuously in Malinowski's diary, that see him agonising about his lack of empathy with his subjects – his loneliness, frustrations, bouts of anger and resentment, and the disobliging stereotypes and profanities he used as a balm for these resentments – had the terrifying weight they had for Malinowski because they bore directly on the infirmities of his conception of anthropological truth. His naive realism led him to hope for a language that would correspond directly to the reality he wished to describe. His psychologism led him to think that to understand Trobriand culture he must penetrate the mysteries of individual Trobrianders' minds. Thus, his claim to authority, and confidence in the rightness of his analyses, seemed to him to depend not on the quality of his observation or argument, but on the personal relationships he was able to maintain with individual Trobrianders and the degree of mental identification he was able to achieve with them; not on professional skills and intelligence, but unexampled sensitivity and unimpeachable moral probity. And of course, by those latter criteria he was bound to fail. The result, says Geertz, was his 'diary disease'. Geertz takes up rather little space in the chapter with direct comment on the anxieties and concerns that fuelled *Writing Culture*, because his point was a simple one: they have landed themselves in the same predicament as Malinowski, because they have a similar naive ambition of gaining unmediated access to the psychic truth of their subjects. Thus, their writings too are smitten with 'diary disease', the symptoms now being a compulsion to try to convince their readers of the depth of their subjective identification with their informants, and the purity of their political sentiments.

And of course, the unspoken further claim of this immodest but impressive book is that to the list of four great anthropologists who have created their own distinctive and discipline-changing forms of anthropological truth, through the singularity and force of their way of writing, a fifth should rightly be added, because Geertz with some justice saw himself as having succeeded in inventing a distinctive new way of writing anthropology. He saw that achievement, ultimately, as untroubled by the objections of these critics. I rather think that, with all his limitations, this is turning out to be history's judgement too.

References

Austin, John L. 1962. *How to Do Things with Words*. Oxford: Clarendon Press.

Bakhtin, Mikhail 1981. *The Dialogic Imagination: Four Essays*, edited by Michael Holquist. Austin, TX: University of Texas Press.

Biehl, João 2005. *Vita: Life in a Zone of Abandonment*. Berkeley, CA: University of California Press.

Bloch, Maurice 1989. *Ritual, History and Power: Selected Papers in Anthropology*. London: Athlone.

——— 2012. *Anthropology and the Cognitive Challenge*. Cambridge: Cambridge University Press.

Clifford, James 1986. Introduction: Partial truths. In James Clifford and George E. Marcus (eds.), *Writing Culture: The Poetics and Politics of Ethnography*, pp. 1–26. Berkeley, CA: University of California Press.

——— 1988. *The Predicament of Culture: Twentieth-Century Ethnography, Literature, and Art*. Cambridge, MA: Harvard University Press.

Clifford, James and George E. Marcus (eds.) 1986. *Writing Culture: The Poetics and Politics of Ethnography*. Berkeley, CA: University of California Press.

Crapanzano, Vincent 1980. *Tuhami: Portrait of a Moroccan*. Chicago, IL University of Chicago Press.

——— 1986. Hermes' dilemma: The masking of subversion in ethnographic description. In James Clifford and George E. Marcus (eds.), *Writing Culture: The Poetics and Politics of Ethnography*. Berkeley, CA: University of California Press.

Darnton, Robert 1984. *The Great Cat Massacre: And Other Episodes in French Cultural History*. Chicago, IL, University of Chicago Press.

Dwyer, Kevin 1982. *Moroccan Dialogues: Anthropology in Question*. Baltimore, MD: Johns Hopkins University Press.

Eagleton, Terry 1983. *Literary Theory: An Introduction*. Oxford: Blackwell.

Geertz, Clifford 1960. *The Religion of Java*. New York, NY: Free Press.

——— 1963a. *Agricultural Involution: The Process of Ecological Change in Indonesia*. Berkeley, CA: University of California Press.

——— 1963b. *Peddlars and Princes: Social Change and Economic Modernisation in an Indonesian Town*. Chicago, IL: University of Chicago Press.

——— 1965. *The Social History of an Indonesian Town*. Cambridge, MA: MIT Press.

——— 1968. *Islam Observed: Religious Development in Morocco and Indonesia*. Chicago, IL: University of Chicago Press.

——— 1973. *The Interpretation of Cultures: Essays in Interpretive Anthropology*. New York, NY: Basic Books.

——— 1980. *Negara: The Theater State in Nineteenth-Century Bali*. Princeton, NJ: Princeton University Press.

——— 1983. *Local Knowledge: Further Essays in Interpretive Anthropology*. New York, NY: Basic Books.

——— 1984. Distinguished Lecture: Anti anti-relativism. *American Anthropologist* 86: 263–278.

——— 1988. *Works and Lives: The Anthropologist as Author*. Palo Alto, CA: Stanford University Press.

——— 1995. *After the Fact: Two Countries, Four Decades, One Anthropologist*. Cambridge, MA: Harvard University Press.

——— 2000. *Available Light: Anthropological Reflections on Philosophical Topics*. Princeton, NJ: Princeton University Press.

——— 2012. *Life Among the Anthros and Other Essays*, edited by Fred Inglis. Princeton, NJ: Princeton University Press.

Geertz, Clifford, Hildred Geertz and Lawrence Rosen 1979. *Meaning and Order in Moroccan Society*. Cambridge: Cambridge University Press.

Geertz, Hildred and Clifford Geertz 1975. *Kinship in Bali*. Chicago, IL: University of Chicago Press.
Gellner, Ernest 1959. *Words and Things: An Examination of, and an Attack on, Linguistic Philosophy*. London: Gollancz.
——— 1981. *Muslim Society*. Cambridge: Cambridge University Press.
——— 1992. *Postmodernism, Reason and Religion*. London: Routledge.
Greenblatt, Stephen 1980. *Renaissance Self-Fashioning: From More to Shakespeare*. Chicago, IL: University of Chicago Press.
Kripke, Saul A. 1982. *Wittgenstein on Rules and Private Language*. Oxford: Basil Blackwell.
Lambek, Michael 1981. *Human Spirits: A Cultural Account of Trance in Mayotte*. Cambridge: Cambridge University Press.
Lévi-Strauss, Claude 1962 [1953]. *Structural Anthropology*. New York, NY: Basic Books.
Malinowski, Bronislaw 1967. *A Diary in the Strict Sense of the Term*. London: Routledge and Kegan Paul.
Meeker, Michael E. 1979. *Literature and Violence in North Arabia*. Cambridge: Cambridge University Press.
Menchu, Rigoberta, with Elisabeth Burgos-Debray 1984. *I, Rigoberta Menchu: An Indian Woman in Guatemala*. London: Verso.
Ortner, Sherry B. 1978. *Sherpas Through their Rituals*. Cambridge: Cambridge University Press.
Rabinow, Paul 1977. *Reflections on Fieldwork in Morocco*. Berkeley, CA: University of California Press.
——— 1986. Representations are social facts: Modernity and post-modernity in anthropology. In James Clifford and George E. Marcus (eds.), *Writing Culture: The Poetics and Politics of Ethnography*, pp. 234–261. Berkeley, CA: University of California Press.
Radcliffe-Brown, A.R. 1952. *Structure and Function in Primitive Society*. London: Cohen and West.
Ricoeur, Paul 1973. The model of the text: Meaningful action considered as a text. *New Literary History* 5: 91–117.
Robbins, Joel 2013. Beyond the suffering subject: Toward an anthropology of the good. *Journal of the Royal Anthropological Institute* 19: 447–462.
Rosaldo, Michelle Z. 1980. *Knowledge and Passion: Ilongot Notions of Self and Social Life*. Cambridge: Cambridge University Press.
Rosen, Lawrence 1984. *Bargaining for Reality: The Construction of Social Relations in a Muslim Community*. Chicago, IL: University of Chicago Press.
Ryle, Gilbert 1949. *The Concept of Mind*. London: Hutchinson.
Said, Edward W. 1978. *Orientalism*. London: Penguin Books.
Sewell, William H. 1980. *Work and Revolution in France: The Language of Labour From the Old Regime to 1848*. Cambridge: Cambridge University Press.
Shostak, Marjorie 1981. *Nisa: The Life and Words of a !Kung Woman*. Cambridge, MA: Harvard University Press.
Taylor, Charles 1985. *Human Agency and Language: Philosophical Papers 1*. Cambridge: Cambridge University Press.
White, Leslie A. 1949. *The Science of Culture*. New York, NY: Grove Press.
Wittgenstein, Ludwig 1953. *Philosophical Investigations*. Oxford: Blackwell.
Zemon Davis, Natalie 1983. *The Return of Martin Guerre*. Cambridge, MA: Harvard University Press.

Chapter 9

The Frankfurt School, critical theory and anthropology

Christos Lynteris

Anthropological theory textbooks and papers often, and in a manner appearing in many cases to be apologetic, mention that the Frankfurt School's 'influence on anthropology has been minimal' (Morris 2014: 298). It may then be surprising to find a chapter on it, and critical theory more generally, in a volume showcasing the main theoretical trends in the discipline. Still, it is perhaps precisely this ability to remain and operate as it were *at the edge of sight* that makes critical theory so crucial for anthropology. This chapter does not aim to review the ways in which anthropologists have employed methods or notions derived from critical theory. Nor is it interested in excavating the 'influence' of the Frankfurt School on anthropology; the very notion being more astrological than analytical, or for that matter historical. Instead, my aim here is to illuminate the *critical theoretical potential* in anthropology, showing both ways in which it has been fruitfully actualised, and, most importantly, how it forms one of the rare inexhaustible undercurrents not of a restricted discipline but of anthropological thinking itself, as a capacity for creating new concepts of, and about, the social world in which we live and act.

In their landmark volume, *Anthropology as Cultural Critique*, George Marcus and Michael Fischer acknowledged the Frankfurt School as, 'perhaps the most important stimulus to the revitalised sense of cultural criticism among the younger generation of American anthropologists during the 1960s and 1970s' (1986: 119). However, they seemed unconvinced about the continuing relevance of critical theory, seeing the fragmentary nature of much of its writings as well as its theoretical, rather than empirical, orientation as severe limitations. In what reads as a reserved appraisal of its legacy, they concluded that, 'while attractive to the temper of the 1970s, however, the contributions of the early Frankfurt School leaves something to be desired now' (p. 122).

In fact, rather than spent, critical theory was poised to be a major force in the reshaping of the intellectual landscape in the social sciences and the humanities of the post-Cold War era. Central to this was the rediscovery of Walter Benjamin, and for the first time, the systematic translation of his works in English. Characteristically, writing two decades after *Anthropology as Cultural Critique*, Fischer would identify his own research as part of 'an increasing flood of work' (2009: 27) in dialogue with Benjamin's opus.

But what is critical theory? This question may be much harder to answer than it first appears. In light of the limited space, I will here relate only to the so-called first generation of the Frankfurt School and its associates, thus leaving out of my discussion the second generation of critical theorists, including philosophers like Jürgen Habermas and Axel Honneth. Also absent from my discussion are the works of Herbert Marcuse and Erich Fromm that, though vital components of the School at different stages of its intellectual and institutional life, fall better within the remit of psychoanalysis and anthropology, or, as the case may be for the former, anthropology and so-called Freudo-Marxism.

Outlining critical theory

Founded in 1923, the University of Frankfurt's Institute for Social Research (in short *Frankfurt School*) began its life as a Marxist-inspired institution, which under Max Horkheimer's direction (1930–1953) and with the help of its main publication, *Zeitschrift für Sozialforschung*, flourished into one of the most influential research institutes of the 20th century. Over the decades, it hosted path-breaking scholars such as Theodor W. Adorno, Erich Fromm, Herbert Marcuse, Jürgen Habermas, Franz Leopold Neumann, Leo Löwenthal and Axel Honneth, whilst associating and interacting with eminent thinkers like Walter Benjamin and Siegfried Kracauer. Banned in Nazi Germany in 1933, the Institute initially moved to Geneva and then to New York two years later, where it conducted, amongst other things, its landmark research on Nazism and on the 'authoritarian personality' (Adorno et al. 1951) before moving back to Germany in 1949. During these developments, different members and associates of the Institute followed diverse but often converging trajectories, whilst conversing vividly with other eminent scholars (many of them also exiles) such as Gerhard Scholem, Martin Buber, Ernst Bloch, Hannah Arendt and Bertolt Brecht. Benjamin, the only associate to remain in Europe, took his own life after an aborted attempt to flee France for Spain in 1940.

In spite of its institutional setting, critical theory never developed into a normative social theory, and hence, following Susan Buck-Morss (1977), the term may be said to lack substansive and historical precision. And yet it would be going too far to say, with Martin Jay (1996: 41), that it was simply a dialogical 'gadfly of other systems'. As Douglas Kellner (1994) has argued, this would be obscuring the fact that critical theory developed in direct relation to pressing social questions of its time; most pointedly the rise of fascism across Europe.

In order to fully understand the scope of this school of social theory, it is conventional and indeed helpful to take as a starting point Max Horkheimer's programmatic 1937 essay 'Traditional and critical theory'. There, critiquing the modelling of the social sciences upon the natural sciences (a hegemonic tendency at the time, see chapter 1), Horkheimer argued that, 'The facts which our senses present to us are socially performed in two ways: through the historical character of the object perceived, and through the historical character of the perceiving organ' (2002 [1944]: 200). Similarly opposed to positivism and to predictive statements in sociology, Horkheimer's close associate and philosopher extraordinaire, Theodor Adorno, maintained that the separation between research methods and the object of research was a mystifying process. Instead, Adorno underlined 'that the research methods employed should always be closely related to, and derived from, the social phenomena being studied' (Morris 2014: 333).

This approach was rooted in Marxian dialectics (see chapter 3), insofar as it drew on Marx and Engels' materialist reconfiguration of Georg Hegel's idea that history proceeded by means of a triadic process of thesis–antithesis–synthesis. Following Marx and Engels, it was only by turning Hegel on his head, and replacing his idealist focus on human consciousness with a materialist focus on human existence that the historical process could be understood as a dialectical process of social transformation 'with one social stage, the thesis, inevitably "sowing the seeds of its own destruction" by harbouring its opposite, the antithesis' (Erickson and Murphy 2008: 44). And yet critical theory was resistant to rendering dialectics into a theory of knowledge as promoted by Soviet-sponsored 'dialectical materialism'. It thus retained at its foundations a radical epistemological and methodological challenge of the relation between subject and object that directly confronted the 'paradigm of knowing as a theoretical representation of a wholly independent object domain' (Bernstein 1994: 1).

Developing an anthropological materialist approach entailed the rejection not only of objectivist ontology but also of the subject as the starting point and pivot 'of bourgeois philosophy' (Horkheimer 2002 [1944]: 211). On the one hand, this led to the recognition of the production of knowledge as a process that required the overcoming of subject/object dichotomies that dominated the social sciences at the time. And on the other hand, it also led to a recognition of the *in process* impact of social scientific research on its subject. In other words, constituting a key conceptual and methodological baseline for crucial developments in anthropology 50 years later (see chapter 8), critical theory ushered in an understanding that social research, in its many forms, does not only impact society through its written outputs, but as a mode of conduct that changes society at the same time as it is studying it; which, in the case of anthropology, means as we are conducting ethnographic fieldwork.

Adorno and Horkheimer's critique of the separation between thought and reality would culminate in their major work, *Dialectic of Enlightenment* (2002 [1944]). Their process of rendering reality into an object of examination or knowledge was linked to the production of rationalism that was, to use Bruce Kapferer's useful turn of phrase, 'instrumental in generating the supposed irrationalism that it encountered and often fought to control' (2007: 86); a dialectic reaching its apex in Nazism. Similarly, as Taussig (1989: 12) notes, in the various fragments composing *Passagen-Werk* or the 'Arcades Project' Benjamin followed the critical-theoretical investigative formula of looking for irrationalism in rationalism in order to expose how 'commodity, in its very modernity and mundaneness, conjured up the archaic and the exotic, the primitive and the mythic' (on Benjamin and myth see Menninghaus 1991; Mali 1999). This was important not only in light of its approach of dialectical reversibility (the process of the thesis containing the seeds of its antithesis, as shown above) as a core element of modern societies (Abélès 2008), but also because, following Adorno (correspondence in Benjamin 2006 [1935]), it allowed an analysis of what Marx had identified as the fetish character of commodities not simply as a fact of human consciousness, but as what, under specific historical conditions, produces and organises consciousness.

This technique of 'identifying archaic elements with the most modern phenomena' (Buck-Morss 1977: 58) was key to critical theory in its many forms, with theoretical affinities and potentials for anthropological analysis being particularly pronounced. In illuminating (rather than 'resolving' in the Hegelian sense) the contradictory character of modernity, it was a method aimed not at revealing some self-contained reality but, to use Benjamin's terms, at awakening us *into* (not *from*) the dream that structures modern life. Central to this project was the employment of a 'microscopic gaze' (a phrase Adorno used for Benjamin): 'a means for making the very particularity of the object release a significance that dissolved its reified appearance and revealed it to be more than a mere tautology, more than simply identical to itself' (p. 74). This was a gaze that retained the particularity of the minute social and cultural phenomena under examination whilst at the same time going beyond their 'given' immediacy. Whereas also key to Adorno's studies of 'cultural production' (a field whose impact on media studies has been immense; see Adorno 2001), no work exemplifies this method more than *The Arcades Project*, which can be seen as 'a historical lexicon of the capitalist origins of modernity, [and] a collection of concrete, factual images of urban experience' (p. 33) whose key operation is, 'telescoping the past through the present' (Benjamin 2002: 588). For it is there that we see emerge the 'dialectical image'; a notion, or indeed method, whose analytical and conceptual potential has fascinated, perplexed and stimulated anthropologists interested in illuminating the constitutive contradictions of modern life (for Adorno's critique of the notion, see Benjamin 2006).

Dialectical image

The dialectical image is an image that, assuming a broader critical-theoretical perspective, we may say operates as a 'switch' insofar as it 'arrests fleeting phenomena' and 'sets reified objects in motion' (Buck-Morss 1977: 106), so as to present us with the non-identity of a historical or ethnographic moment as its unintentional truth:

> It's not that what is past casts its light on what is present, or what is present its light on the past; rather, image is that wherein what has been comes together in a flash with the now to form a constellation. In other words, image is dialectics at a standstill.
>
> (Benjamin 2002: 462/N2a, 3)

'Benjamin's history', Uros Cvoro (2008: 92–93) explains:

> happens through the index of images, whereby the index is both the recognition of a specific historical time to which an image belongs (*Then*) and the recognition of another time in which such an image first became recognizable (*Now*), and the site of their confrontation.

The dialectical image's key function is thus 'to reveal the underlying tensions in history between different conceptions of temporality and difference' (p. 89). Let us look at an example whose impact not only in anthropology but also across the social sciences and the humanities has been far reaching: the flâneur.

In his extensive work on the mid-19th-century French poet, Charles Baudelaire found in *The Arcades Project* (for a systematic edition of these fragments see Agamben et al. 2013), Benjamin explores the figure of the flâneur; a middle-class idler who aimlessly strolls the boulevards and the shopping arcades of the French capital (the 'capital of the 19th century'), lost in the phantasmagoria of merchandise:

> The street becomes a dwelling for the flâneur; he is as much at home among the facades of houses as a citizen is in his four walls. To him the shiny, enameled signs of businesses are at least as good a wall ornament as an oil painting is to the bourgeois in his salon. The walls are the desk against which he presses his notebooks; news-stands are his libraries and the terraces of cafés are the balconies from which he looks down on his household after his work is done.
>
> (2002: 37)

Ushering in not just a new experience of the city but a new anthropological type, for Benjamin, the flâneur 'provide[d] a model for the general relationship between consciousness and experience that became dominant in metropolitan centres in ... the era of high capitalism' (Brand 1991: 7–8). From this perspective, in turn, the flâneur has been approached in anthropology and more broadly the social sciences as a dialectical image of 'our own consumerist mode of being-in-the-world' (Buck-Morss 2006: 35); a figure whose disappearance from the streets of our metropoles has given way, as Adorno (1939) argued, to the 'aural flâneurie' of radio station-switching or indeed, more recently, internet surfing in the form of the 'cyberflâneur' (Manovich 2001; Hartmaan 2004; Hogan 2016). In these terms, as Dana Brand discusses in her acclaimed history of flâneurie in America, through the prism of the dialectical image, this figure is much more than simply a model for what has been or a predecessor of what is; it rather functions, in Benjamin's work, as an exemplification of the anthropological principle (as applied to modernity) that 'every epoch dreams of its successor', but also as a prism for our own society whose 'unquestioned cognitive status' (Lauster 2007: 139) and critical potential is evident from the fact that it visibly continues to excite, puzzle and irritate scholars, critics and commentators (Stephen 2013; Livingstone and Gyarke 2017).

Examining the dialectical image, this famously enigmatic and, as Eduardo Cadava (1997) has shown, 'photographic' notion, that 'never achieved terminological consistency' (Tiedemann 1999: 942), Max Pensky (2006: 117) explains that it 'challenges the familiar Kantian notion of understanding as the capacity to generate knowledge of the world through some rule-governed application of concepts to sensory data'. In this sense, it led not only to a reversal of 'the polarity of the relation between subject and object' (Buck-Morss 1977: 83), but also to an unprecedented turn of analytical attention towards the breaks, cracks and splinters in the social and logical structures under examination. Thus, beyond the relational epistemology carried over by Benjamin's approach (that is also present in Adorno's 'negative dialectics'), what is important is to note that, being simultaneously the *site* and *method* of cultural critique, the dialectical image constitutes 'the scene, space and form of a certain temporal rupture in which time and space are out of joint' (Richter 2006: 148). It is, in other words, to paraphrase Sigrid Weigel (1996), what allows us to see and engage with what lies beyond the ethnographic continuum, or what Vassos Argyrou (2002) has identified as anthropology's 'will to meaning'. 'As an image flashing up in the now of its recognizability [*Erkennbarkeit*]' (Benjamin 2002: 473/ N9,7) the dialectical image emerges out of archival or ethnographic fragments as a constellation that disallows a quick analytical retreat into 'meaning', forcing us instead to take recourse to another method of sense-making: montage.

In its juxtaposition and, at the same time, connection of fragments into hitherto unauthorised constellations, Marcus (1995b) has shown that the method of montage has been key to experiments in ethnographic writing. As variably pioneered and employed by Sergei Eisenstein, the Surrealists and Bertolt Brecht, montage is then not simply a technique but a method and simultaneously a key modernist theoretical concept (p. 37). As Marcus notes, in the field of anthropology a landmark work in the application of this method has been Michael Taussig's pivotal examination of the colonial 'heart of darkness' in the Colombian Amazon, *Shamanism, Colonialism and the Wild Man* (1986), where montage was skilfully employed both 'in its capacity to disrupt the orderly narrative of social science writing and in its capacity as a performative discourse of healing in response to the history of terror and genocide' (Marcus 1995b: 47–48).

To carry over (paraphrasing Benjamin) the principle of montage into ethnography is no less than to forge new, illuminating, and at the same time, Pensky (2004: 186) reminds us, *necessary* relations between the salvaged fragments. By contrast to what we may call a 'systemic' outlook underlining diverse and even opposing anthropological schools over the decades, this method then authorises not only the recognition of usually overlooked or devalued fragments of social life, but also the radical transformation of their use value insofar as it fosters an anti-contemplative, anti-panoramic, and, as a result, counter-reifying, anthropological approach where the ethnographer assumes the guise of the 'ragpicker'; a Benjaminian figure which by contrast to the flâneur (a fundamentally romantic, impulsive and rhapsodic character) is at one and the same time 'methodic, reflexive and implacable' (Berdet 2012: 425):

> Whistling, nose in the air, distracted by illuminated panels, captivated by the latest fashion, a man wanders on the pavestones of his reveries, thoughtfully escaping the capitalist demand to be useful. Grumbling, frowning, scanning the ground, obsessed with nooks, with dark corners, with objects abandoned by society, another man snoops compulsively on the steps of the first, conferring a new utility to everything that is 'no longer of use'. The first is the flâneur, the other the ragpicker.
>
> (p. 425; my translation, as approved by the author)

Whereas advocates of 'postmodern' ethnography (Köpping 2005; Soukup 2012) and some urban anthropologists (Jenks and Neves 2000; Lucas 2004; Kramer and Short 2011; Nas 2012) have, each from a different perspective, endorsed flâneurie as a method of anthropological research (for a review, see Coates 2017), one wonders whether the ragpicker may not present a more fruitful figure for methodological innovation. For what characterises the ragpicker in his or her systematic trajectory is an 'attentive[ness] to the new, without succumbing to madness' (Berdet 2012: 428). Not mentioning the figure of the ragpicker in her *Anthropological Practice*, Judith Okely (2012) comes close to a similar formulation, when she employs André Breton's notion of *disponibilité* to describe the need for the anthropologist's openness to unexpected encounters: '[T]his thirst for wandering in search of everything, I shall be sustained in mysterious communication with other available or *disponible* beings' (Breton 1937 in Okely 2012: 54). As with Berdet's reading of the ragpicker, this proclivity to serendipity is conditioned upon one's attentive receptiveness: 'the seeker must be ready (*guetter*), in a state of attentiveness' (p. 55). By proceeding attentively through the fissures of the social fabric and by assuming its detriti and ruins as its empirical and at the same time conceptual habitus, anthropological thought may then be able, to repeat Adorno, 'to acquire the density of experience without losing any of its rigour' (in Richter 2006: 148; for a comparative perspective on the *historian* as a ragpicker see Wohlfarth 1986).

Mimesis

Critical theory, and Benjamin's work in particular, has found extensive if 'fragmentary' employment in anthropology. Although some applications are inevitably frivolous, seen as a *method*, this fragmentary use of the fragmentary has led to several successful, illuminating anthropological analyses, such as the critical reading of Hurricane Katrina with Benjamin's 'Flooding of the Mississippi 1927' radio broadcast (Fischer 2009), Arjun Appadurai's approach to globalization through the notion of 'mechanical reproduction' (1996) or readings of the artist as ethnographer through the lens of the 'author as producer' (Marcus

1995a: 303). Equally fruitful has been the debate on the application of critical-theoretical notions of fragmentariness and ruination in the discussion of the productive side of destructiveness in several anthropological fields both embracing and critically distancing themselves from 'postmodern' anthropological fascination with the fragmentary (Gordillo 2001; Navaro-Yashin 2009; Lee Dawdy 2010; Ladwig et al. 2012; Stoler 2013).

Still, there is no doubt that anthropological thinking has actualised this critical-theoretical potential most fruitfully with regard to the question of mimesis. Although studies of how humans imitate each other and what lies beyond themselves go back to Plato and Aristotle and were developed extensively by influential modern thinkers like Samuel Taylor Coleridge and Erich Auerbach, the employment of the notion in contemporary anthropology is firmly anchored in a short but conceptually rich work by Benjamin, 'On the mimetic faculty'; a draft that was never published while the author was alive, and was a reworked abbreviated edition of his earlier 'The doctrine of the similar' (1979).

Treating 'the powerful compulsion to become similar and to behave mimetically' (Benjamin 2005: 720) as a key anthropological trait, Benjamin sought to examine the actualisation and transformation of mimesis within modernity. 'On the mimetic faculty' may then be said to represent a key moment in the critical-theoretical examination of the 'change in the structure of experience' (Benjamin 2003: 314) – a milieu of great anthropological importance.

Written in agonistic dialogue with then current anthropological theories (Lévy-Bruhl and Cassirer), Marxist linguistics (Marr and Vigotsky), Freudian psychoanalysis and Kabbalist mysticism (Rabinbach 1979; Hanssen 2004), the essay reflected more widely Benjamin's anthropological materialism in claiming, as its concluding remarks that:

> language may be seen as the highest level of mimetic behavior and the most complete archive of nonsensuous similarity: a medium into which the earlier powers of mimetic production and comprehension have passed without residue, to the point where they have liquidated those of magic.
>
> (2005: 722)

Although the essay is perhaps most faithfully interpreted through the prism of Benjamin's profound and no less complex materialist philosophy of language (Hanssen 2004), its socio/cultural anthropological reading, first attempted to great acclaim by Michael Taussig (1993), relied more on Susan Buck-Morss's reception of the work in tandem with Benjamin's rather less impenetrable writings on photography and more generally mechanical reproduction (Benjamin 2008). 'These technologies', Buck-Morss (1989: 267) claimed, 'provide human beings with unprecedented perceptual acuity, out of which, Benjamin believed, a less magical, more scientific form of mimetic faculty was developing in his own era'. This reading managed to create an unexpected and fruitful synergy between different motions (and eras) in Benjamin's thinking. For Taussig, it provided an opportunity to explore the notion that modern mass culture 'both stimulates and is predicated upon mimetic modes of perception' (1993: 20) and to ask in which ways, in the context of colonialism, 'just as histories enter into the functioning of the mimetic faculty, so the mimetic faculty enters into those histories' (p. xiv).

Although it did not escape criticism by more normative scholars of critical theory (e.g. Jay 1993), Taussig's analysis proceeded by means of a multilayered and entangled approach of mimetic phenomena. These centrally included the employment of figurines carved in the form of Europeans in shamanic healing rituals amongst the Cuna of San Blas (Panama), and Hauka spirit possession rituals in Niger. The question linking these cases was one related to mimesis as a technology of becoming other and, more specifically, the colonial other. Taussig sought to illuminate the implications of this as regards, on the one hand, anti-colonial resistance and, on the other hand, European encounters with these mimetic phenomena; the latter including the anthropologist's own encounter with them, and the way in which, 'The very mimicry corrodes the alterity by which [our] science is nourished' (p. 8). In this way, Taussig ingeniously rendered mimesis into an anthropological dialectical image and at the same time into a dialectical image of anthropology.

For the purposes of this chapter, I will briefly discuss Taussig's Hauka example, as it is the one that has brought his approach of the mimetic faculty in the broadest dialogue with other anthropologists and their work. Examining the Hauka movement amongst the Songhay in the late 1920s, Taussig focused on spirit possession practices. In particular, his attention was caught by the way in which the Hauka engaged in the mimicry of colonial figures; a process in which the possessed copied the posture and mannerisms of colonial officers, or even, with incredible choreographic dexterity, the motion and sound of colonial locomotives, whilst at the same time including bodily traits, such as frothing, body jerking or the bulging of eyes, that were quite removed from a European or colonial body image.

Approaching this contradiction in critical-theoretical terms, insofar as it was seen as a dialectical image of colonialism, Taussig argued that:

> It's the ability to become *possessed*, the ability that signifies to Europeans awesome Otherness if not downright savagery, that allows them to assume the identity of the European, and, at the same time, stand clearly and irrevocably eye-bulgingly apart from it.
>
> (1993: 241)

Still, Taussig was more interested in the 'capture' of this movement by arguably anthropology's greatest auteur, Jean Rouch, in his famous (and for decades banned) documentary *Les maîtres fous*; a film whose powerful effect he traces back to the 'interaction of miming bodies and mimetic machinery' (p. 244); that is, the bodies of the possessed and the motion camera used to capture their movement.

Taussig's understanding of Rouch's film as an instance of 'an era of the borderland where "us" and "them" lose their polarity and swim in and out of focus' (p. 246) borrows from Brecht's notions of dramatic distanciation/estrangement (the so-called *V-effekt*) (see Jameson 1998) and also from Roger Caillois's work on mimicry not as a process of striving to be similar to something, but just to be similar (1984), so as to provocatively proclaim that, 'What's being mimicked is mimicry itself' (Taussig 1993: 241). Following Buck-Morss's synthesis of Benjamin's work, Taussig then goes on to argue that the film:

> borrows from the magical practice of mimesis in its very filming of it . . . In this colonial world where the camera meets those possessed by gods, we can truly point to the Western rebirth of the mimetic faculty by means of modernity's mimetic machinery.
>
> (p. 242)

Although this reading of mimesis, and of Rouch's film in particular, has not remained uncontested (e.g. Kien 2002), Taussig's approach of the mimetic faculty sparked a lasting debate in postcolonial studies and visual anthropology, and at the same time related to core anthropological questions insofar as it postulated that mimesis is a fundamentally sensuous process.

Developed at a time when anthropology and the social sciences as a whole assumed a new and, in some occasions, rigorous interest in the body and embodiment, this provided fertile ground for discussion, leading Paul Stoller (1995) to argue for a mode of 'corporal knowing' that involved processes such as sorcery. Returning to Taussig's reading of Frazer's notion of sympathetic magic, or the idea that 'the magician infers that he can produce any effect he desires merely by imitating it' (1993: 47) via Benjamin's notion of mimesis, Stoller gives us the example of a Songhay sorcerer who crafts copies of magical bows and arrows to which he speaks, 'naming a victim'. Shooting the arrow, Stoller writes, in what may be counted in itself as a powerful moment of mimetic writing in anthropological literature:

> [T]he replica falls harmlessly on the ground in the victim's dwelling or village, but the 'inner' arrow flies through the night air. And if a sorcerer's aim is good . . . the 'inner' arrow strikes its victim. Victims will wake up in the middle of the night, screaming, with pain shooting up their leg. Once struck, they become increasingly weaker. And if they don't seek a cure, they will most certainly die from an invisible (inner) wound.
>
> (1995: 41)

Forming an important part of contemporary anthropological studies of embodiment, the mimetic faculty has more recently also found a novel and challenging application in a prominent example of anthropology's ontological turn (see chapter 14). In his book *Soul Hunters*, Rane Willerslev (2007) explored the way in which moose hunting amongst the Yukaghirs of Siberia is predicated upon the power of mimesis. Arguing for an approach of 'animism as mimesis', Willerslev weaves an intricate ethnography that contests the view that the former is a system whereupon humans treat animals 'as if' they were persons. Going beyond a metaphorical reading of animism requires, Willerslev argues, an approach of it as a mimetic practice through which animals and humans are implicated in a 'paradoxical situation of mutual mimicry' (p. 11). In a move characteristic of anthropological trends in the last decade, Willerslev thus turns away from the hitherto dominant study of mimesis in relation to colonialism, in order to explore the faculty's ontological potential.

Focusing his attention on the way in which 'the copy ha[s] to resemble the original in order for the copy to take power over what it is a copy to' (p. 11), he asks us to rethink mimesis as a process of imperfect doubling. Willerslev thus asserts that it is this imperfection, 'copiedness' or difference, rather than sensuous contact, that is the operative faculty in mimesis; what, in other words, allows the imitator (hunter) to exercise power over the imitated (prey) by being able to maintain a double perspective – that of being not-self and not-not-self, not-animal and not-not-animal, at one and the same time. If we want to talk about 'perspectivism', or the way in which human/non-human relations rely on the difference and exchange of perspectives of themselves and of others (especially as this applies in Amazonia, as shown by Eduardo Viveiros de Castro; see chapter 14), Willerslev argues, in the Yukaghir context this needs to be seen as a process of 'mimetic empathy' based not on 'moving from one point of view to another' but rather on 'not surrendering to a single point of view' (p. 110; see also Bubandt and Willerslev 2015).

Prospects and perils

Moving thus from political resistance to ontological suspension by way of corporeal knowledge and perspectival empathy, the critical-theoretical notion of mimesis may be said to operate not simply as a dialectical image of the societies under examination but at the same time as an image of the critical-theoretical potential of anthropological thinking at different historical moments. What is unique about this potential is the way in which it differs from other anthropological traditions of critiquing settled or so-called master narratives, particularly prevalent in the discipline since the 1980s. Whereas the aim of the former, is to 'deconstruct' and 'demystify' social realities, the primary aim of critical theory is not negation but redemption: in other words, 'to facilitate the construction of new forms of social life from the glimpses provided of alternative futures when otherwise concealed or forgotten connections with the past [are] revealed' (Taussig 1984: 89). To forget this, and to read the anthropological potential of critical theory as part or an appendix of 'postmodern' or deconstructive approaches in the discipline would be to decouple the said dialectical notions both from their history and from their emancipatory potential. Reducing them thus to intellectual fetishes, in other words to abstractions that may be 'applied' to different ethnographic contexts with more or less currency or 'success', poses an acute analytical peril, all the more as critical theory's position outside the social theoretical canon of anthropology makes it particularly vulnerable to misuse. Examples of this range from political anthropological uses of Benjamin's ideas of 'exception' and 'state of emergency' (especially as elaborated upon by the Italian philosopher Giorgio Agamben) that ignore both the complex theoretical genealogy and the dialectical potential of the notions (Jennings 2011), to the pervasive habit to decontextualise and trivialise poetically evocative but also philosophical complex tracts from the former's *Theses on the Philosophy of History* such as Benjamin's commentary on the image of Paul Klee's Angelus Novus, who is propelled by the storm of progress 'into the future to which his back is turned, while the pile of debris before him grows skyward' (1999: 249).

Critical theory and the writings of the Frankfurt School and its associates stand at a paradoxical relation with institutional socio/cultural anthropology. Whilst the critical potential of its anthropological materialist method and analysis pervades socio/cultural anthropological approaches, often becoming visible in the explicit employment of notions such as mimesis, ruination or exception, with the exception of Taussig, and by contrast to urban studies, geography or political theory, anthropology as a discipline has not systematically engaged with the critical-theoretical corpus. Much more than the proverbial reading 'against the grain', as a major European tradition of critical engagement with modernity and its darkest consequences, critical theory remains for anthropology a vast and challenging resource of social theoretical thinking.

References

Abélès, M. 2008. Anthropologie, anticipation, utopie. *Cahiers d'anthropologie sociale 4: Walter Benjamin, la tradition des vaincus*: 13–25.

Adorno, T.W. 1939. *Radio Physiognomik*. Frankfurt-am-Maim: Adorno Estate.

Adorno, T.W. 2001. *The Culture Industry: Selected Essays on Mass Culture*. London: Routledge.

Adorno, T.W., E. Frenkel-Brunswik and D.J. Levinson 1990 [1951]. *The Authoritarian Personality*. London: Norton.

Agamben, G., Chitussi, B. and Härle C.-C. 2013. *Baudelaire, by Walter Benjamin*, translated by Patrick Charbonneau, Martin Rueff, and Étienne Dobenesque. Paris: La Fabrique éditions.

Appadurai, A. 1996. *Modernity at Large: Cultural Dimensions in Globalization*. Minneapolis, MN: University of Minnesota Press.

Argyrou, V. 2002. *Anthropology and the Will to Meaning: A Postcolonial Critique*. London: Pluto Press.

Benjamin, W. 1979 [1935]. Doctrine of the similar, translated by Knut Tarnowski. *New German Critique* 17, Special Walter Benjamin Issue (Spring 1979): 65–69.

——— 1999 [1940]. *Illuminations*, translated by Harry Zorn. London: Pimlico.

——— 2002. *The Arcades Project*, translated by Howard Eiland and Kevin McLaughlin. Cambridge, MA: Harvard University Press.

——— 2003 [1940]. *On Some Motifs in Baudelaire*. In Howard Eiland and Michael W. Jennings (eds) Walter Benjamin Selected Writings, Volume 4, 1938–1940, translated by Edmunt Jephcott and others, pp. 313–355. Cambridge MA: The Belknap Press of Harvard University Press.

——— 2005 [1933]. On the mimetic faculty. In M.W. Jennings, H. Eiland and G. Smith (eds.), *Walter Benjamin Selected Writings, Volume 2, Part 2, 1931–1934*, translated by Rodney Livingston and others, pp. 720–722. Cambridge, MA: The Belknap Press of Harvard University Press.

——— 2006 [1935]. Exchange with Theodor W. Adorno on the essay 'Paris, the capital of the nineteenth century'. In *Walter Benjamin Selected Writings, Volume 3, Part 2, 1935–1936*, translated by Edmund Jephcott, Howard Eiland and others, pp. 50–67. Cambridge, MA: The Belknap Press of Harvard University Press.

——— 2008 [1936]. The Work of Art in the Age of Mechanical Reproduction. London: Penguin.

Berdet, M. 2012. Chiffonnier contre flâneur: Construction et position de la Passagenarbeit de Walter Benjamin. *Archives de Philosophie* 75 (3): 425–447.

Bernstein, J. 1994. Introductory essay: Critical theory – the very idea (reflections on nihilism and domination). In J. Bernstein (ed.), *Frankfurt School, Critical Assessments*, Vol.1, pp. 1–40. London and New York, NY: Routledge.

Brand, D. 1991. *The Spectator and the City in Nineteenth Century American Literature*. Cambridge: Cambridge University Press.

Bubandt, B. and Willerslev, R. 2015. The dark side of empathy: Mimesis, deception, and the magic of alterity, *Comparative Studies in Society and History* 57 (1): 5–34.

Buck-Morss, S. 1977. *The Origin of Negative Dialects. Theodor W. Adorno, Walter Benjamin, and the Frankfurt School*. New York, NY: The Free Press.

——— 1989. The Dialectics of Seeing. *Walter Benjamin and the Arcades Project*. Cambridge, MA: MIT Press.

——— 2006. The flâneur, the sandwichman and the whore: The politics of loitering. In B. Hanseen (ed.), *Walter Benjamin and The Arcades Project*, pp. 33–65. London and New York: Continuum.

Cadava, E. 1997. *Words of Light: Theses on the Photography of History*. Princeton, NJ: Princeton University Press.

Caillois, R. 1984 [1935]. Mimicry and Legendary Psychasthenia, translated by John Shepley. October 31 (Winter): 16–32.

Coates, J. 2017. Key figure of mobility: The flâneur. *Social Anthropology* 25 (1): 28–41.

Cvoro, U. 2008. Dialectical image today. *Continuum, Journal of Media and Cultural Studies* 22 (1): 89–98.

Erickson, P.A. and L.D. Murphy 2008. *A History of Anthropological Theory*, 3rd edition. Ontario: University of Toronto Press.

Fischer, M.M. J. 2009. *Anthropological Futures*. Durham, NC and London: Duke University Press.

Gordillo, G. 2011. Ships stranded in the forest: Debris of progress on a phantom river. *Current Anthropology* 52(2): 141–167.

Hartmaan, M. 2004. Technologies and utopias: The cyberflaneur and the experience of 'being Online' (Doctoral thesis, University of Westminster, London).

Hanssen, B. 2004. Language and mimesis in Walter Benjamin's work. In D.S. Ferris (ed.), *The Cambridge Companion to Walter Benjamin*, pp. 54–72. Cambridge: Cambridge University Press.

Hogan, A. 2016. Network ethnography and the cyberflâneur: Evolving policy sociology in education. *International Journal of Qualitative Studies in Education* 29 (3): 381–398.
Horkheimer, M. 2002 [1937]. Traditional theory and critical theory. In M. Horkheimer, *Critical Theory, Selected Essays*, translated by M.J. O'Connell et al., pp. 188–243. London: Continuum.
Horkheimer, M. and T.W. Adorno 2002 [1944]. *Dialectic of Enlightenment, Philosophical Fragments*, translated by Edmund Jephcott. Palo Alto, CA: Stanford University Press.
Jameson, F. 1998. *Brecht and Method*. London: Verso.
Jay, M. 1993. Unsympathetic magic. *Visual Anthropology Review* 9: 2 (Fall): 79–82.
——— 1996 [1973]. *The Dialectical Imagination: A History of the Frankfurt School and the Institute of Social Research 1923–1950*. Berkeley, CA: University of California Press.
Jenks, C. and T. Neves 2000. A walk on the wild side: Urban ethnography meets the flâneur. *Cultural Values* 4: 1–17.
Jennings, R.C. 2011. Sovereignty and political modernity: A genealogy of Agamben's critique of sovereignty. *Anthropological Theory* 11 (1): 23–61.
Kapferer, B. 2007. Anthropology and the dialectic of enlightenment: A discourse on the definition and ideals of a threatened discipline. *The Australian Journal of Anthropology* 18 (1): 41–62.
Kellner, D. 1994 [1973]. The Frankfurt School revisited: A critique of Martin Jay's *The Dialectical Imagination*. In J. Bernstein (ed.), *Frankfurt School, Critical Assessments*, Vol. 1, pp. 1–40. London and New York, NY: Routledge.
Kien, K. L. 2002. Of mimicry and white man: A psychoanalysis of Jean Rouch's *Les maitres fous*. *Cultural Critique* 51 (Spring 2002): 40–73.
Köpping, K-P. 2005. The fieldworker as performative flâneur: Some thoughts on postmodernism and the transfiguration of doing anthropology. *Zeitschrift für Ethnologie* 130 (1): 1–22.
Kramer, K. and J.R. Short 2011. Flânerie and the globalizing city. *City* 15: 322–342.
Ladwig, P., R. Roque, O. Tappe, C. Kohl and C. Bastos 2012. Fieldwork Between folders: Fragments, traces, and the ruins of colonial archives. *Max Planck Institute for Social Anthropology Working Papers*, Working Paper No. 141.
Lauster, M. 2007. Walter Benjamin's myth of the 'flâneur'. *The Modern Language Review* 102 (1): 139–156.
Lee Dawdy, S. 2010. Clockpunk anthropology and the ruins of modernity. *Current Anthropology* 51 (6): 761–793.
Livingstone, J. and L. Gyarkye 2017. Death to the flâneur. *New Republic* (27 March), https://newrepublic.com/article/141623/death-flaneur.
Lucas, R. 2004. Inscribing the city: A flâneur in Tokyo. *Anthropology Matters* 6 (1): 1–11.
Mali, J. 1999. The reconciliation of myth: Benjamin's homage to Bachofen. *Journal of the History of Ideas* 60 (1): 165–187.
Manovich, L. 2001. *The Language of New Media*. Cambridge, MA: MIT Press.
Menninghaus, W. 1991. Walter Benjamin's theory of myth. In G. Smith (ed.), *On Walter Benjamin: Critical Essays and Recollections*, pp. 292–325. Cambridge, MA: MIT Press.
Marcus, G. 1995a. *The Traffic in Culture: Refiguring Art and Anthropology*. Berkeley, CA: University of California Press.
——— 1995b. The modernist sensibility in recent ethnographic writing and the cinematic metaphor of montage. In L. Devereaux and R. Hillman (eds.), *Fields of Vision: Essays in Film Studies, Visual Anthropology, and Photography*, pp. 35–55. Berkeley, CA: University of California Press.
Marcus, G. and M. M. J. Fischer 1986. *Anthropology as a Cultural Critique: An Experimental Moment in the Human Sciences*. Chicago, IL: Chicago University Press.
Morris, B. 2014. *Anthropology and the Human Subject*. Bloomington: Trafford.
Nas, E. 2012. The urban anthropologist as flâneur: The symbolic pattern of Indonesian cities. *Wacana, Journal of the Humanities of Indonesia* 14: 429–454.
Navaro-Yashin, Y. 2009. Affective spaces, melancholic objects: Ruination and the production of anthropological knowledge. *Journal of the Royal Anthropological Institute* 15(1): 1–18.

Okely, J. 2012. *Anthropological Practice: Fieldwork and the Ethnographic Method.* London: Berg.

Pensky, Max 2004. Method and time: Benjamin's dialectical images. In David S. Ferris (ed.), *The Cambridge Companion to Walter Benjamin*, pp. 177–198. Cambridge: Cambridge University Press.

——— 2006. *Geheimmittel*: Advertising and dialectical images in Benjamin's *Arcades Project*. In B. Hanseen (ed.), *Walter Benjamin and The Arcades Project*, pp. 113–131. London and New York, NY: Continuum.

Rabinbach, A. 1979. Introduction to Walter Benjamin's 'Doctrine of the similar'. *New German Critique*, 17 Special Walter Benjamin Issue (Spring): 60–64.

Richter, G. 2006. A matter of distance: Benjamin's *One-Way Street* through *The Arcades*. In B. Hanseen (ed.), *Walter Benjamin and the Arcades Project*, pp. 132–256. London and New York, NY: Continuum.

Soukup, C. 2012. The postmodern ethnographic flâneur and the study of hyper-mediated everyday life. *Journal of Contemporary Ethnography* 42 (2): 226–254.

Stephen, B. 2013. In praise of the flâneur. *The Paris Review*. October 17, https://www.theparisreview.org/blog/2013/10/17/in-praise-of-the-flaneur/.

Stoler, A.L. 2013. *Imperial Debris: On Ruins and Ruination*. Durham, NC: Duke University Press.

Stoller, P. 1995. *Embodying Colonial Memories: Spirit Possession, Power and the Hauka in West Africa*. New York, NY and London: Routledge.

Taussig, M. 1984. History as sorcery. *Representations* 7 (Summer): 87–109.

——— 1986. *Shamanism, Colonialism and the Wild Man. A Study in Terror and Healing*. Chicago, IL: University of Chicago Press.

——— 1989. History as commodity. In some recent American (anthropological) literature. *Critique of Anthropology* 9 (1): 7–23.

——— 1993. *Mimesis and Alterity: A Particular History of the Senses*. Chicago, IL: Chicago University Press.

Tiedemann, R. 1999. Dialectics at a standstill. In Walter Benjamin *The Arcades Project*, translated by Howard Eiland and Kevin McLaughlin. Cambridge, MA: Harvard University Press

Weigel, S. 1996. *Body- and Image-Space: Re-Reading Walter Benjamin*. London: Routledge.

Willerslev, R. 2007. *Soul Hunters, Hunting, Animism, and Personhood among the Siberian Yukaghirs*. Berkeley, CA: University of California Press.

Wohlfarth, I. 1986. Et cetera? The historian as chiffonnier. *New German Critique*, 39, Second Special Issue on Walter Benjamin (Autumn): 142–168.

Selected further reading

Adorno, T.W. 1981 [1966]. *Negative Dialectics*. London: Continuum.

——— 1999. *Introduction to Sociology*, translated by Edmund Jephcott. Palo Alto, CA: Stanford University Press.

——— 2005 [1951]. *Minima Moralia: Reflections on Damaged Life*. London: Verso.

Arendt, H. 1967. *Men in Dark Times*. New York, NY: Harcourt Publishers.

Benjamin, A. (ed.) 2005. *Walter Benjamin and History*. London: Continuum.

Benjamin, W. 2009. *The Origin of German Tragic Drama*. London: Verso.

Bernstein, J.M. 2010. *Adorno: Disenchantment and Ethics*. Cambridge: Cambridge University Press.

Connerton, P. 1980. *The Tragedy of Enlightenment. An Essay on the Frankfurt School*. Cambridge: Cambridge University Press.

Dews, P. 2007. *Logics of Disintegration: Poststructuralist Thought and the Claims for Critical Theory*. London: Verso.

Gilloch, G. 2002. *Walter Benjamin: Critical Constellations*. London: Polity.

Jameson, F. 2007. *Late Marxism: Adorno: Or, the Persistence of the Dialectic*. London: Verso.

Jennings, M. 1987. *Dialectical Images: Walter Benjamin's Theory of Literary Criticism*. New York, NY: Cornell University Press.

Hanssen, B. 2000. *Walter Benjamin's Other History: Of Stones, Animals, Human Beings, and Angels*. Berkeley, CA: University of California Press.

Gilloch, G. 1999. *Myth and Metropolis: Walter Benjamin and the City*. London: Polity.

Leslie, E. 2000. *Walter Benjamin. Overpowering Conformism*. London: Pluto Press.

Lowy, M. 2016. *Fire Alarm: Reading Walter Benjamin's 'On the Concept of History'*. London: Verso.

O'Connor, B. 2004. *Adorno's Negative Dialectic: Philosophy and the Possibility of Critical Rationality*. Cambridge, MA: MIT Press.

Richter, G. 2016. *Inheriting Walter Benjamin*. London: Bloomsbury.

Rose, G. 2010. *The Melancholy Science: An Introduction to the Thought of Theodor W. Adorno*. London: Verso.

Ross, A. 2016. *Walter Benjamin's Concept of the Image*. London: Routledge.

Savage, M. 2000. Walter Benjamin's urban thought. A critical analysis. In Mike Crang and N.J. Thrift (cds.), *Thinking Space*, pp. 33 53. London and New York, NY: Routledge.

Smith, C. 2006. Resurrecting Benjamin. *Anthropological Quarterly* 79(3): 541–546.

Taussig, M. 1989. Terror as usual: Walter Benjamin's theory of history as a state of siege. *Social Text* 23 (Autumn–Winter): 3–20.

——— 2006. *Walter Benjamin's Grave*. Chicago, IL: Chicago University Press.

Wheatland, T. 2009. *Frankfurt School in Exile*. Minneapolis, MN: University of Minnesota Press.

Wolin, R. 1994. *Walter Benjamin: An Aesthetic of Redemption*. Berkeley, CA: University of California Press.

Chapter 10

The anthropological lives of Michel Foucault

James Laidlaw

Although not himself an anthropologist, the French scholar Michel Foucault (1926–1984), who styled himself a 'historian of systems of thought', has been one of the most powerful influences on anthropological theory since the 1970s. That influence has come in two forms and in two stages, through respectively his earlier and later writings. The former includes the books published in English as *The Birth of the Clinic* (1973 [1963]), *The Order of Things* (1970 [1966]), *Discipline and Punish* (1977 [1975]) and the pivotal work that belongs in both groups, *The History of Sexuality* (1978 [1976]). The latter group includes a number of essays, interviews and lectures along with the second and third volumes of *The History of Sexuality* series, that together constitute the project that Foucault referred to as his 'genealogy of ethics'. A complex debate could be had about the extent of the continuities across Foucault's whole oeuvre. In an exegesis of Foucault's own thinking, it would be a mistake to imply too radical a disjunction in what is a continuous, developing intellectual project; but since my concern here is to give an account of Foucault's writings as they have most influenced anthropology, I shall organize my discussion around this fairly clear and consequential break, because in a sense, and notwithstanding the unity of Foucault's overall intellectual project, he has appeared in the history of anthropological theory as two somewhat different Foucaults.

The earlier Foucault provided anthropologists with an account of European modernity, one that informed their understanding in particular of colonialism and how the societies they studied had been shaped by European colonial rule (e.g. Rabinow 1995; Bayly 1996; Cohn 1996; Inda 2005), and associated with this an understanding of anthropology itself (e.g. Clifford and Marcus 1986; Marcus 1998; Rabinow 2003; Faubion and Marcus 2009), and its place in colonial systems of power (e.g. Mitchell 1988; Thomas 1994; Stoler 2005). This understanding of modernity was underpinned not by any one publication of Foucault's exclusively, but by certain features that were common to all his work during this period. Each of his major works described an aspect of a general, radical discontinuity in the organization of knowledge, as what Foucault called 'the classical age' gave way to the modern era. In each case, new forms of knowledge were embodied in new institutions, that both regulated and policed the emerging new 'disciplines' they presided over, and oversaw the 'disciplining' of those to whom these new forms of knowledge were applied: the mad, the sick, criminals, and increasingly, whole populations. Aspects of this same general change in the mode of governance of European states were visible in the 'great confinement' of the mad into asylums (Foucault 2006 [1961]); in new forms of medical science, including the creation of new institutions to give effect to a distinctively authoritative medical 'gaze' (1973 [1963]); and in a kaleidoscopic reconfiguration of the disciplines of what are now the humanities and social sciences, as 'the human' became constituted as the central organizing category of our knowledge of ourselves in the modern era (1970 [1966]). Of course, the fact that this whole regime of truth could now be grasped, as from the outside, was for Foucault evidence that the era of 'the human' was beginning to draw to a close.

Discipline and Punish (1977 [1975]) was at the same time a recapitulation of the same general narrative as those earlier works, something of an outlier in its mode of explanation, and the most rhetorically effective and encompassing account so far. It was also the most generally influential of his books to date, including in anthropology. It begins with a striking contrast between two modes of punishment: the spectacular multi-stage public torture, mutilation, execution and dismemberment of an unsuccesful regicide; and the regime of a boys' prison, in which physical violence is much less important than a minutely specified timetable of activities, designed to reform and shape the conduct of the inmates, by instilling disciplined habits and dispositions. Only a few decades separate these exemplars of what Foucault proposes are radically different modalities of power and knowledge. Spectacular public display of the force of the sovereign's will, exercised directly on the body, is superseded (though such force is never entirely replaced, of course) by a routine designed progressively to reform inmates, so to speak, from the inside out. This internalization and normalization is exemplified most vividly in Jeremy Bentham's famous design for a 'panopticon' prison, in which the ever-present but always in fact uncertain possibility of surveillance by a prison guard reforms the inmate's conduct by training him or her in self-surveillance. The very form of the building, in conjunction with a daily regimen similar to that of the boys' prison, makes it a machine for reforming the inner dispositions, and hence the external conduct, of the prisoner.

Bentham's prison was never built in exactly the form he envisaged, but it was widely influential on prison design and the design of other more or less total institutions such as schools and clinics; and Foucault suggests that it neatly summarizes the basic workings of the prevailing configuration of power and knowledge in the modern world. In the old regime, knowledge had an external relation to power, as it was characteristically addressed to the sovereign in the form of petition or appeal (or 'speaking truth to power'), whereas in a regime of disciplinary power truth is *constitutive* of the power that acts upon and reconstitutes the subject as an object of reform. This *internal* relation of truth to power lies behind the use of the expression 'power/knowledge'. Attention in criminal justice shifts from the material form of forbidden and prohibited acts, to the 'passions, anomalies, infirmities' and other internal states that reveal the real nature of an act. Insanity stops being a reason to put a case outside the judicial system, and becomes instead a factor to be weighed within it, as psychiatric expertise is included in the creation of 'medico-judicial treatment'. The line between criminality and society at large is blurred as similar modes of knowledge govern and seek to reform the population as a whole, in population management, city planning and the modern 'social sciences', and the mature realization of this mode of governance, which Foucault refers to as 'biopower', is found in the 20th-century European welfare state.

One of the reasons reading *Discipline and Punish* was such a revelatory experience for many, was that it invited radical reconsideration and moral re-evaluation of what had been accepted as a simple story of enlightened reform and increasing humanity. Foucault invites us to see the new stress on 'rehabilitation' not as simply more lenient and humane than its punitive predecessors, but also as a more uncompromising and intrusive intervention, a thoroughly mechanical re-engineering of the subject, a more thoroughgoing and total control than external coercion alone could ever achieve. By the end of this process, asserts Foucault in a joking inversion of Plato, 'the soul is the prison of the body'.

It had been a notable feature of Foucault's earlier works that profound change in the configuration of power/knowledge was described, without a causal account being offered of how or why that change came about. While in some respects acutely observant sociologically, he largely did without notions of grand historical forces, and recognized no obligation to explain ideas or how they changed in reductive or materialist terms: so Foucault's use of 'discourse' in his analyses of systems of power/knowledge conspicuously eschews two ideas inherent in Marxist concepts of 'ideology': that the ideas in question were the distortion or inversion of a truth to which our (historical–materialist) theory would independently give us objective and certain access, unavailable to those within the system in question; and that those distortions and inversions were invariably and necessarily in the service of identifiable interests (to whose identity, again, our theory would give us access). So it is no surprise that Foucault frequently distanced himself from Marxism, and also from the other all-encompassing meta-narrative and totalizing explanatory framework of his day, Freudian psychoanalysis. His mode of thought and scholarship was indeed profoundly inimical to both these systems, so there is no reason he should have felt any need of them, but the strength of his antipathy was no doubt due in part to the pathologizing of homosexuality in the Freudian thought of his time, and to his experience of the totalitarian Marxism (totalitarian both in form and in content) that Louis Althusser imposed on so much of French intellectual life, Foucault having experienced it at close quarters in his youth at the Ecole Normale Supérieure. This latter experience was reinforced by his brief experience of 'actually existing socialism' in Poland in 1958 to 1959 (on Foucault's life, see Eribon 1991; Macey 1993; Miller 1993; and on Foucault in intellectual context in France see Paras 2006; Bourg 2007). *Discipline and Punish* is a partial exception to all of this. Influenced at the time by his friendship with Gilles Deleuze, that book admits of being read (e.g. Hoy 1986) as arguing the conventional Marxist case that the changes it describes occurred because they were functional for a capitalist political economy. And many, including anthropologists, whose acquaintance with Foucault's thought has been principally with this book, have thought it natural to mix Foucauldian and Marxist ideas. However, the book Foucault began writing the very day *Discipline and Punish* was finished, the first volume of a projected *History of Sexuality*, if its arguments are accepted, decisively rules this out by rejecting the specific Marxist–functionalist argument that can be seen in the earlier book, and with it also both Marxist historical materialism and Freudian psychoanalysis, and indeed the synthesis of those two systems that dominated French Leftist intellectual circles at that time. If later authors, including anthropologists, have missed the depth and significance of the move Foucault made between these two books, Deleuze himself did not: he never spoke again to Foucault after reading the book that would be published in English as *The History of Sexuality*.

Like its predecessor, *The History of Sexuality* proceeds by debunking a widely accepted narrative, but instead of the liberal-progressive account of an increasingly humane penal system, this time what is repudiated is a Marxist–Freudian construction that Foucault calls 'the repressive hypothesis'. This was the idea, virtually unchallenged in Leftist thought of the time, and articulated by thinkers as otherwise different as Marcuse and Deleuze, that capitalism requires for its functioning, and therefore has brought about, pervasive sexual repression, as a sort of monstrously paranoid mechanism of labour discipline. It followed, according to this 'hypothesis', that the liberation of sexual desire – anything and everything that defies bourgeois sexual repression – is a step towards bringing down the capitalist system. Foucault has four decisive objections to the repressive hypothesis. Far from the development of capitalism being accompanied by a repressive stifling of all talk of sexual desire, the period saw instead a 'discursive explosion' in relation to sex: a proliferation of new disciplines that claimed to know the truth about sex and its significance for education, fertility and mental and physical health. Power has therefore come to be exercised by prohibiting the expression of sexual desire, but instead through techniques for the *elicitation* of expressions of desire: rituals for speaking the truth about the self as the truth of one's desires. Thus 'sexuality' was not an eternal good requiring liberation, but an invention of this period, as persons with different sexual desires had become defined as distinct human types, whose essential truth is revealed by those desires. This means that the Marxist–Freudian idea of liberating 'sexuality' is not a form of effective resistance to the prevailing forms of power/knowledge, but an unwitting and complicit intensification of them. The prophets of the sexual revolution were extending and completing the work of the therapists and other specialists: requiring us to speak to them of our desires, and to accept the truth so revealed as our essential nature, so that our freedom can consist of nothing but the obligation to be faithful to that nature. If sexual repression had really been essential to the capitalist system, quipped Foucault, it would scarcely have given way to permissiveness so readily.

The analysis of the importance of forms of knowledge about sex, reproduction, health and populations in modern forms of governance (biopolitics) developed in *The History of Sexuality* left Foucault with two new questions. First, if the understanding of freedom offered by the prophets of 'liberation' was so flawed (and a dangerous trap), how then is 'freedom' to be understood? How might a freedom worth exercising be conceived? And second, how was it that European societies had come to seek the truth of the self – such that 'it' might be thought to need liberation – in sexual desire? Where did the subject of 'sexuality' come from? Answering these questions is what turned the 'history of sexuality', as envisaged in *The History of Sexuality*, into a rather different project, as it in fact appeared in the two subsequent volumes of the series, and that Foucault entitled his 'genealogy of ethics'.

The History of Sexuality contained a programme for uncovering 'the historical ontology of ourselves', the process whereby the modern desiring subject had come into being and come to be experienced as natural and necessary. Foucault's genealogical method, discovering the contingency of the apparently necessary by tracing the points at which its components had come together, was adapted from Nietzsche (1994 [1887]). Foucault initially assumed that this genealogy could be accomplished, as had all his previous works, largely within the chronological span of post-Renaissance Europe. But he found that in order to find the point at which the question of the subject was posed in a coherent and consistently different way, he had to go much further back in time. And the same was true of the question of power and freedom.

In one of his most important later essays, 'The subject and power' (in Foucault 2006), Foucault lays the groundwork for his genealogical investigation of freedom by announcing the necessity of developing what he called a 'non-juridical concept of power': that is to say, a conception of power not limited by the imagination of sovereignty, in which the exercise of power takes the form of prohibition, and in which it makes sense to think of knowledge as exterior to power. This then was a direct development of a central theme in earlier works, such as the *History of Madness* and *Discipline and Punish*, and a reason not to see a radical discontinuity between the early and late Foucault. He distinguishes what he calls power, as typically exercised in social relations, from 'capacity', or sheer physical force. The latter does of course play some part in human relations, but for the most part, we do not act on other people and cause them to act as we wish by the exercise of such brute capacity (as we might push a cart or cage an animal). Instead, we influence others *as subjects* with their own intentions and capacities, so that *they* take the action we wish of them. This Foucault referred to as the conduct of conduct; it is a matter of orchestrating, or conducting, as with a musical ensemble; and therefore the exercise of power among persons involves reckoning with the freedom of each other. It means influencing what others do, *by means* of their freedom. It is incoherent to imagine, therefore, that there could ever be a society without relations of power, or that we become 'more free' by the removal of power. Freedom cannot be an absence of power: 'power exists only over free subjects, and only insofar as they are "free"'. Unfortunately, our understanding of power, including the ideas of the prophets of 'liberation' but also those of most thinkers in the liberal tradition, continued to be dominated by the now largely anachronistic problem of resisting the coercive power of the sovereign state. Was there a time when our thought had not been so constrained? When had the question of freedom been posed other than as the absence of power? Not coincidentally, Foucault concluded that the answer to this question was the same as the answer to the question of when the subject had been understood other than as the subject of desires. Both the subject and freedom had been systematically understood in fundamentally different ways in classical Athens.

This conclusion delayed the completion of the succeeding volumes of the *History*, because they now had to take a shape, and concern an historical period, quite different from what was envisaged in the first volume, and Foucault had to master whole new periods and bodies of texts. The 'later Foucault', who is the second decisive influence on anthropology, is a remarkably unified project, referred to by Foucault as his 'genealogy of ethics'. The writings cover roughly a millennium and consist of two further volumes of *History of Sexuality*, one on classical Athens and one on the Hellenistic period – an incomplete fourth volume on early Christianity remains unpublished, according to Foucault's own instructions – together with a number of associated essays and lecture series.

In terms of the influence of this project on anthropology, we should begin with the set of concepts and analytics Foucault developed in order to carry it out, because in addition to being influenced by his substantive conclusions about the history of European ethical life, anthropologists have adopted and adapted major components of this apparatus as they have approached the ethical life of quite other times and places. There are three principal components.

The first is the notion of subjectivation (*assujetissement*) to describe the processes by which a subject is constituted. The important point here is that these processes are located within, but not reducible to, social structures and institutional contexts. The subject also, in Foucault's view, actively participates in its own self-constitution, notably through its capacity for reflective thought ('freedom in relation to what one does'): the capacity, that is, to establish some reflective distance from oneself, to constitute oneself and one's own conduct as an object of knowledge, and to act in such a way as to modify it. The recognition of this possibility is a major point of divergence from materialist reductionism, and from the structural Marxism of Althusser and others. It is in this context that we must see Foucault's remarks on his own earlier studies of asylums and prisons, that those works had put too much stress on techniques of domination at the expense of other dimensions of power relations. There are, he wrote, techniques of production, techniques of domination and techniques of signification, but there are also techniques of the self. His earlier works had tended to emphasize the first two, or in some cases (*The Order of Things* and *The Archaeology of Knowledge*), the third. But the fourth, the techniques by which we actively participate in our self-constitution as a subject, are equally a part of the material of social life, and a rounded picture of any historical era or social system must include them.

The second component of Foucault's method in his genealogy of ethics is a distinction between moral codes and ethics. By 'moral codes', Foucault means to refer to rules as to what one should and should not do, together with the questions of how these are defined, codified and enforced in institutions, and together also with the questions of how people variously obey, resist, challenge or evade them. All of this is an undeniably important part of any form of moral life. But in addition, it follows from Foucault's account of subjectivation and reflective thought, there is also what he calls 'ethics', which refers to the ways in which people respond to injunctions or embrace projects to make themselves a certain sort of person. They do not, of course, do so *ab initio*; they find in their culture ideals and values and exemplars, and in some cases, well laid out projects of self-fashioning or self-transformation, but it is always to some extent a matter for them how they respond to these ideals. In any historical epoch or cultural setting, the field of moral discourse and practice will consist of both moral codes and ethics in this sense. They are not entirely separate matters, but in all cases 'two sides of the same coin'. Nevertheless, some forms of moral life are more dominated by moral codes; others give a larger place to ethics. And the variation historically and culturally between forms of moral life is mostly a variation in ethics. Wherever you look, the moral codes (do not kill; do not steal) show very great similarities; it is in their ethics, including indeed quite subtle differences in their ethics, that societies most profoundly differ. Moral codes and ethics may change independently of each other, and changes in moral life are mostly changes in ethics, as is the case through the millennium described in Foucault's genealogy of ethics, in which, he claims, moral codes differ hardly at all but ethics shifts, slowly but decisively.

That this is so explains the need for the third component of Foucault's method, which is his analytic for the understanding of ethical projects. Ethical projects may be compared and contrasted, writes Foucault, by asking four questions in relation to them. First, what is their *ontology*? What are the parts or elements of the self that are of ethical significance? Shall I focus on, and try to improve, my actions, my desires, my heart, my soul, my virtues? Second, there is the question of *deontology*. What is the mode of subjectivation by which I recognize an ethical injunction or standard? As what does it apply to me? Is it because I am rational, or in response to divine command, or because I am a warrior or a king or a mother, or as a human being? Third, there is the question of *ascetics*. What are the specific techniques and practices used to carry out the necessary work to effect whatever changes I am aiming at? Must I fast, keep a diary, make confession, practise meditation, exercise the body? And finally, there is the *teleology*. What are the qualities or state I seek to realize, the *telos* towards which I am working? How do I conceive the ideal that guides my conduct?

Insofar as there are one or more reasonably coherent projects of self-formation in existence, or being advocated, in a certain social context, applying this analytic and answering these four questions will enable us to draw out their specificity, to detect the subtle shifts in orientation that might take place even if overt rules and regulations remain unchanged, and to draw comparisons and contrasts between even radically different cultural contexts. Indeed, it enables us to give the vague idea of 'cultural context' fairly precise content. Although Foucault took to calling his new project a 'genealogy of ethics', a term that better captures its scope and comparative range, he retained 'The History of Sexuality' for the subtitles of the two books that were published together, very shortly before his death, in 1984. These volumes present an analysis of the projects of ethical self-fashioning that Foucault finds articulated – not often all together in one place, to be sure – in a range of polemical and practical texts providing advice, largely to elite males (as it was to them that written advice was almost exclusively addressed) in the ancient world. *The Use of Pleasure* does this for classical Athens, *The Care of the Self* for the Hellenistic period. And it is vital to notice that these books are contributions to 'the history of sexuality' in the specific sense only that they describe a changing world in which 'sexuality' *does not yet exist.*

The texts that feature in *The Use of Pleasure* and *The Care of the Self* concern what we would now see as a range of topics that overlap but do not even approximately coincide with our category of 'sexuality': the general topic they have in common, and therefore in Foucault's terms the *field of problematization* with which they are concerned, is *aphrodisia*: how to conduct oneself in relation to pleasures. Foucault divides *aphrodisia* into four broad fields: dietetics and health, including advice on exercise and how to respond to changes in the weather; marital relations and household management, including how to exercise authority over one's wife, children and slaves; 'erotics', which here refers specifically to passionate relations with male youths; and the idea that abstinence gave some kind of special access to truth. For Foucault, these four were structured similarly in the ancient world because they were a single field of problematization, and constituted the general area of ethical concern elite males were invited to attend to in their self-cultivation. The uncompleted *History of Sexuality* had become, then, a project for describing how this form of ethics was slowly transformed over a millennium and more, and how *aphrodisia* slowly gave way to sexuality.

All this quite elaborate exegetical apparatus is required to convey what Foucault saw as the hugely consequential difference between the ethics of the ancient world and our own, beneath its apparent familiarity especially in terms of moral codes. This was a world, Foucault claimed, in which the self was not understood as defined by its desires, and therefore the privileged way to know and to reform the self was not through knowing and altering its desires: hence it was 'before sexuality' (see Halperin et al. 1990). And it was, at the same time, a world in which the question of what it was to be free was not posed as a matter of the absence of power. These two facts are closely connected, a connection that is brought out by Foucault's description of the change between the ethics of *aphrodisia* in classical Athens and the ethics that first takes shape in late antiquity, the latter being the predecessor of modern 'sexuality', as a transformation of an *aesthetics of existence* into a *hermeneutics of desire.*

The latter term, although it is not used there, sums up much of what Foucault had been saying almost a decade before, in *The History of Sexuality*: even those Leftist intellectuals who claim a critical purchase on our social and political structures nevertheless take for granted the idea that the essential inner truth of the self is given by our desires; that because those desires are not immediately transparent to us, we may be guided to self-knowledge by experts in their decipherment; and that liberation consists in our being 'true to' the nature they reveal for us, and requires therefore only the removal of constraints on its expression. But for Foucault, this 'liberation' was a kind of servitude and an evasion of our responsibility for ourselves, because it involves accepting an always-already given fact of the matter of who we essentially and necessarily are.

There is, of course, no reason why anthropologists, any more than anyone else, should necessarily agree with Foucault's personal and political preferences here, on the value of self-creation as opposed to what he saw as the fictions of self-discovery and 'authenticity'. Where his scepticism does speak directly to anthropologists, as anthropologists, is in his impetus to consult the full range of human social experience, before accepting any aspect of our contemporary self-understanding as necessary and immutable, and in his impulse to look for times and places in which profoundly different understandings held sway, in order to secure a better appreciation of our limits and possibilities. And it is this impulse that motivates the profoundly ethnographic exercise Foucault undertakes, in *The Use of Pleasure* in particular, in reconstructing imaginatively the deeply different kind of ethical project that was the ancient aesthetics of existence.

The aesthetics of existence recommended to Athenian citizens differed from the later hermeneutics of desire, according to Foucault, across the range of domains from dietetics, through domestic relations, erotic relations with youths and questions of abstinence and truth. Importantly, the advent of Christianity was a point in a gradual and complex reconfiguration, and not a dramatic discontinuity. The gradual change in the core ethical substance (ontology) from pleasures to desires was a change from a *dynamic* concern with what are the effects of pleasurable activities and how these respond to considerations of timeliness, frequency, need and the relative status of participants, to a *morphological* knowledge of permitted and prohibited acts. The aesthetics of existence required knowledge in the form of a *savoir-faire* and the effortful *use* of pleasures aimed at achieving freedom, conceived not as absence of constraint but as self-rule. It involved an *exercise* rather than an absence of power, because just as a ruler who was not in control of his own pleasures would be unable to care for those in his power, and was therefore the epitome of a tyrant, so a man who was not in command of himself in the exercise of his pleasures would become a slave to them instead.

Nothing brings out the distinctiveness of an aesthetics of existence, and its distance from an ethics of sexuality, more clearly than Foucault's account of just why and in what ways same-sex relations were problematized in the ancient system. Erotic relations between adult men and youths (those just on the cusp of adult status) were not, as had sometimes been claimed, regarded in a relaxed way or as not a moral matter. Far from being morally neutral or 'tolerated', they were highly problematized, regarded by some as the most elevated and beautiful relationships of all, by others as the corruption of youth. The reason they were thus debated, furthermore, was not a matter of sexuality: same-sex relations were not, as such, either admired or condemned. The fact of the relations being same-sex was not in itself ethically significant. What was significant was not *what you desired*, but *whether you could exercise self-command* in your enjoyment of whatever it was that you desired. So, for example, in discussion of the terms on which a man might enjoy his slaves sexually, the fact of whether they were men or women was not significant. There, as in a man's relations with his wife and in his government of his household, what virtue was generally argued to consist in was moderation in the exercise of one's own freedom *in relation to those over whom one had power*. What was different about relations between a citizen and a youth, however, was that here right conduct consisted in *accommodating the freedom of the other*, because, unlike a wife or a slave, he was himself destined to be a citizen. As such, he must learn to be free and to exercise his freedom in relation to those over whom he had power. He must be fitted to take his place in the assembly with his fellow citizens. Ethical concern about pederastic friendships was about whether they aided or inhibited the preparation of a youth to assume these responsibilities. Were they, as maintained by those who idealized such relations, the crowning stage of his education, or did they risk fixing him in immaturity, as was claimed by those who disapproved? And this was a matter of whether, in the relation, the youth learned to be master of his pleasures, rather than becoming the object of someone else's. Any citizen who was so much not his own master that he was the slave to another's wishes was unfit to act as a citizen, and in a direct democracy such as Athens, such a man was a danger to the integrity and the safety of the state. What was morally hazardous about such relations was nothing to do with sexuality, therefore, in the modern sense. It was to do with the fundamentally political matter of who is fit to exercise freedom.

Now, obviously Foucault had a number of reasons for finding this whole subject fascinating, and the care with which he explicates the specificities of ancient Athenian elite male erotics was motivated partly by his interest in the experimental possibilities of male–male erotic relations in his own time. But Foucault's principal claim for his ethnography of ancient Greece, and the contribution he claimed it could make to the thought of his time, was of a different order: it had something important to teach us about how we should think about freedom. Of course, he insisted repeatedly and as forcefully as he could, ancient Greece could in no sense be a model for a modern society. The reasons for this are as numerous as they are obvious. But, as anthropologists ought especially to be aware, the ethnography of other times and places may inform and enrich our thinking on a wide range of grounds and bases other than that. The principal point Foucault makes is rather a focused one. The ancient Greeks had asked themselves the question of what it was to be free, and what you need to do in order to be able to be free, not as a matter of how one might remove power from human relations, or how one might remove oneself from relations of power, neither of which he thought possible, but as a matter instead of how to *exercise* power. He did not claim that the ancient Greeks provided a usable answer to that question, or indeed that there ever could be a definitive answer. His claim was simply that we had something to learn from the very possibility of *asking* that question.

The genealogy of ethics developed in Foucault's later works has been formative for the developing anthropology of ethics and morality in a number of ways, some more obvious than others. His broad conceptual framework was an important starting point for some early prospectuses for the enterprise of an anthropology of ethics (e.g. Faubion 2001a; 2011; Laidlaw 1995; 2002; 2014), and the general theme of how subjects are constituted in an interplay of 'relationships of self to self' and 'relationships of self to others', including, in his last two series of Collège de France lectures, in practices of truth-telling (*parrēsia*) in relation to procedures of government (2010; 2011), together with his analytic for the analysis of projects of self-fashioning: all these rich intellectual resources have informed anthropological analyses of ethical life, in contexts as diverse as religious traditions and piety movements (Asad 1993; Laidlaw 1995; Faubion 2001b; Mahmood 2004; Robbins 2004; Hirschkind 2006; Cook 2010), everyday rural life in the face of state developmentalism (Pandian 2009) or human rights NGOs (Englund 2006), parenthood (Paxson 2004; Clarke 2009; Kuan 2015), artistic vocations (Faubion 2014) and activism (Dave 2012; Heywood 2015a; 2015b; Lazar 2017), as well as in reflections on the ethics of the anthropological enterprise itself (Faubion 2009; Laidlaw 2014: Ch. 6).

The apparent divergence between two different ways in which Foucault's ideas have been adopted and adapted by anthropologists owes more to the interests the latter have had in reading him, and to the other intellectual traditions they have been engaging with as they did so, than to any profound discontinuity in Foucault's own thought, that proceeded in general by incremental steps, taking up new problems and questions as they came into view, correcting what he thought were over-emphases and blind-spots in earlier studies, and broadening the historical and cultural range of his enquiries, to address, ever more searchingly, ever more fundamental questions.

Bibliography

Foucault's principal books

[1961] Folie et déraison: histoire de la folie à l'âge classique. (Translated first as *Madness and Civilization*, 1967; and then *History of Madness*, 2006.)

1963. *Naissance de la clinique: une archéologie du regard medical.* (*The Birth of the Clinic* 1973.)

1966. *Les mots et les choses: une archéologie des sciences humaines.* (*The Order of Things* 1970.)

1969. *L'archéologie du savoir.* (The *Archaeology of Knowledge* 1972.)

1975. *Surveiller et punir: naissance de la prison.* (*Discipline and Punish* 1977.)

1976. *La volonté de savoir: histoire de la sexualité, I.* (*The History of Sexuality. Vol 1. An Introduction*, 1978.)

1984. *L'usage des plaisirs: histoire de la sexualité, II.* (*The Use of Pleasure: The History of Sexuality, Vol. 2*, 1985.)

1984. *Le souci de soi: histoire de la sexualité, III.* (*The Care of the Self: History of Sexuality, Vol. 3*, 1986.)

Essay collections

The best English-language collections are the three volumes of *Essential Works*:

1997. *Essential Works of Michel Foucault,* Vol 1: *Ethics, Subjectivity and Truth*, edited by Paul Rabinow.

1998. *Essential Works of Michel Foucault,* Vol 2: *Aesthetics, Method, Epistemology*, edited by James D. Faubion.

2000. *Essential Works of Michel Foucault,* Vol 3: *Power*, edited by Paul Rabinow.

Posthumously published

The most important posthumously published works are transcripts of Foucault's lectures at the Collège de France. The most significant of these are:

Society Must be Defended: Lectures at the Collège de France, 1975–1976 (2003).

The Birth of Biopolitics: Lectures at the Collège de France, 1978–1979 (2008).

The Hermeneutics of the Subject: Lectures at the Collège de France, 1981–1982 (2005).

The Government of Self and Others: Lectures at the Collège de France, 1982–1983 (2010).

The Government of Self and Others II: The Courage of Truth. Lectures at the Collège de France, 1983–1984. (2011).

Other references

Asad, Talal 1993. *Genealogies of Religion: Discipline and Reason of Power in Christianity and Islam.* Baltimore, MD: Johns Hopkins University Press.

Bayly, Christopher A. 1996. *Empire and Information: Intelligence Gathering and Social Communication.* Cambridge: Cambridge University Press.

Bourg, Julien 2007. *From Revolution to Ethics: May 1968 and Contemporary French Thought.* Montreal: McGill-Queens University Press.

Clarke, Morgan 2009. *Islam and the New Kinship: Reproductive Technology and the Shariah in Lebanon.* Oxford: Berghahn.

Clifford, James and George E. Marcus (eds.) 1986. *Writing Culture: The Poetics and Politics of Ethnography.* Berkeley, CA: University of California Press.

Cohn, Bernard S. 1996. *Colonialism and its Forms of Knowledge: The British in India.* Princeton, NJ: Princeton University Press.

Cook, Joanna 2010. *Meditation in Modern Buddhism.* Cambridge: Cambridge University Press.

Dave, Naisargi N. 2012. *Queer Activism in India: A Story in the Anthropology of Ethics.* Durham, NC: Duke University Press.

Englund, Harri 2006. *Prisoners of Freedom: Human Rights and the African Poor.* Berkeley, CA: University of California Press.

Eribon, Didier 1991. *Michel Foucault.* Cambridge, MA: Harvard University Press.

Faubion, James D. 2001a. Toward an anthropology of ethics: Foucault and the pedagogies of autopoiesis. *Representations* 74: 83–104.

——— 2001b. *The Shadows and Lights of Waco: Millennialism Today.* Princeton, NJ: Princeton University Press.

——— 2009. The ethics of fieldwork as an ethics of connectivity, Or the good anthropologist (isn't what she used to be). In James D. Faubion and George E. Marcus (eds.), *Fieldwork is Not What It Used to Be.* Ithaca, NY: Cornell University Press.

——— 2011. *An Anthropology of Ethics.* Cambridge: Cambridge University Press.

——— 2014. Constantive Cavafy: A parrhesiast for the cynic of the future. In James D. Faubion (ed.), *Foucault Now: Current Perspectives in Foucault Studies.* Cambridge: Polity Press.

Faubion, James D. and George E. Marcus (eds.) 2009. *Fieldwork is Not What It Used to Be.* Ithaca, NY: Cornell University Press.

Foucault, Michel. 2006 [1961]. *The History of Madness*, translated by Jean Khalfa. London: Routledge.

Halperin, David M., John J. Winkler and Froma I. Zeitlin (eds.) 1990. *Before Sexuality: The Construction of Erotic Experience in the Ancient Greek World.* Princeton, NJ: Princeton University Press.

Heywood, Paolo 2015a. Freedom in the code: The anthropology of (double) morality. *Anthropological Theory* 15: 200–217.

——— 2015b. Agreeing to disagree: LGBTQ activism and the Church in Italy. HAU: *Journal of Ethnographic Theory* 5: 59–78.

Hirschkind, Charles 2006. *The Ethical Soundscape: Cassette Sermons and Islamic Counterpublics.* New York, NY: Columbia University Press.

Hoy, David Couzens (ed.) 1986. *Foucault: A Critical Reader.* Oxford: Blackwell.

Inda, Jonathan Xavier (ed.) 2005. *Anthropologies of Modernity: Foucault, Governmentality and Life Politics.* Oxford: Blackwell.

Kuan, Teresa 2015. *Love's Uncertainty: The Politics and Ethics of Child Rearing in Contemporary China.* Berkeley, CA: University of California Press.

Laidlaw, James 1995. *Riches and Renunciation: Religion, Economy, and Society among the Jains.* Oxford: Clarendon Press.

——— 2002. For an anthropology of ethics and freedom. *Journal of the Royal Anthropological Institute* 8: 311–332.

——— 2014. *The Subject of Virtue: An Anthropology of Ethics and Freedom.* Cambridge: Cambridge University Press.

Lazar, Sian 2017. *The Social Life of Politics: Ethics, Kinship and Union Activism in Argentina.* Palo Alto, CA: Stanford University Press.

Macey, David 1993. *The Lives of Michel Foucault.* London: Hutchinson.

Mahmood, Saba 2004. *Politics of Piety: The Islamic Revival and the Feminist Subject*. Princeton, NJ: Princeton University Press.

Marcus, George E. 1998. *Ethnography Through Thick and Thin*. Princeton, NJ: Princeton University Press.

Miller, James 1993. *The Passion of Michel Foucault*. Cambridge, MA: Harvard University Pres.

Mitchell, Timothy 1988. *Colonising Egypt*. Cambridge: Cambridge University Press.

Nietzsche, Friedrich 1994 [1887]. *On the Genealogy of Morality*. Cambridge: Cambridge University Press.

Pandian, Anand 2009. *Crooked Stalks: Cultivating Virtue in South India*. Durham, NC: Duke University Press.

Paras, Eric 2006. *Foucault 2.0: Beyond Power and Knowledge*. New York, NY: The Other Press.

Paxson, Heather 2004. *Making Modern Mothers: Ethics and Family Planning in Urban Greece*. Berkeley, CA: University of California Press.

Rabinow, Paul 1995. *French Modern: Norms and Forms of the Social Environment*. Cambridge, MA: MIT Press.

——— 2003. *Anthropos Today: Reflections on Modern Equipment*. Princeton, NJ: Princeton University Press.

Robbins, Joel 2004. *Becoming Sinners: Christianity and Moral Torment in a Papua New Guinea Society*. Berkeley, CA: University of California Press.

Stoler, Ann Laura 1995. *Race and the Education of Desire: Foucault's History of Sexuality and the Colonial Order of Things*. Durham, NC: Duke University Press.

Thomas, Nicholas 1994. *Colonialism's Culture: Anthropology, Travel, and Government*. Cambridge: Polity Press.

Chapter 11

From 'the body' to 'embodiment', with help from phenomenology

Maryon McDonald

Introduction

In the late 1980s and early 1990s, structuralism (see chapter 2) seemed to have come and gone but an embarrassing party celebrating ideas, interpretations, myth, symbolism and other 'representations' was still going on in the main anthropology lecture theatres in the UK. New theories had already tried to get anthropology's feet back on the ground – whether through Marxism (e.g. Bloch 1975; see chapter 3), for example, or the definitional realities of 'semantic anthropology' (e.g. Ardener 1982). In some ways, semantic anthropology prefigured the very much more recent 'ontological turn' (see chapter 14) but much of anthropology simply ignored or condemned it at the time. Where Marxist anthropologies asserted their own analytical realities – their own language of context, we might say – semantic anthropology sought to take seriously the realities of those that anthropologists studied. Such an ambition was held by other moves against structuralism in this period, particularly those taking their inspiration from phenomenology. Although all these approaches were very different from each other, they shared aspects suggestive of a sweeping search for 'reality talk' in this period – a period of anthropology that could otherwise seem to be afloat on cosmic balloons of conceptual structures. Anti-idealist and anti-objectivist battles ensued, and it was out of this general post-structuralist pursuit that the analytical language of 'embodiment' came. Bodies were real.

This chapter points to some of the innovative aspects of 'embodiment', an innovation that became so popular that its pervasive, self-evident banality would later cause Bruno Latour (2004: 209; see also chapter 13) to proclaim that 'the opposite of embodied is dead'.

From 'the body' to embodiment

'The body' had long been a taken-for-granted object in much social analysis – and was often assumed to be a universal, natural object best left to the biologist and medical doctors to deal with. A social science interest in the body was encouraged from the 1960s and 1970s onwards by several factors, amongst which the consumer culture of late capitalism, plus feminism (see chapter 12), Foucault's work (see chapter 10) and the growing 'bio' industry are among those most commonly cited (Lambert and McDonald 2009). 'The body' – along with the realm of the natural – was now problematized and, in keeping with new modes of flexibility in work and production, objectifications of 'the body' began to favour a flexible, changing entity (Martin 1994). It is in this more general context of body-consciousness that part of the appeal of embodiment can be situated.

One of the best known introductions to embodiment in social anthropology came through the work of Thomas Csordas, with his 1994 edited volume entitled *Embodiment and Experience* becoming a landmark work in this respect. We will turn to this in a moment. Csordas was not the first to draw our attention to the analytical importance of the corporeal. Mauss (1979[1935]) and then, much later on, Bourdieu (1977), were amongst those who preceded him (see chapter 2). 'Embodiment' is generally attributed to the influence in the social sciences of phenomenology. The philosophers usually drawn on have included Husserl, Merleau-Ponty and Heidegger. Several anthropologists took up the insights of phenomenology and combined them with earlier insights concerning the body that they found in the work of Marcel Mauss. Mauss had written about 'techniques of the body' by which he meant 'the ways in which from society to society men [sic] know how to use their bodies' (1979: 97). These 'techniques' involve the ways in which we walk, run, dance, swim, ride a bicycle and more. All these techniques were 'social' for Mauss in that they are learnt and become habitual or customary, together forming what he termed a *habitus*. Bodies are socialized, in his terms, into a *habitus*.

A well-known social theorist who later took up Mauss's ideas – including the notion of *habitus* – was Bourdieu (1977; see also chapter 2). *Habitus* for Bourdieu was 'a system of lasting, transposable dispositions which, integrating past experiences, functions at every moment as a matrix of perceptions, appreciations and actions' (1990: 70). He talked of 'socialization', which was the inculcation of dispositions and generative schemes. Bourdieu, like others at that time, was trying to move beyond the apparent idealism and objectivist abstractions of structuralism. Using phenomenology to put sentient bodies back into the picture felt like one important way of doing this. In the work of Bourdieu, bodies developed 'automatisms' through 'socialization' and social order was thus naturalized. You talk, eat, sit, write in particular ways at particular times, without necessarily thinking too much about it: it just starts to feel natural. Bourdieu refined Mauss's notion of *habitus* through his work, and Csordas suggests how he also refined Merleau-Ponty: whereas the latter stressed 'being toward-the-world', Bourdieu injected a 'reciprocity of body-world' (Csordas 2011: 138). However, even those initially keen on the idea of *habitus* felt it was still without historicity (Toren 2012: 71–72). Instead, once acquired, *habitus* seemed to imply a fixity of dispositions, a fixity of custom or tradition. Critics of Bourdieu's work have suggested that a seeming fixity of the *habitus* was perhaps due to a romanticism in Bourdieu's construction of the world of the Kabyle where he did his first fieldwork (Roodenburg 2004).

Csordas's own work was partly in the field of ritual – part of that exotic triad of myth, ritual and symbolism that structuralism had made its own. Such a focus could easily appear to leave empirical realities – whether framed as the political, economic or as the concern of the natural sciences – to others. However, Csordas was not content with this situation. He was keen to re-inject what he saw as missing. For him, this was the body but he did not introduce it as a 'brute fact of nature' (1994: 1). He quite explicitly wanted the body to be seen not as object but as agent – as 'an experiencing agent' (p. 2). Structuralism, he felt, had merely encouraged a study of 'culture from the neck up' (p. 2). The mind/body dichotomy was one central duality that embodiment wanted to go beyond. It was no help that, in the face of accusations of idealism, Lévi-Strauss's cosmic objectivity had ultimately come to rest in some universal 'human brain'. An embodiment approach moved away from the idealism of structuralism, its concern with ideas, and away from the objectivism that had led us to a universal human brain. Csordas's body was not a universal, naturalized or essentialized body of this kind: instead, it was a body that was always 'located'.

Perceptions are always embodied, grounded in a particular embodied standpoint. This approach carried caveats for other anthropologies of the time, including interpretivism (see chapter 8): it meant accepting always the 'interpretive consequences' of partiality, imperfect communication or what Csordas termed the 'relatedness' of our being (p. 2). There was – to cite a paraphrase from elsewhere – no unlocated interpretation, no 'god's eye view' (e.g. Haraway 1991).

Anthropological work has long been cited that suggests that 'the body' as a distinct entity was introduced in some parts of the world by Christianity, medicine and colonialism (see, for example, Leenhardt (1979[1947]) on the Canaques of New Caledonia, cited in Csordas 1994: 6–8). The objectification of 'the body' with which we are now so familiar has been shaped and encouraged by Christianity, by consumerism and by the biological and medical sciences. Amongst the consequences of this objectification has been a further encouragement to individuation, and to dualities that have posed analytical hurdles for anthropology – dualities that Csordas lists as 'culture and biology, the mental and the material, culture and practical reason, gender and sex' (p. 7). These are dualities that slide into a distinction of mind/subject/culture on the one hand and body/object/biology on the other, with the former seeming to be derivative or mere 'representation'. 'Embodiment' seemed to be a language and approach that offered a means of getting outside this framework, potentially both avoiding and collapsing the dualities involved. For Csordas and others, 'embodiment' meant changing gear analytically to take up phenomenological notions of being-in the-world as prior. Paying attention to the priority of being in the world could turn embodiment into historically informed engagement with the world of which bodies are part. In this approach, dualities such as subject and object or meaning and the material world (evoking mind/body) can be collapsed. They are collapsed in the world of analysis – but may take on an ethnographic reality instead, as the experiential realities of those whom we study.

For Csordas, 'embodiment' is a 'paradigm' and a 'methodological perspective' rather than a route to any distinct theory (2011). Within this framework, we must redirect our ethnographic attention and stop any loose talk of 'the body' as if it were a given object that we can take for granted. Critiques of biology at this time – critiques that we might see as having come mainly from the two 'fs', Foucault and feminism – together with studies of illness, pain and his own work on healing, gave Csordas the confidence to see biology as no longer a 'monolithic objectivity' (p. 3). Instead, 'the body' is transformed in this approach, and we turn to a prior 'embodiment' that the 'anthropology of the body' had previously taken for granted. It can only be through embodiment that 'the body' can be objectified in the first place. It is bodies that make objectifications possible. If we are to accept a methodological framework of embodiment, then the body is not invoked analytically as a given empirical thing or interesting theme but as 'the existential ground of culture and self' (1990; 2011: 4). Acknowledging his debt to Merleau-Ponty (1962), Csordas emphatically moves bodies in this way from being objects to being 'the ground of perceptual processes that *end* in objectification' (2011: 7) It is this process of objectification that has involved the ontologization of those same analytically problematic dualities such as mind/body or culture/biology (or culture/nature), gender/sex, and so on (1990; 2011: 7). 'Embodiment' approaches struggled to make clear that such dualities become matters of ethnographic interest, distinctions that we watch being made, rather than being a taken-for-granted part of any analytical framework imposed by the anthropologist. Ethnographically, we can note where and when they both structure and are confirmed by experience.

This is an important point that Csordas's work, and that of others who then took up embodiment, have made in various ways. Versions of a conceptual/material or mind/body dichotomy that structuralism had appeared to encourage, if only through the debates that it generated, had to be overcome. Work by the philosopher and medical doctor Drew Leder, entitled *The Absent Body* (1990), offered a useful critique of mind/body as an analytical framework, and Csordas made explicit use of Leder's work. Leder's book makes the point that, in everyday life, the body seems to disappear from awareness. In instances of disease or dysfunction, however, there is an unwanted appearance of the body, a kind of 'dys-appearance' and alienation. This can only arise, Csordas argues (1994: 7), through the embodied nature of mind. The unwanted consciousness of the body when ill encourages attention to 'the body' as an objective reality – and a correlative sense of the immateriality of mind and thought. Such points underline that the mind/body duality is an experiential reality, not a natural given. Csordas (p. 8) points out that in different contexts, either mind or body can appear to be objects – the mind objectified in cognitive science or the body in biological science, for example – but anthropology can no longer see the body as 'biological raw material' on which either 'the mind' or culture operates. Making the body a 'precultural substrate' in that way would be simply to reproduce those same dualities, with mind as subject and body as object. Putting this another way, we can note once again that what advocates of 'embodiment' seemed to want to do here was not only to go beyond the objectivism of structuralism but also to encourage us to treat ethnographically dualities that anthropology might otherwise have been tempted at this time to take for granted.

The chapters gathered in Csordas's edited volume *Embodiment and Experience* (1994) exemplify some of his aims. For example, Setha Low's chapter entitled 'Embodied metaphors: Nerves as lived experience' (1994) makes it clear that we are trying to pay attention to the embodied realities of those studied. She discusses the condition of *nervios* in Costa Rica, Guatemala and elsewhere. In situations of poverty, migration and discrimination, for example, the illness of *nervios* occurs – with sufferers shaking, feeling dizzy, weak and feeling unable to sleep. Setha Low stresses that talking of *nervios* or nerves is not a metaphor about the body, or a metaphor imposed on the body. It is real, it is embodied, it is a condition in which what we might want to see as the 'social' emerges in bodily experience. We note here that talk of 'embodiment' does not preclude other analytical languages that in some way describe the 'social context' in which bodies are situated. If bodies are 'located', it will often require an analytical language from the ethnographer to situate or 'locate' them, and embodiment does not dictate what that language should be. We might introduce a language of context such as 'exploitation' or 'alienation', for instance – but, at the same time, there is an ever-present danger hovering here, and in several other chapters of the same volume, of rendering embodiment secondary to this 'context'. *Nervios* still risks appearing to be a condition that might all too easily be dismissed as somehow a metaphor for, or as 'representing', something else that is more real. The way this matter of context is presented is important, and has gradually been refined in anthropology since the 1990s. What Csordas's 1994 volume seems to be trying to do is to move us towards grasping that it is, in priority, people's own context that they embody, and Low's chapter gives some material that is successful in that respect, with some sufferers speaking of their *nervios* in a context of disruptions in their family relationships, for instance (pp. 147–148). There are nevertheless lingering tensions throughout the volume between 'the body' and embodiment, and between 'embodiment' and how to talk about what it is that is being embodied such that embodiment is still taken seriously. And there are frequent lapses into divisions of 'the body' and its 'cultural construction'. Social or cultural constructionism were among the other products of phenomenological influence that had taken hold at this time, and they haunt many of the pioneering 'embodiment' approaches. Evoking the very dualities that embodiment wanted to go beyond, this constructionism assumed an object (e.g. the body) that is then culturally or socially constructed (see chapter 13).

Embodiments

There have been many enlightening uptakes and refinements of the 'embodiment' paradigm that Csordas, above all others, was successful in launching. One of the subsequent refinements of both Bourdieu and Csordas is found in the work of Christina Toren. Advocates of 'embodiment' have sometimes used notions of 'intentionality' and 'intersubjectivity' (ideas important in the work of phenomenologists). Csordas cites these ideas in his early work and, within the same decade, Toren (1999) subsequently gave them prominence. Toren's 1999 book shows in some detail how hierarchical relations are progressively embodied in Fiji from a young age. This ethnography was an unusual work for its time in that it very clearly did not take anything called 'social structure' or 'structure' as ready-made. She injected historicity, and 'child development' took on a new ethnographic life. Toren draws on phenomenology, Bourdieu and embodiment. Through bodies as 'embodied minds' (or 'enminded bodies'), she rethinks her own earlier training in psychology and, above all, she insists on the historicity of embodiment, on bodies as process.

'Intentionality' had been important for Merleau-Ponty, and Toren cites it frequently. For those who advocated embodiment theory, intentionality came to mean that all consciousness is consciousness of something and that the things that we are conscious of and might take for granted are brought into being as a function of our lived experience in the world. There are no given mind/body or subject/object distinctions here and phenomenology is seen to have inverted Descartes in some respects. We are, in priority, embodied: we are in and of the world and it is only through this that we can think and reflect, and know who we are. The world is fundamentally part of the constitution of what it means to be human, and inherent to the constitution of any 'self'. Being in the world necessarily has a historicity and we embody the spatiotemporal dimensions of the world in which we live. Toren stresses that it is not just a question of being in time but of embodying the time that we have lived (p. 13). Subjectivity presupposes intersubjectivity – the relations that produce a self. These two ideas, of intentionality and intersubjectivity, are most explicitly brought together in anthropology by Toren who cites Merleau-Ponty to clarify: 'I am installed on a pyramid of time which has been me' (Merleau-Ponty 1964: 14; cited Toren 1999: 13). From within this phenomenological framework, Toren summarizes the production of a body-self, a relational embodiment, in the following terms:

> [Y]ou were born into a set of relations with others and the ideas held by those others and the practices with which they are associated have informed, and continue to inform, the process of your becoming.
>
> (p. 7–8)

Toren pushed hard to build in historicity and to take on board what a collapse of dualities such as biology and culture might mean. It is not just that biology is itself culturally produced as a discourse. Rather, if we are to use that language at all, then she felt we were culturally biological and biologically cultural. Such insights pointed us firmly in the direction of thinking that, if we are to talk of embodiment, then we are talking of bodies (in the plural) and bodies that are constituted and reconstituted over time (2012). Another anthropologist who had long been working in a similar direction is Tim Ingold (e.g. 1998, 2000, 2004). Ingold, however, used phenomenology and embodiment in a slightly different way – and introduced 'enskilment' and 'dwelling' (Heidegger 2001 [1971])). What might otherwise have been loosely talked of as cultural variation, and linked to language, becomes for Ingold a matter of a variation in embodied skills. He was, like others, partly arguing against structuralism and other contemporary language-dominated models of 'culture' – but also against their associated models of 'knowledge transmission'. Skills, he argued, become 'incorporated into the *modus operandi* of the developing human organism through training and experience in the performance of particular tasks' (2000: 5). The tasks that people in any situation get involved in depends on their modes of 'dwelling' – that might, for example, be their modes of travelling, of recreational activity, or their means of subsistence or earning a living and so on (Ingold 2000: Ch 3; 2004; Ingold and Kurttila 2001). Enskilment is not a simple matter of being on the receiving end of the 'transmission' of knowledge but it is like being an apprentice who has progressively to fine-tune their perception. This is done through 'observation and imitation', which means paying active attention to others' orientation to their environment (observation) and 'aligning that attention to the movement of one's own practical orientation towards the environment' (Ingold 2000: 37). The 'apprentice', the newborn, the novice – or perhaps the anthropologist in the field – thus learns 'to feel this, taste that, or watch out for the other thing' (p. 22).

Since the late 1990s, work has proliferated that has developed embodiment approaches in this way (one example would be Gieser 2008) – but also in other ways, too. This is especially the case in medical anthropology, but also in anthropological studies that have focused on specific topics such as a comparative focus on persons (e.g. Lambek and Strathern 1998; Strathern and Stewart 2011), and there are large compendia displaying a great variety of embodied wares, from neoliberalism to taste (e.g. Mascia-Lees 2011). Refinements have certainly taken place but not all anthropologists have found 'embodiment' either useful or appropriate.

Moving on

One critic of embodiment is Aparecida Vilaça, who has worked in the Amazon. Her mentor was Viveiros de Castro (see chapter 14). Vilaça takes embodiment to task from the point of view of her Amazonian ethnographic material (2009). Embodiment, she says, looks to bodies as the seat of perception and as the substrate of intersubjective relations. Of course, we can follow this up by saying, well, differences between bodies would imply differences in perception, constituting objects in a particular way. But generally, she feels, there seems – in Csordas particularly – to be an assumption of the individual and of the human as starting points.

In Amazonia, however, the starting point would have to be different: it would have to be an extended notion of the human, including various types of animal. The human is a position, a temporary outcome of a complex play of perspectives. This is, as Viveiros de Castro (1998) pointed out, a perspectivist world of multinaturalism as opposed to multiculturalism (see chapter 14 for a more detailed account of this general argument). Every being has *jam*, the capacity to *jamu* – to change affects and adopt other habits – enabling the person to be seen as similar by other types of beings. Beings have a metamorphic capacity, a capacity to transform, a 'soul' – and prophylactic measures are taken to control the capacity for transformation and fix the 'soul': for example, tattooing to keep a person distinct from animals. Whereas, Vilaca says, embodiment often assumes subjects constituted in advance, anterior to the relations that produce them. In Amazonia, different relational contexts produce not only distinct objectifications of phenomena but different bodily constitutions of the subject. The result of a relational context in Amazonia may therefore be an Indian or a jaguar or a tapir (2005). This is a world of unstable bodies and there is no stable substrate. Vilaça (2009) therefore rejects 'embodiment' as inappropriate: it carries specific assumptions from different and broadly 'Western' contexts that would traduce Amazonian (amongst other) modes of being.

Through Amazonian ethnography, some of the interesting but problematic assumptions that 'embodiment' carried were highlighted. In common with others, Vilaça looked for help to reflections by Latour (2004) on the body. Latour's work extends the human in a different way (for an introduction to Latour's work more generally, see chapter 13). Otherwise, he evokes Toren and Ingold in some respects, stressing historicity, change and learning – but he writes of a process of 'learning to be affected'. We can summarize Latour's approach by saying that he speaks not of embodiment or enskilment but of *bodies that are acquired* (see McDonald 2014). He uses the example of the odour kit used in the training of 'a nose' in the perfume industry in France. In this training process, the kit becomes co-extensive with bodies; bodies are dynamic trajectories, and the kit produces a transformation in the body of the person and in the universe of which they are a part, one element articulating another, one affecting the other. It is not a question, Latour stresses, of an anterior subject (is my nose accurate?) representing an object (is it an accurate representation of odours already existing in world?). It is instead a question of 'articulation' – or indeed a process of 'mutual articulation' (McDonald 2014). The odour kit has itself already been set up through the publications, conferences, documentation, training, conventional materials and practices of the chemists and engineers who made it. In using this kit, the trainees in the perfume industry learn to be affected by ever more subtle distinctions. They move from simple, sharp contrasts to fine distinctions. The 'dumb nose' that initially can only smell a nice/nasty difference becomes a nose in a world of ever finer distinctions. Pupils learn to be affected by chemicals that previously might have 'bombarded them to no avail'. 'Body parts' are acquired at the same time as 'world counterparts' are being registered in a new way (Latour 2004). It is not a question of a more and more accurate 'representation' of odours that are really already there in the world – but rather a process of mutual articulation. The more sensitive the bodies the trainees acquire, the more the world affords them – in this instance, the subtler the fragrances they smell.

In this framework then, bodies are readily changeable and are acquired through an environing world that is itself constituted and reconstituted in the process.

Conclusion

A focus on embodied experiential realities became an anthropology that posed from its early days as another critique of 'representation' (Csordas 1994: 9; see chapters 13 and 14). All too often in the humanities, the body was previously interpreted as a 'text', a 'sign' or as the passive bearer of various political and social meanings, including gender or ethnicity, for example. Structuralist (see chapter 2) and interpretative (see chapter 8) theories encouraged this. The body was of interest in its symbolism, or the attitudes towards it and discourses about it. There was always an 'it' – the body. The work of Bourdieu (see chapter 4) and then the increasingly dynamic correctives of 'embodiment' theory brought a new attention to the performing and understanding body, the active body, the lived body. Gradually, instead of any fixed, habitual dispositions, and instead of 'the body', there now emerged a world of bodies that are constituted and reconstituted through an inherence in their environing world – and then bodies acquired through an environing world itself constituted and reconstituted in the process.

More than any other anthropologist, Csordas brought 'embodiment' to anthropological attention. Toren injected historicity and relationality, and rethought aspects of psychology. Ingold introduced enskilment and dwelling and rethought older ideas of learning. Vilaça and Latour, each in their own way, looked beyond a stable substrate and beyond the human. This story is necessarily brief, simplified and highly selective, suggesting just a few of the approaches that 'embodiment' introduced. Since the 1990s, embodiment and its successors have been progressively 'mainstreamed' in social anthropology and become pervasive enough to be unspoken. An interest in embodiment has also encouraged a great deal of new and critical ethnographic work in related fields such as 'affect' and the senses (see, for just a few such examples amongst many: Clough and Halley 2007; Pink 2009; Howes 2011; Navaro-Yashin 2012). We have come a long way now from that natural object, the body, that was left to medicine and the natural sciences. Anthropologists have been able to engage in dialogue with the natural sciences, and to rethink both 'the body' and its 'environment' through ideas such as 'local biologies', and they have pointed to epigenetics and microbiomes as interesting ways of collapsing old ideas of the body/environment distinctions (see e.g. Lock and Nguyen 2010; Lock 2011). In an important sense, bodies are environments – and they embody their environments (see e.g. Hamdy [2011] for poignant ethnographic illustration). Embodiment may mean that you are what you eat or how you dwell, or the person your relatives and environments make you, and that you make, too. There is no universal 'human body' on which relational embodiments depend. That 'human body' is an objectification common in arenas such as medicine or fine art – and in the training involved in either of these fields, particular circumstances are embodied, and particular bodies have to be acquired, and then an objectified 'body' can be reproduced on canvas or notebook.

References

Ardener, E. 1982. Social anthropology, language and reality. In D. Parkin (ed.), *Semantic Anthropology*. London: Academic Press.

Bloch, M. (ed.) 1975. *Marxist Analyses and Social Anthropology*. London: Routledge.

Bourdieu, P. 1977. *Outline of a Theory of Practice*. Cambridge: Cambridge University Press.

——— 1990. *The Logic of Practice*. Cambridge: Polity Press.

Clough, P.T. and J. Halley (eds.) 2007. *The Affective Turn: Theorizing the Social*. Durham, NC: Duke University Press.

Csordas, T 1990. Embodiment as a paradigm for anthropology. *Ethos* 18(1): 5–47.

——— (ed.) 1994. *Embodiment and Experience*. Cambridge: Cambridge University Press.

——— 2011. Cultural Phenomenology Embodiment: Agency, Sexual Difference, and Illness. In F.E. Mascia-Lees (ed.), *A Companion to the Anthropology of the Body and Embodiment*, pp.137–156. Oxford: Wiley-Blackwell.

Gieser, T. 2008. Embodiment, emotion and empathy. *Anthropological Theory* 8(3): 299–318.

Hamdy, S. 2011. When the state and your kidneys fail: Political etiologies in an Egyptian dialysis unit. *American Ethnologist* 35(4): 553–569.

Haraway, D. 1991. *Simians, Cyborgs, and Women: The Reinvention of Nature*. New York, NY: Routledge.

Heidegger, M. 2001 [1971]. Building dwelling thinking. In *Poetry, Language, Thought*, pp. 143–159. New York, NY: Harper Perennial Modern Classics.

Howes, D. 2011. The senses. Polysensoriality. In F.E. Mascia-Lees (ed.), *A Companion to the Anthropology of the Body and Embodiment*, pp. 451–460. Oxford: Wiley-Blackwell.

Ingold, T. 1998. From complementary to obviation: On dissolving the boundaries between social and biological anthropology, archaeology and psychology. *Zeitschrift für Ethnologie* 123: 21–52.

——— 2000. *The Perception of the Environment: Essays on Livelihood, Dwelling and Skill*. London and New York, NY: Routledge.

——— 2004. Culture on the ground: The world perceived through the feet. *Journal of Material Culture* 9 (3): 315–340.

Ingold, T. and T. Kurttilla 2001. Perceiving the environment in Finish Lapland. In P. Macnaghten and J. Urry (eds.), *Bodies of Nature*, pp. 183–196. London: Sage.

Lambek, M. and A. Strathern (eds.) 1998. *Bodies and Persons. Comparative Perspectives from Africa and Melanesia*. Cambridge: Cambridge University Press.

Lambert, H. and M. McDonald 2009. Introduction. In Lambert, H. and McDonald, M. (eds.), *Social Bodies*, pp. 1–16. Oxford and New York, NY: Berghahn.

Latour, B. 2004. How to talk about the body? The normative dimension of science studies. *Body & Society* 10: 205–229.

Leder, D. 1990. *The Absent Body*. Chicago, IL: University of Chicago Press.

Leenhardt, M. 1979 [1947]. *Do Kamo: Person and Myth in the Melanesian World*, translated by B. Miller Gulati. Chicago, IL: University of Chicago Press.

Lock, M. 2011. Genomics. Embodying molecular genomics. In F.E. Mascia-Lees (ed.), *A Companion to the Anthropology of the Body and Embodiment*. Oxford: Wiley-Blackwell.

Lock, M. and V.-K. Nguyen (eds.) 2010. *An Anthropology of Biomedicine*. Oxford: Wiley-Blackwell.

Low, S. 1994. Embodied Metaphors: Nerves as Lived Experience. In T. J. Csordas (ed.) *Embodiment and Experience*. Cambridge: Cambridge University Press.

Martin, E. 1994. *Flexible Bodies. Tracking Immunity in American Culture from the Days of Polio to the Age of AIDS*. Boston, MA: Beacon Press.

Mascia-Lees, F.E. (ed.) 2011. *A Companion to the Anthropology of the Body and Embodiment*. Oxford: Wiley-Blackwell.

Maus, M. Mauss, Marcel 1979 [1935]. Body techniques. *Sociology and Psychology: Essays*, translated by Ben Brewster. London: RKP.

McDonald, M. 2014. Bodies and cadavers. In P. Harvey, E. Conlin Casells, G. Evans, H. Knox, E.B. Silva, N. Thoburn and K. Woodward,. (eds.), *Objects and Material*, pp. 128-143. *A Routledge Companion*. Abingdon: Routledge.

Merleau-Ponty, M. 1962. *Phenomenology of Perception*, translated by Colin Smith. London: Routledge & Kegan Paul.

——— 1964. *The Primacy of Perception*, translated by James Edie. Evanston, IL: Northwestern University Press.

Navaro-Yashin, Y. 2012. *The Make-Believe Space: Affective Geography in a Post-War Polity*. Durham, NC: Duke University Press.

Pink, S. 2009. *Doing Sensory Ethnography*. London: Sage.

Roodenburg, H. 2004. Bourdieu: Issues of embodiment and authenticity. *Etnofor* 17 (1/2): 215–226.

Rutherford, D. 2016 Affect theory and the empirical. *Annual Review of Anthropology* 45: 285–300.

Strathern, A. and P. Stewart 2011. Embodiment and Personhood. In F. Mascia Lees (ed.), *A Companion to the Anthropology of the Body and Embodiment*. Oxford: Wiley-Blackwell.

Toren, C. 1999 *Mind, Materiality and History Explorations in Fijian Ethnography*. London and New York, NY: Routledge.

——— 2012. Imagining the world that warrants our imagination. The revelation of ontogeny. *Cambridge Anthropology* 30 (1): 64–79.

Vilaça, A. 2005. Chronically unstable bodies: Reflections on Amazonian corporalities. *Journal of the Royal Anthropological Institute* (N.S.) 11(3): 445–464.

——— 2009. Bodies in perspective: A critique of the embodiment paradigm from the point of view of Amazonian ethnography. In Lambert, H. and M. McDonald (eds.), *Social Bodies*, pp. 129-147. New York, NY and Oxford: Berghahn.

Viveiros de Castro, E. 1998. Cosmological perspectivism in Amazonia and elsewhere. Four Lectures given in the Department of Social Anthropology, University of Cambridge, February–March. *Hau Masterclass Series Volume 1*. online at http://www.haujournal.org/index.php/masterclass/issue/view/Masterclass%20Volume%201 (accessed 20 September 2014).

Chapter 12

Feminist anthropology and the question of gender

Jessica Johnson

Bringing women in

In 1974 it was possible, and necessary, for Michelle Rosaldo and Louise Lamphere to argue that anthropologists needed 'to begin to think about women' (1974: vi). Indeed, until that time, very little had been published in anthropology that looked in detail at the lives and perspectives of women (see chapter 1). There were a few exceptions, of course, notably the work of Audrey Richards (1982 [1956]), for example, and a number of impressive studies of women that had been published by the wives of male anthropologists, generally untrained in anthropology themselves, whose work received little attention from professional anthropologists.[1] From the 1970s, inspired by the women's liberation movement, women in anthropology began to place women at the centre of their research. They were convinced that anthropology as a discipline was impoverished by its seeming lack of interest in women, and argued that paying attention to women's lives would lead to richer ethnographic accounts and more reliable theory.

One of the few anthropological texts that these early pioneers of what we might call at this stage 'the anthropology of women' (Moore 1988: 6) rather than the 'anthropology of gender', had in their arsenal was a chapter by Edwin Ardener, 'Belief and the problem of women' (1972). Perhaps unwittingly, Ardener had provided a key reference for these early debates. He argued that it was not the intrinsic significance (or otherwise) of women's activities that led to their relative absence from anthropological texts, but rather the fact that they tended to offer visiting ethnographers less satisfying models of society than male informants did, and less overarching accounts of social life. He suggested that women were less likely to offer insights and explanations that met male anthropologists' expectations; as a result, women seemed inarticulate and made frustrating interviewees.

Ardener's article was taken up by early anthropologists of women as a seminal text, at the time one of few available texts about women. It was certainly influential for Rosaldo and Lamphere, whose own volume marked the arrival of the anthropology of women. They began from the assumption of generalised inequality between men and women, seeing women as universally subordinated to men.[2] In their view: 'all contemporary societies are to some extent male-dominated, and although the degree and expression of female subordination vary greatly, sexual asymmetry is presently a universal fact of human social life' (1974: 3). For her part, Rosaldo argued that women's subordination could be explained by the fact that, generally speaking, 'a good part of a woman's adult life is spent giving birth to and raising children [and this] leads to a differentiation of domestic and public spheres of activity' (1974: 23). Confined to the domestic sphere, generally the home, where they are socially isolated and occupied by the tasks of household labour and child care, women lack access to the kinds of status, respect or cultural value that men enjoy.

Writing in the same volume, Sherry Ortner (1974) offered a more symbolic (see chapter 8), or structuralist (see chapter 2) analysis. She asked: 'What could there be in the generalised structure and conditions of existence, common to every culture, that would lead every culture to place a lower value upon women?' (p. 71). Her response was that women were seen as closer to nature than men, who were considered more cultural. She did not see these associations as necessary in the sense that they could never be different, but she did argue that they were universal. The association of women with nature, she suggested, provided a ready rationale for female subordination. For Ortner, culture was 'the notion of human consciousness (i.e. systems of thought and technology), by means of which humanity attempts to assert control over nature' (p. 72). Culture was thus, by definition, transcendent over nature.

There are three aspects to women's relative closeness to nature in Ortner's account: women's physiology, as expressed in reproduction and child rearing; women's social roles as mothers and in child care; and what she calls 'woman's psyche', which she relates to the ways in which young girls and women are socialised to engage in more personalised modes of relating to others while men engage in more abstract or general modes of relating. Thus, for both Rosaldo and Ortner, it was largely women's roles in child care and reproduction that led to their secondary status. Both authors were looking for universal answers to the question of seemingly universal female subordination.

Although contributors to Rosaldo and Lamphere's volume began from the assumption of women's universal subordination, they sought to show that sexual inequality was 'not a necessary condition of human societies but a cultural product [and therefore] accessible to change' (1974: 13). Taken together, the chapters highlight the ethnographic range of differences between men and women, the wide array of social roles women play, differences in their public status, cultural definitions and daily activities. They thus show women's lives to be more varied and interesting than had previously been assumed. Despite relating women's subordinate status to their biological role in reproduction, they did not see women's position as biologically determined. They recognised the significance of cultural 'interpretations of biology' (p. 5); differences in the degree to which societies were male dominated; and the possibility that 'with changes in technology, population size, ideas, and aspirations, our social order can change' (p. 7). 'Surely', they say, 'the diversity of human cultures and the evidence of societies in which women have achieved considerable recognition and social status might make us optimistic about the possibility of realising sexual equality in our world today' (p. 14). The politics of feminism, and the urgency of women's struggles in the 1970s, were clearly reflected in this early work that sought to bring women into the anthropological fold.

However, it did not take long before the idea of the universal association of women with nature came under scrutiny. Carol MacCormack and Marilyn Strathern (1980) soon transformed the terms of debate by arguing that the concepts of nature and culture, and the meanings associated with male and female persons, were not 'given' and they could not be 'free from the biases of the culture in which the concepts were constructed' (MacCormack 1980: 6). MacCormack thus described Ortner's argument as 'remarkable for its ethnocentricity' (p. 16). The significance of their volume was two-fold. They argued that women were not universally associated with nature, and, more radically, that the nature/culture distinction itself was not universal. Rather, they suggested that this dichotomy was better understood as a product of a particular (cultural and Western) mode of interpreting the world.

This latter position was most clearly articulated by Strathern (1980), who offered an ethnographic examination of the distinctions Hageners of Highland Papua New Guinea make between categories of domestic and wild, and how these relate to their distinctions between male and female. She argued against the straightforward mapping of 'our' nature–culture distinction onto Hageners' wild–domestic distinction, showing in some detail how they differed. The wild–domestic distinction did not relate in a direct manner to Hageners' gender categories. In some contexts, women were considered wild and men domesticated, in other contexts it was the other way around. The domestic and the wild were also differently related to each other than nature and culture were in Ortner's framework. They were opposites but they were not hierarchical or processual in the sense that nature might either be transformed into culture or dominated by it. The domestic domain was not seen as colonising the wild, and the human socialisation process was not represented as the transcendence of nature by culture. Such critiques served to undermine the universalist confidence of the early anthropology of women, suggesting that more care needed to be taken in generalising about women's status or the meanings attached to male and female identities.[3]

MacCormack and Strathern's volume might be said to have marked a key point of transition from the anthropology of women to the anthropology of gender. This transition continued apace throughout the 1980s and beyond, as ethnographic and theoretical work continued to unsettle easy distinctions between male and female social roles, revealed vast differences in cultural interpretations of social reproduction and experiences of gendered identities, and exposed the impossibility of generalising about 'men' and 'women' in a meaningful sense. It has also become clear that gender is relational, and that studies of masculinity, as well as femininity, have a great deal to offer the discipline (e.g. Ortner and Whitehead 1981; Shapiro 1981; Collier and Yanagisako 1987; Cornwall and Lindisfarne 1994; Moore 1994). However, as we shall see below, the deconstruction of gender that such studies entailed has not always been easy to reconcile with feminist scholarship dedicated to a focus on women.

Can there be a feminist ethnography?

Following on the heels of these developments, Lila Abu-Lughod (1990), writing explicitly as a feminist anthropologist, posed the question 'Can there be a feminist ethnography?' Abu-Lughod, while cognisant of developments in the anthropology of gender, maintained a primary interest in the lives of women, in all their social and cultural diversity. Her article represented a different kind of movement beyond the anthropology of women, critiquing that early work for not having questioned the social scientific goal of objectivity. As we have seen, early interventions argued that leaving women out of ethnographic texts distorted anthropological accounts. The suggestion was that anthropologists would gain more objective, or complete, knowledge of the societies they sought to understand if they included women more thoroughly in their ethnographic studies.

In the years since the publication of Rosaldo and Lamphere's volume, feminist anthropologists had begun to question the goal of objectivity. They no longer saw objectivity as a neutral aim. This view built on work by reflexive anthropologists who had stressed that so-called ethnographic facts were actually the outcome of particular interactions in specific contexts. Their work led to the acceptance that no two anthropological accounts of the same people and place could be expected to look the same. By this point, anthropologists had also become much more concerned with the processes and effects of writing and representation (Clifford and Marcus 1986; see chapter 8).

In tandem with these developments in anthropology, feminist scholars in other disciplines also attacked ideas of objectivity. Abu-Lughod charts these debates, and shows that there were several different angles that feminists took. In the history and philosophy of science for example, Evelyn Fox Keller (1982; 1985) argued that distinctions made in science between subjectivity and objectivity were deeply gendered, and that science valorised qualities considered masculine (e.g. reason, detachment, impersonality). Keller thus distinguished between 'objectivism' – the ideology of science – and true 'objectivity' (Abu-Lughod 1990: 13). The feminist legal scholar Catherine MacKinnon (1982) went further. She argued that objectivity was 'a strategy of male power' by which 'men, always dominant, create the world from their point of view, particularly in objectifying women, and then adopt an epistemological stance – namely objectivity – that corresponds to the world they have created' (Abu-Lughod 1990: 14). While Abu-Lughod sounded a word of caution against the totalising nature of MacKinnon's position, she highlighted the crucial point that if we believe that there is inequality between men and women and that perspective affects understandings of the world, then 'there is no ungendered reality or ungendered perspective' (MacKinnon 1982: 636, cited in Abu-Lughod 1990: 15).

Reflecting on these advances in 1990, Abu-Lughod laments: Where have the feminist anthropologists been? Where were the studies that might 'bring to life what it means to be a woman in other places and under different conditions'? (p. 27). She did not see much feminist ethnography, despite the fact that changing understandings of the discipline of anthropology might well have opened the way for it. Notably, for example, no feminist anthropologists were included in the *Writing Culture* project that rethought the way in which anthropologists wrote and the kinds of texts they produced (Clifford and Marcus 1986). One part of the answer as to why feminist anthropologists had not been more visible, according to Abu-Lughod, is to do with the politics of academia. Job security greatly improves the likelihood of innovative work, and academic job security (then and now) is a privilege disproportionately enjoyed by men (1990: 17).

Another part of the problem is that women's earlier experimental texts, which were often written by the untrained wives of anthropologists, had been ignored. Instead of looking to these texts for inspiration when they wanted to rethink genres of anthropological writing, anthropologists involved in the writing-culture project had turned to the elite, professional disciplines of philosophy and literary studies. In a context in which women's contributions were dismissed as unprofessional, descriptive rather than theoretical, based on personal experience rather than authoritative literature and so on, Abu-Lughod argued that early feminist anthropologists were so busy trying to 'establish their credibility, gain acceptance and further their intellectual and political aims' (1990: 19) as feminists that they were not in a position to take risks or experiment in their work.

Nevertheless, Abu-Lughod ended her piece on a hopeful note. She identified a growing awareness that the category of 'women' was diverse and inflected by economic factors, race, sexuality, nationality, religion and so on, and noted that recognition of this diversity was occurring at the same time that anthropologists themselves were also becoming more diverse in terms of their origins as well as their gender. In turn, clear-cut distinctions between 'home' and 'field' were being unsettled, and explicitly debated. Abu-Lughod argued that:

> feminist ethnographies, . . . ethnographies that explore what work, marriage, motherhood, sexuality, education, poetry, television, poverty, or illness mean to other women, can offer feminists a way of replacing their presumptions of *a* female experience with a grounded sense of our commonalities and differences.
>
> (1990: 27)

Indeed, in an explicit effort to respond to feminist critiques of objectivity and representation, Abu-Lughod (1993) went on to publish a book that deliberately eschewed generalisations in order to stress the particularity of experience and the inevitable situatedness of storytelling, which were two important messages of feminist theory.

Engaging with theories of performativity

When anthropologists working on gender have taken inspiration from theorists, Judith Butler, and in particular her book *Gender Trouble* (1990), have figured prominently. The book emphasises the importance of performance to the creation of gender identities and it can be seen as presenting a radical constructionist position. Butler's aim in *Gender Trouble* was to 'to trace the way in which gender fables establish and circulate the misnomer of natural facts' (p. xiii). According to Butler, sex categories (male and female) constitute 'a discursive formation that acts as a naturalised foundation' (p. 37), and gender is defined as 'the repeated stylization of the body . . . within a highly rigid regulatory frame that congeal[s] over time to produce the appearance of substance, of a natural sort of being' (p. 33, cited by Franklin 2002: 310). She terms these processes 'performance'. In part, Butler's argument is reminiscent of MacCormack and Strathern's emphasis on the contextual specificity of what is considered natural. She posits that a binary difference between male and female does not exist 'as a presocial fact' (Franklin 2002: 310), but rather is achieved or produced through discursive and embodied performance. Thus, in a radical move, she questions the notion of sex as a non-cultural category that can be separated out and distinguished from gender.

Work by anthropologists that engages Butler's ideas and pushes against the edges of gender theory has often appeared in studies of kinship. In fact, the study of kinship and the study of gender have a great deal in common. Both are fields of anthropological enquiry that have emphasised distinctions between what is natural and what is cultural, what is biologically determined and what is the result of human creativity (Yanagisako and Collier 1987).[4] A key aspect of early work on gender was an emphasis on the distinction between gender – as a social role – and sex – as the physical body (e.g. Ortner 1974; Rosaldo and Lamphere 1974; Ortner and Whitehead 1981). By demonstrating that ideas about bodies were distinct from the bodies themselves, feminist scholars posited that the cultural values attached to bodies might shift, and they questioned the seeming inevitability of women's association with the domestic realm.

Janet Carsten (2004) has suggested that distinguishing between sex and gender led quite quickly to a 'theoretical impasse' (p. 62). The distinction rested on a separation between biology and culture and the studies that followed largely focused on variations in cultural understandings and symbols of gender. What was left unexamined and untheorised was 'the question of the analytic significance of the actual physical attributes of the body' (p. 62).

Subsequently, anthropologists were accused of having 'assumed the existence of natural differences between women and men when they should have sought to explain how such differences were conceived' (p. 62). They had focused on culture (gender) and neglected biology (sex). Butler's position challenged these distinctions by arguing that sex was as much of a discursive construction as gender.

Carsten examines constructionist arguments, including those of Butler and Michel Foucault (1978), who also set great store by culturally and historically situated discourse, and argues that these kinds of approach are, like the arguments they sought to overcome, too totalising, with discourse about sex and gender becoming determining in place of biological sex differences (Carsten 2004: 66). Ultimately, she argues that the framing of the debate in terms of the distinction between sex and gender has restricted anthropological work, limiting the possibilities for research. Building on her own Malay ethnography, Carsten shows how gender differentiation varies over the life course with reference to events such as birth, marriage and death. In this way, gender differentiation interacts with kinship, and the two need to be examined together.

Carsten does not advocate doing away with biology, or ideas of what is fixed and given. She says it is not only Western understandings that contain a sense of the difference between sex and gender, between what is given and what is acquired or learned, and subject to change throughout the life course. She suggests that instead of assuming how these distinctions are made and how they relate to other aspects of people's thinking about gender, kinship, relationships and so on, we should enquire about these categories and processes. For example, her Malay informants consider that people become kin through living and eating together, and as Carsten points out, 'it is difficult to know whether this should be considered as a "social" or a "biological" process' since it incorporates aspects of both (p. 81). For Carsten, then, kinship brings empirical substance and ethnographic possibilities to the study of gender and there is much to be gained by 'bringing together sex and gender as a way of retaining biology as part of what anthropologists have to understand' (p. 82). Other scholars who have done just that, incorporating and moving beyond Butler's ideas of performativity in the process, include Middleton (2000), Ramberg (2014), Weston (1997 [1991]) and Franklin (2002).

Anthropology and feminism: An awkward relationship?

It is worth pausing at this point to reflect on the relationship between anthropology and feminism, discussions of which bear the imprint of developments in the broader anthropology of gender.

In a widely read article, Rosaldo (2006 [1980]) noted that, although by 1980 there was 'plenty of data "on women"', when it came to writing about them, few anthropologists knew where to begin. She argued that the problem was no longer a lack of data, but of research questions (p. 108). Although, in general terms, questions about human origins had gone out of favour in anthropology, those kinds of questions still dominated feminist writings, where 'were things always as they are today?' was still commonly asked (p. 108). Non-Western societies were taken as sources for answers to questions about the distant past and this seemed to be the kind of information feminists expected anthropologists to supply. Rosaldo's concern was that such universalist questions lent themselves to universalist answers (of the kind she herself had provided in 1974). By 1980, she argued that feminist anthropologists needed new questions, rather than accounts of 'how it all began'.

By posing different questions she hoped anthropologists would have more than raw data to contribute to feminist debates. In her view, they needed to be able to analyse relationships between men and women 'as aspects of a wider social context' (p. 120). By this point, she wanted to foreground ethnographic specificities rather than abstract theorising. While Rosaldo was clear that she did not want to deny 'that biological facts like reproduction leave their mark on women's lives', she wanted to place greater emphasis on sociocultural context and focus on the ways in which 'men and women both participate in and help to reproduce the institutional forms that may oppress, liberate, join or divide them' (p. 122).[5] In addition to concern over the development of the relationship between anthropology and feminism, we can also see here a shift towards incorporating men into the study of gender rather than advocating an anthropology of women.

Twenty years later, Ana Maria Alonso (2000) published a response to Rosaldo's piece. In the meantime, feminist scholarship had moved on significantly and the field of gender studies had become more firmly established. Scholars such as Judith Butler had come to prominence and many of the questions being asked had in fact changed. Alonso opened her paper by describing a moment during a feminist reading group when a new member mentioned menstruation as a possible '"point of departure" for a feeling of commonality among women' (p. 221). She was immediately accused of being an 'essentialist' – someone who believes that gender is fixed and determined – to which she replied that currently dominant approaches in feminist scholarship were 'engaged in a denial of the body' (pp. 221–222).

Reflecting on this incident, Alonso asked whether 'essentialist' had become a 'gatekeeping label' that prevented the recognition of 'biocultural bodily processes' that had largely been put to one side in the theoretical shift towards discourse (p. 222). She recognised many academic feminists' fear that acknowledging a bodily basis to gender would imply that gender inequality is inevitable (p. 222) Indeed, many of Alonso's own students defined Rosaldo's work as 'essentialist'. Most of these students identified as liberal feminists; for them, 'freedom meant choice, and gender could not be chosen if it was in any way anchored in human biology – particularly in sexual reproduction' (p. 223). Alonso was struck by what she saw as the privilege underpinning this position. These were women who had benefitted from advances such as:

> contraception, mammograms, breast reconstructive surgery, infant formulas, breast pumps, child care alternatives, sanitary towels and so on; for them it was possible to forget that their access to these things was the result of 'earlier generations' struggles against men's control of women's [bodies], against gender bias in the medical establishment, [and] against state regulation of women's sexuality.
>
> (p. 223)

By contrast, she argued that less privileged women, whether in the US or elsewhere, tended to be 'more aware of the materiality of the body because they have to struggle with the changes brought on by their cycles, pregnancies, and illnesses in ways that carry more sensory immediacy' (p. 223).

Alonso posited that anthropological work done in the 1970s had actually been more holistic than later work on gender, which she considered to have placed too much emphasis on discourse, seeing things as constantly in flux and inequalities as negotiated. In her view, more recent work that deconstructs the idea of 'women' as the subject of feminism, 'has little to offer women outside the academy or . . . activists working in battered women's shelters, legal aid, and numerous women's health, social and political organisations' (p. 229). She concluded that a debilitating division had emerged 'between feminists who believe in women and those who do not' (p. 229). Evident here is the status of feminism and gender studies as contested fields, characterised by debate, disagreement and tension between feminist scholarship and feminist political action.

Picking up on such frictions, Strathern memorably characterised the relationship between anthropology and feminism as 'awkward' (1987). Her intervention was motivated in part by her observation of the limited degree to which feminism and feminist anthropology had penetrated or transformed the discipline of anthropology as a whole. She used the term 'awkward' to signal a tension between feminist scholarship and anthropology that is, in her words 'a doorstep hesitation rather than barricades' (p. 286). Her thesis is that anthropology and feminism have certain things in common, notably a shared emphasis on experience, but that they are different in ways that mean that they do not confront or challenge each other head on, but rather strike tangential blows and sideways, mocking glances, because of their contrasting perspectives.

Central to her discussion is a consideration of the ways in which anthropology and feminism relate to their respective others. Anthropology, she says, deals with an other that is socially or culturally distinct; while for feminism, the other is patriarchy: 'the institutions and persons who represent male domination', or in its short hand form: men (p. 288). While feminism operates in opposition to its other, anthropology is committed to a stance of openness towards others' experiences and perspectives. The anthropological aim is to 'create a relation with the Other' (p. 289). Strathern suggests that feminism mocks anthropologists' efforts to create collaboration, dialogue or even co-authorship with their informants by pointing out the inevitable asymmetries of power between anthropologists and the people they work with and study. Anthropologists are aware of this, but collaboration remains their 'ideal ethical situation' (p. 290), an idea that feminism challenges because, 'from a feminist perspective ... there can be no collaboration with the Other' (p. 290). Although this feminist critique is based upon different premises, it hits anthropology where it hurts. In its turn, anthropology mocks feminism: feminism contains the idea that feminists come to know the self through its difference from the other. For feminists 'It is an achievement to perceive the gulf [between women and patriarchy], and in turn, an ethical position, for this is what validates women's commitment to one another' (p. 290). Anthropology mocks the idea that feminists can achieve this separation because anthropologists' ethnographic work indicates just how much is shared by feminists with the society they critique: feminists can be seen to operate 'within the sociocultural constraints of their own society' (p. 291).

The idea of the awkward relationship encapsulates, for many feminist anthropologists, the tensions they feel personally as they strive to be both feminists and anthropologists. It also gets to the heart of the question of how feminist anthropologists deal with difference. Strathern's thesis has nevertheless elicited critique. Thus, Abu-Lughod has suggested that Strathern is guilty of retreating 'from the problem of power' (Abu-Lughod 2006 [1991]: 154). She points out the contrast between the relative position, in terms of power, that anthropologists and feminists have tended to hold in relation to their respective others. Anthropologists generally stand on the side of the more dominant West, in the West–non-West encounter (whether they are Western born, Western educated or simply schooled in the discourse and practice of the discipline of anthropology), whereas feminists are associated with the subordinate position in their confrontation with patriarchy.

Moreover, according to Abu-Lughod, feminism has struggled more openly and successfully with difference within its own ranks than anthropology has. Nevertheless, she recognised increasing diversity within anthropology by the 1990s. She thus speaks of 'halfie' anthropologists, herself included, 'whose national or cultural identity is mixed by virtue of migration, overseas education, or parentage' (p. 153).[6] Abu-Lughod argues that feminist and halfie anthropologists share an inability to 'comfortably assume the self of anthropology' (p. 155). This gives them a greater awareness of issues of power and positionality, as well as the differences between the various audiences they write for. Unlike earlier generations of anthropologists, they have to ask themselves 'what happens when the other that the anthropologist is studying is simultaneously constructed as, at least partially, a self?' (p. 155). What Abu-Lughod points us to here, in part, are debates within feminism and feminist anthropology centring on differences of gender, race and class (e.g. Sacks 1989).

Postcolonial critiques

Other scholars have not looked as favourably on feminism as Abu-Lughod (e.g. Amadiume 1987; Oyěwùmí 2003a).[7] Aihwa Ong (2001 [1988]), for example, echoing Strathern's point that feminists operate within the same sociocultural world as the patriarchy they oppose, accuses feminist anthropologists of inappropriately applying the Western standards and goals of 'rationality and individualism' (p. 108) to other societies. Moreover, she suggests that 'when feminists look overseas, they frequently seek to establish *their* [own] authority on the backs of non-Western women, determining for them the meanings and goals of their lives' (p. 108). In this sense, she argues that for feminists working outside the West, the other is not men, as Strathern had suggested, but non-Western women, so that feminists and the non-Western women they study stand in an oppositional relationship to one another.

Ong's concern is with the 'intersections between colonial discourse and feminist representations of non-Western women' (p. 108) and she focuses on books written about 'women in development' to make her point. She suggests that there is a neo-colonial quality to the ways in which feminist scholars write about non-Western women, who are 'taken as an unproblematic universal category' in much of this writing, and whose status is judged against 'a set of legal, political, and social benchmarks that Western feminists consider critical in achieving a power balance between men and women' (p. 110). In other words, non-Western women are measured by Western standards and compared with Western feminists' own goals and ideals. As a result, feminists reinforce their 'belief in their own cultural superiority' (p. 113). She urges feminist anthropologists not to give up, but rather to alter their approach in order to understand the lives and ambitions of women in other parts of the world on their own terms and to 'accept that others often choose to conduct their lives separately from our particular vision of the future' (p. 116)

Ong's critique resonates with a highly influential article published by Chandra Mohanty the same year. Mohanty argued that much Western feminist writing obliterates the 'material and historical' variety of women's lives in the third world and ends up presenting a 'composite [and] singular "third world woman"' (1988: 62), characterised by her subordination.[8] She sees in this creation of the generic third world woman the workings of power in feminist writing. Ong and Mohanty's critiques also chime with the work of a number of African feminist scholars, which constitutes a rich resource for those seeking to engage with other ways of thinking about gender both historically and in relation to contemporary concerns. In different ways, Ifi Amadiume (1987; 1997), Oyèrónkẹ́ Oyěwùmí (1997; 2003b) and Nkiru Uwechia Nzegwu (2006), for example, challenge us to rethink deeply held assumptions about the significance of motherhood for women's social status and highlight the detrimental effects of colonisation upon gender relations in Africa. They emphasise the contrasting historical experiences of European and African women, and argue that the tools of Euro-American feminist analysis are insufficient to the task of understanding African women's lives.

In a similar vein, Saba Mahmood's (2005) study of Egyptian Muslim women members of the piety movement in Cairo presents us with a profound challenge. The women Mahmood worked with did not necessarily share the goals of Western liberal feminists – namely freedom and equality – and thus Mahmood asks: 'does a commitment to the idea of equality in our own lives endow us with the capacity to know that this ideal captures what is or should be fulfilling for everyone else?' If not, she continues, 'as is surely the case, then I think we need to rethink, with far more humility than we are accustomed to, what feminist politics really means' (2005: 38). Indeed, this challenge can be said to lie at the heart of the tension between feminism and anthropology.

Reflecting on the ways in which the lives of Muslim women had come under the scrutiny of European and American media and social commentators, including feminists, in the wake of the atrocities of 11th September 2001, Abu-Lughod issued two warnings to feminist anthropologists. One is that we must not succumb to the polarising trap of placing 'feminism on the side of the West' (2002: 788). We need to be able to think about feminisms in the plural. The second is against vocations of 'saving' women in other parts of the world. She suggests that the very idea of saving other women is patronising and insensitive to other ways of understanding their lives (2002; 2013). Instead of saving Muslim women she asks that:

> where we seek to be active in the affairs of distant places ... we do so in the spirit of support for those within those communities whose goals are to make women's (and men's) lives better ... [and that we] use a more egalitarian language of alliances, coalitions, and solidarity, instead of salvation.
>
> (2002: 789)

Conclusion

Beginning with the formation of feminist anthropology from the 1970s, this chapter has examined the development of the anthropology of gender. Recognising the exclusion of women from anthropological texts, pioneers in this field sought to establish an anthropology of women, which soon grew and diversified into the broader study of gender relations and gendered identities. In particular, we have considered advances in the field of kinship studies, where the mutual influence of anthropological work and the writings of gender theorists is particularly evident. These debates continue to foreground ideas about nature and culture similar to those that animated earlier studies. However, the move to interrogate nature/biology (rather than simply focusing on culture/gender) has unsettled the distinction between sex and gender that was central to earlier work. Other important developments in the anthropology of gender that I have not been able to consider in any detail in this chapter include the emergence of studies of masculinity (e.g. Gutmann 1997; Osella and Osella 2000; Connell 2005 [1995]; Simpson 2009) and sexuality (e.g. Kulick 1997; Boellstorff 2005; 2007; Cornwall, et al. 2011; Lyons and Lyons 2011; Tamale 2011; Spronk 2014).

Over time, as we have seen, scholars have wrestled with the relationship between feminism as a political commitment to gender equality, and the anthropological imperative not to assume the content or meaning of gendered identities and gender relations elsewhere. The seemingly inevitable tension or 'awkwardness' between anthropology and feminism raises the possibility that 'feminist anthropology' is a contradiction in terms, and yet, somehow, sensitive feminist anthropological enquiry continues to provide provocative ethnographic and theoretical interventions (e.g. Arnfred 2011; Hodgson 2011).

In conclusion, it is worth reflecting on just how successful feminist anthropology and the anthropology of gender have been. In the 21st century it is hard to imagine an anthropologist, male or female, explicitly feminist or not, proceeding to write about a society or a social phenomenon without explicitly considering the gendered dimensions – without talking to both men and women, thinking carefully about how their perspectives differ and why that might be, looking at their various activities and roles, or the ways in which men and women cooperate in the same tasks. Nor would contemporary anthropologists necessarily assume the relevance of binary gender categories in a particular ethnographic setting. Whatever they were most interested in and whatever their resulting ethnography was about, these things would make their way into the text. They would be built into the methods of the study during fieldwork in order to find ways of accessing male and female voices, perspectives and activities. And in situations where that was not possible, these limitations would be discussed, as would the implications of the ethnographer's own gendered identity. The ground really has shifted.

Notes

1 Perhaps the best known of these texts is *Nisa* (Shostak 1981). For an overview of this genre, see the introduction to Lewin (2006) and Abu-Lughod (1990).
2 It is worth noting that another volume published the following year (Reiter 1975) did not share this approach. See especially the chapter by Gayle Rubin.
3 As Heywood (chapter 14) demonstrates, Strathern's argument has also had a profound and lasting influence beyond anthropological debates about gender, in particular, in relation to the development of what has come to be known as 'the ontological turn'.
4 See Carsten (2004) for a critical discussion of Yanagisako and Collier's position.
5 See Nicholson (1982) for an early (and critical) response to Rosaldo's argument.
6 'Halfie' is not a term that has taken off in anthropology, but it is one that Abu-Lughod finds useful; I employ it here by way of reference to her work.
7 I refer here to the comparison, discussed above, that she made in the early 1990s between anthropology and feminism, and her sense at that time that feminists had been struggling more successfully with their own internal differences than anthropologists had. Nevertheless, as we shall see below, Abu-Lughod is by no means uncritical of feminism, as is evident in her condemnation of the ways in which many feminists have depicted Muslim women since 2001 (Abu-Lughod 2002; 2013).
8 By 'Western feminism' Mohanty refers to feminism that takes its 'primary point of reference [as] feminist interests as they have been articulated in the US and Western Europe' (1988: 61) and she suggested that middle-class urban African and Asian scholars are often also implicated in the hierarchical approach to feminist writing that she critiqued.

References

Abu-Lughod, L. 1990. Can there be a feminist ethnography? *Women & Performance: A Journal of Feminist Theory* 5 (1): 7–27.

——— 1993. *Writing Women's Worlds: Bedouin Stories*. Berkeley, CA: University of California Press.
——— 2002. Do Muslim women really need saving? Anthropological reflections on cultural relativism and its others. *American Anthropologist* 104 (3): 783–790.
——— 2006 [1991]. Writing against culture. In E. Lewin (ed.), *Feminist Anthropology: A Reader*, pp. 153–169. Oxford: Blackwell.
——— 2013. *Do Muslim Women need Saving?* Cambridge, MA: Harvard University Press.
Alonso, A.M. 2000. The use and abuse of feminist theory: Fear, dust, and commensality. In A. Lugo and B. Maurer (eds.), *Gender Matters: Rereading Michelle Z Rosaldo*, pp. 221–231. Ann Arbor, MI: University of Michigan Press.
Amadiume, I. 1987. *Male Daughters, Female Husbands* London: Zed Books.
——— 1997. *Reinventing Africa: Matriarchy, Religion and Culture*. London: Zed Books.
Ardener, E. 1972. Belief and the problem of women. In J.S. La Fontaine (ed.), *The Interpretation of Ritual*, pp. 135–158. London: Tavistock.
Arnfred, S. 2011. *Sexuality and Gender Politics in Mozambique: Rethinking Gender in Africa*. Woodbridge: James Currey.
Boellstorff, T. 2005. *The Gay Archipelago: Sexuality and Nation in Indonesia*. Princeton, NJ: Princeton University Press.
——— 2007. *A Coincidence of Desires: Anthropology, Queer Studies, Indonesia*. Durham, NC: Duke University Press.
Butler, J. 1990. *Gender Trouble: Feminism and the Subversion of Identity*. London: Routledge.
Carsten, J. 2004. *After Kinship*. Cambridge: Cambridge University Press.
Clifford, J. and G. Marcus (eds.) 1986. *Writing Culture: The Poetics and Politics of Ethnography*. Berkeley, CA: University of California Press.
Collier, J.F. and S.J. Yanagisako (eds.) 1987. *Gender and Kinship: Essays Toward a Unified Analysis*. Palo Alto, CA: Stanford University Press.
Connell, R.W. 2005 [1995]. *Masculinities*. Berkeley, CA: University of California Press.
Cornwall, A. and N. Lindisfarne (eds.) 1994. *Dislocating Masculinity: Comparative Ethnographies*. London: Routledge.
Cornwall, A., J. Edstrom and A. Greig (eds.) 2011. *Men and Development: Politicising Masculinities*. London: Zed Books.
Foucault, M. 1978. *The History of Sexuality Volume 1: The Will to Knowledge*. London: Penguin Books.
Franklin, S. 2002. Biologization revisited: Kinship theory in the context of the new biologies. In S. Franklin and S. McKinnon (eds.), *Relative Values: Reconfiguring Kinship Studies*, pp. 302–325. Durham, NC: Duke University Press.
Gutmann, M.C. 1997. Trafficking in men: The anthropology of masculinity. *Annual Review of Anthropology* 26: 385–409.
Hodgson, D.L. (ed) 2011. *Gender and Culture at the Limit of Rights*. Philadelphia, PA: University of Pennsylvania Press.
Keller, E.F. 1982. Feminism and science. *Signs: Journal of Women in Culture and Society* 7 (3): 589–602.
——— 1985. *Reflections on Gender and Science*. New Haven, CT: Yale University Press.
Kulick, D. 1997. The gender of Brazilian transgendered prostitutes. *American Anthropologist* 99 (3): 574–585.
Lewin, E. (ed.) 2006. *Feminist Anthropology: A Reader*. Oxford: Blackwell.
Lyons, A.P. and H.D. Lyons (eds.) 2011. *Sexualities in Anthropology: A Reader*. Oxford: Wiley-Blackwell.
MacCormack, C.P. 1980. Nature, culture and gender: A critique. In C.P. MacCormack and M. Strathern (eds.), *Nature, Culture and Gender*, pp. 1–24. Cambridge: Cambridge University Press.
MacCormack, C.P. and M. Strathern (eds.) 1980. *Nature, Culture and Gender*. Cambridge: Cambridge University Press.
MacKinnon, C. 1982. Feminism, Marxism, method and the state: An agenda for theory. *Signs: Journal of Women in Culture and Society* 7 (3): 515–544.

Mahmood, S. 2005. *Politics of Piety: The Islamic revival and the Feminist Subject*. Princeton, NJ: Princeton University Press.

Middleton, K. 2000. How Karembola men become mothers. In J. Carsten (ed.), *Cultures of Relatedness: New Approaches to the Study of Kinship*, pp. 104–127. Cambridge: Cambridge University Press.

Mohanty, C.T. 1988. Under Western eyes: Feminist scholarship and colonial discourses. *Feminist Review* 30: 61–88.

Moore, H.L. 1988. *Feminism and Anthropology*. Cambridge: Polity.

——— 1994. *A Passion for Difference: Essays in Anthropology and Gender*. Bloomington, IN: Indiana University Press.

Nicholson, L.J. 1982. Comment on Rosaldo's 'The use and abuse of anthropology'. *Signs* 7 (3): 732–735.

Nzegwu, N.U. 2006. *Family Matters: Feminist Concepts in African Philosophy of Culture*. Albany, NY: State University of New York Press.

Ong, A. 2001 [1988]. Colonialism and modernity: Feminist re-presentations of women in non-Western societies. In K. Bhavnani (ed.), *Feminism and Race*, pp. 108–118. Oxford: Oxford University Press.

Ortner, S.B. 1974. Is female to male as nature is to culture? In M.Z. Rosaldo and L. Lamphere (eds.), *Woman, Culture, and Society*, pp. 67–88. Palo Alto, CA: Stanford University Press.

Ortner, S.B. and H. Whitehead (eds.) 1981. *Sexual Meanings: The Cultural Construction of Gender and Sexuality*. Cambridge: Cambridge University Press.

Osella, F. and C. Osella 2000. Migration, money and masculinity in Kerala. *Journal of the Royal Anthropological Institute* 6 (1): 117–133.

Oyěwùmí, O. 1997. *The Invention of Women: Making an African Sense of Western Gender Discourses*. Minneapolis, MN: University of Minnesota Press.

——— (ed.) 2003a. *African Women and Feminism: Reflecting on the Politics of Sisterhood*. Trenton, NJ: Africa World Press.

——— 2003b. The white woman's burden: African women in Western feminist discourse. In O. Oyěwùmí (ed.), *African Women and Feminism: Reflecting on the Politics of Sisterhood*, pp. 25–43. Trenton, NJ: Africa World Press.

Ramberg, L. 2014. *Given to the Goddess: South Indian Devadasis and the Sexuality of Religion*. Durham, NC: Duke University Press.

Reiter, R.R. (ed.) 1975. *Toward an Anthropology of Women*. London: Monthly Review Press.

Richards, A.I. 1982 [1956]. *Chisungu: A Girl's Initiation Ceremony among the Bemba of Zambia*. London: Routledge.

Rosaldo, M.Z. 1974. Woman, culture, and society: A theoretical overview. In M.Z. Rosaldo and L. Lamphere (eds.), *Woman, Culture, and Society*, pp. 17–42. Palo Alto, CA: Stanford University Press.

——— 2006 [1980]. The use and abuse of anthropology: Reflections on feminism in cross-cultural understanding. In E. Lewin (ed.), *Feminist Anthropology: A Reader*, pp. 107–128. Oxford: Blackwell.

Rosaldo, M.Z. and L. Lamphere 1974. Introduction. In M.Z. Rosaldo and L. Lamphere (eds), *Woman, Culture, and Society*, pp. 1–15. Palo Alto, CA: Stanford University Press.

Sacks, K.B. 1989. Toward a unified theory of class, race, and gender, *American Ethnologist* 16 (3): 534–550.

Shapiro, J. 1981. Anthropology and the study of gender, *Soundings: An International Journal* 64 (4): 446–65.

Shostak, M. 1981. *Nisa: The Life and Words of a !Kung woman*. Cambridge, MA: Harvard University Press.

Simpson, A. 2009. *Boys to Men in the Shadow of AIDS: Masculinities and HIV Risk in Zambia*. Basingstoke: Palgrave Macmillan.

Spronk, R. 2014. Sexuality and subjectivity: Erotic practices and the question of bodily sensations. *Social Anthropology* 22 (1): 3–21.

Strathern, M. 1980. No nature, no culture: The Hagen case. In C.P. MacCormack and M. Strathern (eds.), *Nature, Culture and Gender*, pp. 174–222. Cambridge: Cambridge University Press.

——— 1987. An awkward relationship: The case of feminism and anthropology. *Signs* 12 (2): 276–292.

Tamale, S. (ed) 2011. *African Sexualities: A Reader*. Oxford: Pambazuka Press.

Weston, K. 1997 [1991]. *Families We Choose: Lesbians, Gays, Kinship and the New Reproductive Technologies*. New York, NY: Columbia University Press.

Yanagisako, S.J. and J.F. Collier 1987. Toward a unified analysis of kinship and gender. In J.F. Collier and S.J. Yanagisako (eds.), *Gender and Kinship: Essays toward a Unified Analysis*, pp. 14–50. Palo Alto, CA: Stanford University Press.

Chapter 13

No actor, no network, no theory

Bruno Latour's anthropology of the moderns

Matei Candea

Introduction: Identifying ANT

This chapter explores the theoretical movement known as 'actor-network theory', which originated in science studies in the 1980s, and became influential in the social sciences more broadly in the late 1990s. Actor-Network Theory (henceforth ANT) has been invoked approvingly or critically by anthropologists studying not only scientists at work (Rabinow 1996; Houdart 2008; Helmreich 2009; Candea 2013) but also a wide a range of other topics, from finance (Zaloom 2003), archaeology (Abu El-Haj 1998) or indigenous rights (Kaplan and Kelly 1999) to forest fires (Candea 2008; for a review of some of this work, and a reflection on the influence of ANT on anthropology, see Oppenheim 2007).

ANT has enthralled and infuriated commentators in equal measure, in part due to the iconoclastic style of argument and seemingly outrageous metaphysical claims of some of its main proponents, chief amongst them the French sociologist/anthropologist/philosopher Bruno Latour. Latour (2000) famously argued, for instance, that when archaeologists claim Egyptian pharaoh Ramses II died of tuberculosis, they are engaging in a kind of time travel, as tuberculosis was only discovered in 1882. In a monograph on the failure of a Parisian transport system project, Latour included transcripts of imagined dialogues between different mechanical elements of a railway carriage (1996). And in a retrospective paper on the growing success of ANT, he helpfully observed that 'there are four things that do not work with actor-network theory; the word actor, the word network, the word theory and the hyphen!' (1999: 15).

This chapter follows the lead of Latour's gnomic utterance. After an initial extended example of the ANT approach in practice, the chapter will develop the negative points above. We will explain why ANT is not in fact about networks or actors in the usual sense, and why its potentially most productive claim – and actor-network theorists' best response to their critics – is that it is not a theory at all.

This negative approach might seem facetious, but it is required because a more frontal description of ANT, outlining what it is about or what it argues, is actually quite difficult. In a very general way, one can identify some key themes and concerns associated with ANT: an interest in materiality and the active properties of non-humans, a concern with tracing the effects of persons, objects and ideas over multiple locations, a suspicion of settled social scientific explanatory frameworks and of standard dichotomies such as science versus politics, nature versus culture or human versus non-human. These themes and concerns are part of a broader theoretical atmosphere in which anthropology and social science have been steeped since the 1990s (see, for instance, Strathern 1988; Gell 1998;

Viveiros de Castro 1998; Ingold 2000; Miller 2005 for very different approaches to some or all of these themes).

ANT stands out quite markedly from these other approaches, but in trying to specify in what way it does so, one begins to run into a number of key problems. The first is that, as suggested above, ANT has been deployed to describe an enormous range of phenomena, and it is thus difficult to circumscribe it by its topic or object. Latour alone (or in some cases with others) has written books on science (1987), technology (1996), modernity (1993), Paris (1998), politics (2004a), law (2009), religion (2013b) and the varieties of existence in general (2013a). This eclecticism is widespread amongst actor-network theorists. As John Law – himself a key proponent of ANT – has noted:

> Truth and falsehood. Large and Small. Agency and structure. Human and non-human. Before and after. Knowledge and power. Context and content. Materiality and sociality. Activity and passivity. In one way or another all of these divides have been rubbished in work undertaken in the name of actor–network theory.
>
> (1999: 3)

The second difficulty is that the concepts used by actor-network theorists are themselves constantly shifting, not simply from one author to the next, but even within the work of single authors. Novice readers will soon be confronted with a bewildering array of terms and distinctions: actor-networks, quasi-actants, hybrids, immutable mobiles, matters of fact and matters of concern, intermediaries and mediators, regions, networks, fluids, translation, trail, alliance, the sociology of translation, the sociology of association as opposed to the sociology of the social… Confusing and irritating as this may be, this constantly shifting terminology is, as we shall see below, not a bug but a feature of ANT. It embodies a general aversion to setting out a stable 'framework' or conceptual dogma. For its proponents, to quote Law again:

> [A]ttempts to convert actor-network theory into a fixed point, a specific series of claims, of rules, a creed, or a territory with fixed attributes also strain to turn it into a single location … But this is a nonsense for, to the extent that it is alive, to the extent that it does work, to the extent to which it is inserted in intellectual practice, this thing we call actor-network theory also transforms itself. This means that there is no *credo*. Only dead theories and dead practices celebrate their self-identity. Only dead theories and dead practices hang on to their names, insist upon their perfect reproduction. Only dead theories and dead practices seek to reflect in every detail, the practices which came before.
>
> (1999: 10)

For these reasons, it is in fact easier, as I will do in the rest of this chapter, to identify ANT by outlining what it is not, rather than what it is. I will also be focusing here – again for simplicity's sake – on one particular version of ANT, namely that proposed by Bruno Latour. In light of Law's statement above, however, the reader will not be surprised to hear that a number of other authors associated with ANT – Law himself, but also Annemarie Mol (2002), Michel Callon (1986), Steve Woolgar (1991), Helen Verran (2001), Antoine Hennion (2015) – produced distinctly different versions or understandings of it.

ANT might seem to be more obviously a contribution to sociology or even to philosophy than to anthropology. In many ways, this is so. Indeed, Mol, a prominent proponent of ANT, has characterised her own work as 'empirical philosophy' (2002: 1). That works as an apt characterisation of ANT more generally. And yet, the impulse behind ANT was anthropological in one key sense: it rested on a very particular reading of the anthropological commitment to 'making the familiar strange'. Bruno Latour outlines this commitment at length in his now classic book, *We Have Never Been Modern* (1993). There he proposes that an 'anthropology of the moderns' is required in order to see the most central assumptions of Western modernity – notably, all of the 'divides' enumerated by Law above – as if from the outside.[1] The counterintuitive pronouncements of ANT – including its claim that it is not a theory – can all be seen as an attempt to craft the conceptual language for such an anthropology.

I will give in this chapter only one detailed ethnographic exemplification of ANT, and this is a slightly paradoxical one for two reasons. First, because the book is not everyone's idea of an ethnography. Second, because the term ANT had not in fact been coined at the time at which the book was written. And yet, this now classic monograph – which is both one of the first and still arguably one of the most incisive ethnographies of science in action – first established some of the key principles of what would later be called ANT. The book is *Laboratory Life* (Latour and Woolgar 1979): an ethnographic study of a biology lab at the Salk institute in La Jolla, California. The next section examines this case study in some detail, after which the chapter turns to a broader delineation of ANT, concluding with a consideration of some of the limits of this approach.

The first intimations of ANT: *Laboratory Life*

Two elements of context are important in considering Latour and Woolgar's *Laboratory Life*. The first is that this book was written at a time when anthropologists were increasingly committed to bringing the discipline 'back home' (Cole 1977). Latour and Woolgar's study was thus part of a handful of attempts to bring anthropology 'all the way home' by focusing on the very centres of the production of Western power and narratives, such as scientific laboratories or Western political institutions and ideologies. *Laboratory Life* was thus part of a broader move towards putting 'the West' itself into an ethnographic perspective.

The second element of context concerns discussions in the sociology and history of science, to which this book made a transformative and critical contribution.[2] While examining these arguments might seem to take us out of our focus on ANT as a general 'theoretical' resource for anthropologists, many of the identifying features, strengths and weaknesses of ANT were developed in this context so it pays to examine them in some detail.

In the 1960s and 1970s, sociologists and historians who studied science were increasingly boldly claiming that scientific facts were 'social constructions', and that the content of scientific discoveries could be to some extent explained by, or at least correlated to, factors such as scientists' political interests, social class, gender, cultural imaginations of 'nature' and ideologies more broadly, structural inequalities in the society scientists belonged to and so on. Even if one remained agnostic about whether a particular body of scientific knowledge was true, the fact that it was generally believed to be so was, surely, a social fact. As sociologist David Bloor put it: 'the objectivity of knowledge resides in its being the set of accepted beliefs of a social group' (1974: 76).

This 'social constructionism' in science studies drew directly on a tradition of anthropological analysis of the relationship between knowledge and society, ultimately rooted in the sociology of knowledge launched by Emile Durkheim at the beginning of the 20th century (see chapter 1). Bloor again:

> Social organization can indeed ensure that a given theory is perceived as true. Evans-Pritchard showed as much when he examined the way in which Zande society is organized around the premise that the natural world is replete with the forces of witchcraft. Philosophers as well as anthropologists have increasingly investigated the social processes whereby theories are sustained and anomalies absorbed ... Social organization, then, is the crucial variable determining the perception of the truth and falsity of any given theory.
>
> (p. 76)

This approach led to a focus on the detail and practice of scientific work, to examine sociologically the day-to-day processes whereby 'theories are sustained and anomalies absorbed' (p. 76). In anthropology, Sharon Traweek's ethnography of American and Japanese physicists (1988) sought, in this vein, to treat physics as a cultural emanation of a particular social context, inherently informed by the socially structured and gendered practices of the physicists who produced it (for a critique, see Latour 1990).

The first stirrings of ANT emerged within, but also against, this social constructionist context. *Laboratory Life* (1979), subtitled *The Social Construction of Scientific Facts*, gave an account of biologists' practices that sought to show how facts were – literally – constructed in laboratory conditions, through a series of practices of testing, measuring and writing. The philosophical implications were, in one sense, in line with those of the social constructionist sociology above: facts were not 'discovered', they were made, and the processes that went into making them were (in part) human social processes and practices. For instance, the authors looked in detail at the way the language of scientific papers invokes different levels of certainty about particular statements – from controversial, hypothetical or merely possible statements, to ones solidly grounded through references to other papers, all the way to obvious truths that 'everyone knows' and can be stated without even requiring any reference at all. Establishing a fact, they argued, is partly a struggle to convince readers through the accumulated production of statements tied together through intertextual citational practices. Facts, examined empirically, are nothing other than these interconnected sets of statement and references, inscribed on pieces of paper.

However, Latour and Woolgar took a distinctive turn away from classic social constructionism in their close focus on the role machines and artefacts played in this process of construction. The construction of stable scientific facts – the writing of those papers – would not be possible, they pointed out, without an extensive array of 'things' and non-human entities: the mice whose blood can be extracted, the centrifuges that separate the blood into constituents, the array of machines that allow various other kinds of test to be done on those constituents, the computers that enable the results to be mapped into graphs, that in turn can be printed on paper and integrated into articles in which they support statements. At each of these steps, ideas, questions and answers are given a material solidity. Key to this process are the various technical apparatuses that measure and extract data from biological materials. Latour and Woolgar describe these as 'inscription devices' whose role is to 'transform pieces of matter into written documents' (1979: 51). The lab as a whole in this portrayal emerges as a 'system of literary inscription' (p. 52) – a kind of factory in which facts are built through a very physical, material process of accumulated 'writing'. Its aim:

> was to persuade readers of papers ... that its statements should be accepted as fact. To this end rats had been bled and beheaded, frogs had been flayed, chemicals consumed, time spent, careers had been made or broken, and inscription devices had been manufactured and accumulated within the laboratory.
>
> (p. 88)

Laboratory Life nowhere invoked the term ANT. This would come later as a retrospective label, and one can argue as to whether the book was a work of ANT 'proper'. But a number of the key elements of the approach were already implied in this early work. I will now tease four of these out, and show how they were elaborated more formally in later writings under the heading of ANT. Here we find again that list of things that ANT is *not* about. The first is a vision of entities as networks of other entities (why ANT is not about 'networks' in the usual sense). The second is a concern with the way action is distributed between humans and non-humans (why ANT is not about 'actors' in the usual sense either). The third is a general account of ANT as a project, that Bruno Latour has dubbed 'an anthropology of the Moderns', and this leads to the fourth, namely a critique of classic sociological modes of explanation. Together these last two points show why ANT might claim not to be a 'theory' in the usual sense.

No 'networks': A material semiotics

Since the 1960s and in some quarters still today, the notion of networks has been a popular alternative for sociologists seeking to characterise social organisation in ways more fluid than those allowed by the notion of structure (see chapters 1 and 2). The image of social actors linked to each other through complex and shifting social networks became a key figure of the sociological imagination. This is emphatically *not* the sense in which the word network is used in ANT. To understand the role that term plays in ANT, one has to start somewhere rather different, namely at the interface between concepts and things.

One of the key effects of Latour and Woolgar's descriptions of scientific work in *Laboratory Life* was to blur the boundary between conceptual and material processes. In classic philosophy of science, science is frequently portrayed as a mental activity, a process of rational enquiry. Arguably, the social constructionist approach to science, with its emphasis on collective representations and cultural relativity, retained this approach to science as an essentially mental activity – a question of thoughts, perspectives, interpretations and symbols. Latour and Woolgar's account, by contrast, bracketed the question of what happened in scientists' minds, and focused instead on the material practices of cutting, mixing, comparing, talking and writing that they observed taking place in the lab. From this perspective, the scientific papers that issued forth from the lab could no longer be seen as simply a condensation of abstract thought happening in the minds of scientists (influenced or not by their 'culture' or 'collective consciousness'). Rather, papers and the facts they contained were simply the final stage in the stabilisation of a process that was both conceptual and material through and through.

Crucially, facts, in this view, are no longer seen as more or less accurate representations of a 'reality out there'. They are simultaneously conceptual and material entities that are produced, and can only continue to exist, within a certain kind of *network* made of social practices, technical devices and statements. The question of whether they are true or false can only be asked within these networks. The authors put the point very clearly in the following passage that deserves extended quotation:

> Let us consider one particular statement: 'somatostatin blocks the release of growth hormones as measured by radioimmunoassay'. If we ask whether this statement works outside science, the answer is that the statement holds true in every place where the radioimmunoassay has been reliably set up. This does not imply that the statement holds true everywhere, even where the radioimmunoassay has not been set up. If one takes a blood sample of a hospital patient in order to determine whether or not somatostatin lowers the level of the patient's growth hormone, there is no way of answering this question without access to a radioimmunoassay for somatostatin. One can believe that somatostatin has this effect and even claim by induction that the statement holds true absolutely, but this amounts to a belief and a claim, rather than to a proof. Proof of the statement necessitates the extension of the network in which the radioimmunoassay is valid, to make part of the hospital ward into a laboratory annex in order to set up the same assay. We are not arguing that somatostatin does not exist, nor that it does not work, but that it cannot jump outside of the very network of social practice which makes possible its existence.
>
> (Latour and Woolgar 1979: 183)

Facts cannot 'jump outside' the networks they exist in. Pushing the point, one might say that facts *are* these networks. That insight is borrowed from the relational logic of semiotics (see chapter 2), in which the meaning of any term can only be understood because of its place in a broader network of other terms. Imagine trying to learn a language from a dictionary: each word is defined by a set of other words (**cow**: 'a fully grown female animal of a domesticated breed of ox, kept to produce milk or beef'). You can look these up in turn, and you will find they too are defined by a set of further words (**beef**: 'the flesh of a cow, bull, or ox, used as food'), and so forth. The meaning of each term *is* that network of other terms.

ANT takes that semiotic insight and applies it not only to words or ideas, but also to material entities in the world (Law 1999: 4).[3] Thus, facts are not simply meanings consisting of a network of other meanings, but rather entities that are simultaneously material and conceptual (inscriptions of meaning on paper) consisting of a network of other things, persons, practices and meanings (ink, paper, ideas, human habits of writing, evidence produced by machines, etc.).

The same could be said, in turn, of the machines and other objects in the lab. Just like the facts they help to construct, these material devices are themselves the effects of conceptual and material activities, some of which have taken place elsewhere. A centrifuge, for instance, is made of machined bits of metal and plastic, to a blueprint designed by technicians, who were themselves applying mechanical principles derived from other scientific studies, of force, velocity and so forth. In the same way, facts 'constructed' at the Salk institute (for instance about the effects of hormones) would eventually find their way, for instance, into medical technologies. In other words, science and technology flow into and out of one another. But more profoundly, material and conceptual processes are flip sides of each other too: machines are stabilised, materialised facts and theories. They in turn allow new facts and theories to be stabilised and materialised. A spectrometer or a hormonal treatment, is, in this particular sense, a network of other material and conceptual entities, in the same way that a fact is.

Most provocatively, this extends not simply to ideas or to machines, but also to people themselves. In this view, 'scientists' *as scientists*, are nothing other than networks, each composed of their bodies, of course, habits, abilities and proclivities, but also their CVs (lists of papers), the techniques they have learned, the facts they know, their relations to other experts and non-experts, the machines and lab equipment they can leverage and so forth. This – perhaps the most counterintuitive – extension of the notion of networks to persons themselves, has been made less strange perhaps by technological developments in the intervening time, such as social media and new techniques of online visualisation (Latour et al. 2012). Consider the Facebook page: that profoundly personal, individual, partly public and partly intimate, partly honest and partly contrived, account of who its owner is. What is it made of, apart from links to other people and things, each of which leads elsewhere: friends, family members, photos of events, things once said, music once liked, news that raised hackles or confirmed opinions? An individual Facebook page is not 'in' that broader network, it *is* that network – all of it is entirely constituted by those other entities, elsewhere. An actor-network theorist would simply add that what is true for that online facet of your person is true more generally also of: your physical clothes and possessions, that mark you out as individual by linking you to specific and generic others elsewhere; your habits, aversions and desires, that were themselves learned, inherited or produced wilfully by contrast to other persons, ideas and things; and indeed your physical body – that is relationally constituted by genetic material from your parents and their parents before them, by the food you eat, the exercise you do with others and so forth... Of course, as we shall see below, critics have suggested that a number of crucial things might be missing from this radically relational view of human persons.

Hence ANT is not about 'networks' in the usual sociological sense. Indeed, the distance between ANT and the – as actor-network theorists might see it – bland, monochromatic sociological vision of 'social networks', should now be starkly visible. Actors are not *in* networks – they *are* networks. And these networks are not simply 'social' in the classic Dukheimian sense (made up of social connections to other human actors – see for instance chapter 1): they are *heterogeneous*. Actor-networks assemble together humans, objects, ideas and materials of varied and multiple kinds.

No 'actors': Action is distributed

The counterpart of this point is also that it is networks, rather than self-contained, bounded human actors, who act: action is distributed. A lone naked scientist contemplating a mouse in a bare room would never, with all the power of her intellect, be able to produce scientific facts about somatostatin. The distinctive activity of science is possible only through and with other scientists, machines, benches, scalpels, paper and pens. All of these other entities therefore participate in the activity of the scientist. It is the combined activity of this network of entities, rather than the naked human individual, that produces the outcome that is a well-established scientific fact.

Laboratory Life thus shows the first stirrings of what became the most famous and counter intuitive move in ANT: extending agency beyond the human. In the story of how scientific facts are constructed and stabilised, human scientists (with their culture, social structures, interests, assumptions, beliefs and so forth) emerge as only one part of the picture. The machines, the objects and materials, are just as important in the story. Without the machines' ability to hold things in place, to document, inscribe, map and measure, no amount of rational discussion and enquiry, no amount of social structure or ideology, would suffice to bring forth stable facts. But equally, the objects of study – the blood, cells, chemical elements and so forth – have to behave themselves. It is not always in the scientists' power (or even in that of their machines) to make those elements do what is required of them (for a radical exemplification of this point, in which bacteria play a leading role, see Latour 1988a). What is true of the human scientists is true also of their machines, cells or facts: insofar as each of these entities can 'act', it is through its relation to other entities in the network (see Callon [1986] for a famous scallop-based example). If actors are networks, it follows that networks are actors.

This vision of networks as actors takes the counterpoint of classic sociological models of agency in which archetypal agents are imagined as, as it were, 'naked' human beings with their intentions and interests, confronted with, entangled in, or even empowered by, broader social, cultural, cognitive or political 'structures' (for which sociologists have designed a number of convenient labels – class, gender, age, culture, etc.). In this classic vision, human agency is exerted against, within or even through those broader structures, with endless debates raging about the respective balance and interplay between those two imponderable quantities (see chapters 1, 2, 5 and 8). And yet the difficulties with this pair of concepts are well known: considered properly, 'structures' are themselves nothing other than the effect of the activity of other persons. As for the agency, intentionality and 'interests' seemingly deeply lodged within individual human actors, they too already seem to come from elsewhere, from somewhere external and structural. Successive scholars sought to resolve this problem by fusing or collapsing the distinction, for instance through Bourdieu's notion of habitus (see chapters 5 and 11), or through the study of the social and discursive contexts that enable (rather than simply limit) particular kinds of reflective freedom (as in the work of Foucault, see chapter 10). Actor-network theorists, instead, simply choose to ignore the distinction altogether and look elsewhere: to the objects and material things that scaffold our everyday lives.

Your alarm woke you up this morning, because you set it to do so last night, and because someone somewhere programmed your phone to have an alarm app on it, because that person works for a company whose market research leads it to expect that buyers of phones such as yours will expect to find alarm programs on them. One might of course seek to parse this set of networked actions and expectations into structure and agency: your compliance with a particular culture of clock-time (see Thompson 1967)? A neoliberal global order driven by transnational electronics companies whose products are shaping you into a self-disciplining subject? These are possible readings, of course, but in trying to tease out where agency splits off from structure or individual initiative from context, they require the invocation of mysterious conceptual entities ('culture', 'neoliberalism' on the one hand; 'free will', 'the self' on the other) that take on an active role in the story, despite being theoretical productions of sociology's (and philosophy's) own making. By contrast, an actor-network theorist might trace the ways in which your meaningful, intentional action (your decision to wake up this morning), already comes mediated and scaffolded by your actions yesterday, by the distinctive propensities, flexibilities and resistances of electronic components put together in a factory in China, by the shortcuts, the memory and path-dependencies of software written in California, and by the actions (some intentional, some thoughtless and routine) of all the other humans involved in those assemblages.

The point of ANT then, is not that machines, bacteria or indeed mice are endowed with a human-like intentionality. Rather, the point is once again negative: grand dichotomies between human action and non-human passivity are misleading. Initiative and control, intentionality and mechanical repetition, activity and passivity, should rather be seen as the ends of a spectrum, with multiple stopping-points in between, because:

> there might exist many metaphysical shades between full causality and sheer inexistence. In addition to 'determining' and serving as a 'backdrop for human action', things might authorize, allow, afford, encourage, permit, suggest, influence, block, render possible, forbid, and so on.
>
> (Latour 2005: 72)

These kinds of action are just as available to a range of non-human entities as they are to human beings. By reintegrating these non-human entities into our accounts of human action, Latour suggests, the false opposition between agency and structure is dissolved.

This approach recasts classic sociological visions of power. As the title of one of Latour's papers puts it, 'Technology is society made durable' (1991). That is to say, the stable and resilient technological artefacts that scaffold our lives are themselves the effect of stabilised networks of social actions (the expectations of designers, the skills of builders, etc.). Conversely, one might say, technology makes society durable – social bonds, norms, promises, habits, interests, prejudices, beliefs and expectations (the classic panoply for explaining social stability in sociology) are by themselves a weak kind of glue. It is our technical apparatuses (from writing or alarm clocks through to iPhones or Facebook) that stiffen these social patterns and make our behaviour regular and – up to a point – expectable.[4] Or to put it more bluntly, if society seems to have 'structure' it is in no small part because it is full of the activity of guns, uniforms, passports, credit cards, prison walls, money, clothes, motorways, borders and the like.

No 'theory': ANT as a critique of social theory

Before moving on to our final negative assertion – that ANT is not a theory – let us pause for a moment to examine some of the critical objections that have been raised against the above account. In the main, these follow two key strands. The first attacks ANT for its lack of concern with politics and inequality. The second, to which I will return in the conclusion, attacks ANT for its failure to engage ethnographically with what matters to the (human) participants in the research.

Early instances of the first strand of criticism – that ANT ignores politics – came from critical sociologists of science of what was known as the Edinburgh school (Shapin 1988; Sturdy 1991; Bloor 1999) who saw in ANT a refusal to acknowledge the human interests, social classes or economic forces that sociologists had traditionally used to explain scientific knowledge. Ian Hacking sums up the difference in perspective between the Edinburgh School and Latour with characteristic clarity:

> [There is] a vivid contrast between Latour's project and work deriving from the Edinburgh strong program in the sociology of knowledge. He shares with them

> the idea of facts being made, not discovered, but has abandoned any thought that it is human interests, social classes and economic forces that do the making. He calls all that, somewhat dismissively, 'interest theory'. By now the (mostly British) sociology of scientific knowledge scholars suspect that Latour is a dangerous reactionary. This is a reminder of different national histories. French intellectuals of a previous generation simply lived, spoke, breathed Marxism in a way that was never true in the United Kingdom. People of Latour's ilk and age have grown past that, while for his contemporary British thinkers that is still something to be fascinated by because it has not yet been fully experienced.
>
> (1992: 511)

Analogous concerns to those of the Edinburgh School were raised by science studies scholars writing in traditions informed by feminism, postcolonialism and queer theory, who found Latour's work productive but occasionally expressed discomfort about its perceived lack of interest in inequality (Haraway 1989; Barad 2007).[5] Anthropologists, in close conversation with these other critiques, complained about the lack of 'a larger historical or cultural context' (Martin 1998: 27). The concern in essence turned on the perceived lack of critical bite of ANT accounts. To these scholars, critique was the proper duty of the sociological commentator, particularly on such a power-laden subject as Western science, and critique had to involve some engagement with explanatory or contextualising frames, some ability to step outside the description itself in order to trace responsibilities, effects and inequalities.[6] By refusing to draw on such explanatory resources, at best, the stories ANT told about networked entities and distributed agency seemed to provide little beyond a restatement, in abstruse terms, of a given situation. At worst, ANT's picture of entities vying for the extension of their networked reach was suspected of a kind of neoliberal sympathy, 'an affinity for demiurgic, agonistically self-constituting entrepreneurial selves' (Oppenheim 2007: 473).

To these sorts of challenge, Latour responded head-on with a critique of explanation (1988b) and a critique of critique (2004b). In the second edition of *Laboratory Life* (Latour and Woolgar 1986) – in which, subtly but significantly, the subtitle had been changed from 'The social construction of scientific facts', to 'The construction of scientific facts' – Latour and Woolgar had already added a postface that foregrounded an aspect of that book's account that many readers had not seemed to take sufficient notice of. *Laboratory Life,* they pointed out, was a reflection on the construction of scientific facts in biology, but it was also a reflection on the construction of scientific facts in sociology. Class, interests, gender, culture or society were tools of the sociologist and anthropologist, and they were constructed in much the same way as the pipettes, spectrometers, graphs and sentences used by the biologists at the Salk institute to stabilise their facts. If one is prepared to examine the construction of the latter, one ought surely also to examine the construction of the former. The challenge to sociological explanations of science could not be more direct. As Callon wrote around the same time:

> For [sociologists of science] Nature is uncertain but Society is not . . . Sometimes the effect can be so devastating that the reader has the impression of attending a trial of natural science presided over by a privileged scientific knowledge (sociology) which has been judged to be indisputable and above criticism.
>
> (1986: 197–198)

Latour later hammered in these points and expanded them beyond the sociology of science through a relentless attack on 'social constructionism' and 'critical sociology' (2005). Critical sociologists, Latour argued, claimed for themselves precisely the kind of power to know the real that they refused to allow other actors might have access to. This held beyond the realm of the sociology of science, and Latour excoriated the sociology of art or the sociology of belief in much the same terms. Everywhere, he claimed, the critical sociologist denies the existence and power of those non-human entities that actors are claiming make them act (molecules in the lab, paintings in the gallery, God in heaven), and replaces these non-human powers with explanatory devices of his own construction, made out of that mysterious thing called 'the social'.[7]

This brings us finally to the sense in which ANT might claim not to be a theory. Theory in this ANT view is imagined as a set of stable conceptual resources that can be used to frame, contextualise and explain any given situation; a framework that already includes pre-loaded assumptions about what the relevant categories and distinctions are (class, gender, etc.), what the likely problems and tensions will turn out to be (domination, false consciousness, inequality, silencing) and who the cast of likely characters is, including likely suspects; a framework that already includes a set of assumptions about the relative power and scale of these actors (individuals, groups, corporations, etc.). If that is what theory is, then ANT, its proponents claim, is quite the opposite. Little more than a set of negative injunctions that a researcher ought to bear in mind upon setting off to give an account of a particular topic: don't assume you know who or what is acting, what the 'kinds' of entity in presence are going to be and what their relative power is. Starting from this methodological injunction to keep the world 'flat', an actor-network theorist should painstakingly 'trace' the relations as they appear to him or her in the course of the research. This in turn explains the profusion of terms and distinctions with which this chapter began. This bewildering profusion is the mark of a particular way of thinking about concepts: not as the building blocks of a progressively growing theoretical edifice, but as a kind of theoretical scaffolding that will enable a description of a particular case and can be set aside once that description is complete. 'Description' using ad hoc terminology thus takes the place of 'explanation' in view of an established theoretical framework. It is in this sense that ANT is characteristic of what I have described in the introduction as a 'heuristic turn' in social science (see chapter 14 for a further discussion of heuristics).

So far, it might sound like Latour's response to those who accuse ANT of being apolitical or acritical is merely to relinquish critique in favour of a kind of superior objectivity or more careful empiricism. And surely critics would be right to respond that there is no reason to assume that the playing field in any given situation will be level – so why should we assume methodologically that it is? To this explicitly political rejoinder one might add two methodological ones. First, that the methodological injunction outlined above is surely impossible – who can really claim to approach a situation without any presuppositions? And second, that the proposed 'method' would seem to invite the relatively absurd result of an endless description of heterogeneous chains of co-activity. As Steve Shapin put it in an early critique of ANT: 'there is little to be said from within a seamless web' (1988: 547).

But Latour's final – and boldest – rejoinder against his critics is to claim that ANT is in fact even more political and more engaged than critical sociology. Indeed, Latour claims, critical sociology may rile against the status quo, make accusatory gestures and point to collective and individual injustices and responsibilities. By contrast, the very same logic of material–semiotic 'production' of facts as real entities in the world, that Latour and

Woolgar long ago described for scientists at the Salk institute, applies to the construction of ANT accounts. ANT accounts are not presented as more or less correct representations or interpretations, or denunciations or critiques of a situation. Rather, they are new things in the world – empirical interventions on their readers: they reconfigure relationships and entities in such a way as to make new ways of seeing and new ways of being possible. ANT should thus be seen, not as a set of descriptions, but as a particular type of performative or 'ontological politics' (Mol 1999). In these latter (and later) developments, one might see the partial integration into ANT of some of the critiques and perspectives coming from the tradition of feminist technoscience (Haraway 1989; Stengers 2000).

Conclusion: On taking seriously

I wrote previously that there were two main strands to the critiques of ANT, the first of which I have now examined at some length. The second is more distinctly anthropological, and it has been adumbrated above. This is the sense that ANT accounts tend to suffer from a certain kind of ethnographic deficit. Somewhat unkindly, but not entirely unfairly, one might characterise many ANT accounts as tacking back and forth between very close – and occasionally tedious – descriptions of practice and technical detail, and some fairly far-reaching philosophical conclusions, often with little of what is conventionally known as sociology or history in between. More specifically, a number of anthropologists have noted that ANT accounts are relatively unconcerned with the perspectives and subjectivities of scientists at work (Rabinow 1996). Beyond science, ANT's account of action distributed seamlessly across humans and non-humans has been critiqued for evacuating that crucial aspect of meaningful human life, namely the way in which we ascribe responsibility to each other (Laidlaw 2014).

This is the case, despite the frequent claim that ANT is better than critical sociology at taking into account those entities that truly matter to its subjects of study (God to believers, facts to scientists, artworks to aesthetes, etc.), and despite the injunction to aspiring actor-network theorists that their prime commitment should be to 'follow the actors'. Indeed, this is the case constitutively, I would argue, because of the nature of the project of an 'anthropology of the moderns' itself.

There is, of course, the basic problem of identifying, for empirical purposes, 'the moderns' in the first place (see note 1). In practice, ANT tends to bracket that problem by seeing modernity as unlocated and pervasive, a matter of global assemblages (Collier and Ong 2005), that contrasts with anthropologists' traditional concern with identifying particular contexts of knowledge and practice (Tsing 2010). To some, doing away with contexts in this way will seem a refreshing move. But it also does away with a particular conceptual device central to many anthropological arguments; namely that of presenting an alternative perspective or conceptual possibility as a way to challenge our own assumptions – a move most recently reinvented by the so-called 'ontological turn' (see chapter 14).

Another way of putting it is that in the 'anthropology of the moderns', as conceived by ANT, the perspective that is being challenged is also the perspective of those being studied. From the very outset, ANT was committed to producing an account that did not replicate or take for granted the core divisions that are so often invoked in the kinds of everyday Western settings in which ANT descriptions often take place. This concern was explicitly articulated as early as *Laboratory Life* in the worry about 'going native' and the concern not to 'reiterate the accounts of scientists themselves' (Latour and Woolgar 1979: 39, 44). In Latour's work, in particular, this concern became increasingly cast as a desire not merely to understand 'the moderns' but to challenge and reform them, to give them (us?) a different account of them(our?)selves (Candea 2016). Whether this is still a recognisably anthropological project is a matter of perspective.

Notes

1 The whole point of Latour's book was to challenge the classic narrative of modernity and the 'great divide' (Latour 1993: 11), which is often imagined to exist between 'us, the moderns' and 'others'. And yet in a number of ways, that book has also been seen as reinstating that distinction (for a sophisticated critique along those lines, see Strathern 1999).
2 For a broader outline of relation between Science and Technology Studies and anthropology, see Candea (2017).
3 In this and other ways, ANT owes more than a passing debt to the work of Michel Foucault (see chapter 10).
4 For an intriguing comparison between human social life (scaffolded by multiple objects) and primate social life (comparatively free of technology), see Strum and Latour (1987).
5 '[O]ne can't simply bracket (or ignore) certain issues without taking responsibility for the constitutive effects of these exclusions . . . I want to emphasize in the strongest terms possible that it would be a mistake to think that the main point is simply a question of whether or not gender, race, sexuality and other variables are included in one's analysis. The issue is not simply a matter of inclusion. The main point has to do with power. How is power understood? How are the social and the political theorized? Some science studies researchers are endorsing Bruno Latour's proposal for a new parliamentary governmental structure that invites nonhumans as well as humans, but what, if anything, does this proposal do to address the kinds of concerns that feminist, queer, postcolonial, (post-)Marxist, and critical race theorists and activists have brought to the table? . . . their presence has barely been acknowledged' (Barad 2007: 58).
6 A concern very similar to that raised against Foucault by some Marxist writers. Here again, we see the family resemblance between Foucault's work and that of ANT.
7 This commitment to 'taking seriously' what matters to the actors, is another sense in which ANT arguments show an affinity with those of the ontological turn (see chapter 14). As I point out in the introduction, however, there are a number of important discrepancies between the two approaches.

Bibliography

Abu El-Haj, Nadia 1998. Translating truths: Nationalism, the practice of archaeology and the remaking of past and present in contemporary Jerusalem. *American Ethnologist* 25 (2): 166–188.
Barad, Karen Michelle 2007. *Meeting the Universe Halfway: Quantum Physics and the Entanglement of Matter and Meaning*. Durham, NC: Duke University Press.
Bloor, David 1974. Essay review: Popper's mystification of objective knowledge. *Science Studies* 4 (1): 65–76.
——— 1999. Anti-Latour. *Studies in History and Philosophy of Science* 30 (1): 81–112.
Callon, Michel 1986. Some elements of a sociology of translation: Domestication of the scallops and the fishermen of St Brieuc Bay. In John Law (ed.), *Power, Action and Belief: A New Sociology of Knowledge?* London: Routledge & Kegan Paul.
Candea, Matei 2008. Fire and identity as matters of concern in Corsica. *Anthropological Theory* 8 (2): 201–216.
——— 2013. Objects made out of action. In P. Harvey (ed.), *Objects and Materials Companion*. London: Routledge.

——— 2016. We have never been pluralist: On lateral and frontal comparisons in the ontological turn. In Pierre Charbonnier, Gildas Salmon and Pierre Skafish (eds.), *Comparative Metaphysics: Ontology after Anthropology*. London: Rowman & Littlefield.

——— 2017. Science. In Felix Stein, Sian Lazar, Matei Candea, Hildegard Diemberger, Joel Robbins, Andrew Sanchez and Rupert Stasch (eds.), *The Cambridge Encyclopedia of Anthropology*. Cambridge: Cambridge University Press. Cambridge: Cambridge University Press. http://www.anthroencyclopedia.com/entry/science.

Cole, John 1977. Anthropology comes part way home: Community studies in Europe. *Annual Review of Anthropology* 6: 349–378.

Collier, Stephen J. and Aihwa Ong (eds.) 2005. *Global Assemblages: Technology, Politics and Ethics as Anthropological Problems*. London: Blackwell.

Gell, Alfred 1998. *Art and Agency: An Anthropological Theory*. Oxford, Oxford University Press.

Hacking, Ian 1992. Review: Science in action. *Philosophy of Science* 59 (3): 510–512.

Haraway, Donna Jeanne 1989. *Primate Visions: Gender, Race and Nature in the World of Modern Science*. London: Routledge.

Helmreich, Stefan 2009. *Alien Ocean: Anthropological Voyages in Microbial Seas*. Berkeley, CA: University of California Press.

Hennion, Antoine 2015. *The Passion for Music: A Sociology of Mediation*. London: Routledge.

Houdart, Sophie 2008. *La cour des miracles: Ethnologie d'un laboratoire japonais*. Paris: CNRS.

Ingold, Tim 2000. *The Perception of the Environment: Essays on Livelihood, Dwelling & Skill*. London and New York, NY: Routledge.

Kaplan, Martha, and John Kelly 1999. On discourse and power: 'Cults' and 'orientals' in Fiji. *American Ethnologist* 26 (4): 843–863.

Laidlaw, James 2014. *The Subject of Virtue: An Anthropology of Ethics and Freedom*. New York, NY: Cambridge University Press.

Latour, Bruno 1987. *Science in Action: How to Follow Scientists and Engineers through Society*. Milton Keynes: Open University Press.

——— 1988a. *The Pasteurization of France*. Cambridge, MA: Harvard University Press.

——— 1988b. The politics of explanation: An alternative. In Steve Woolgar (ed.), *Knowledge and Reflexivity: New Frontiers in the Sociology of Knowledge*, London: Sage.

——— 1990. Postmodern? No simply amodern; Steps towards an anthropology of science. *Studies in History and Philosophy of Science* 21 (1): 145–171.

——— 1991. Technology is society made durable. In J. Law (ed.), *Technology is Society Made Durable. Vol. A Sociology of Monsters: Essays in Power, Technology and Domination*. London: Routledge.

——— 1993. *We Have Never Been Modern*. London: Harvester Wheatsheaf.

——— 1996. *Aramis, Or, The Love of Technology*. Cambridge, MA: Harvard University Press.

——— 1998. *Paris ville invisible*. Paris: Institut Synthélabo.

——— 1999. On recalling ANT. *The Sociological Review* 47 (S1): 15–25.

——— 2000. On the partial existence of existing and nonexisting objects. In Lorraine Daston (ed.), *Biographies of Scientific Objects*, Chicago, IL: Chicago University Press.

——— 2004a. *Politics of Nature: How to Bring the Sciences into Democracy*. Cambridge, MA: Harvard University Press.

——— 2004b. Why has critique run out of steam? From matters of fact to matters of concern. *Critical Inquiry* 30: 225–248.

——— 2005. *Reassembling the Social: An Introduction to Actor-Network-Theory*. Oxford: Oxford University Press.

——— 2009. *The Making of Law: An Ethnography of the Conseil D'Etat*. Revised edition. Cambridge and Malden, MA: Polity Press.

——— 2013a. *An Inquiry into Modes of Existence: An Anthropology of the Moderns*. Cambridge, MA: Harvard University Press.

——— 2013b. *Rejoicing: Or the Torments of Religious Speech*, Cambridge: Polity Press.

Latour, Bruno, Pablo Jensen, Tommaso Venturini, Sébastian Grauwin, and Dominique Boullier 2012. 'The whole is always smaller than its parts' – A digital test of Gabriel Tardes' Monads. *The British Journal of Sociology* 63 (4): 590–615.

Latour, Bruno, and Steve Woolgar 1979. *Laboratory Life: The Social Construction of Scientific Facts*. Beverly Hills, CA and London: Sage Publications.

——— 1986. *Laboratory Life: The Construction of Scientific Facts*, edited by Jonas Salk. 2nd edition. Princeton, NJ: Princeton University Press.

Law, John 1999. After ANT: Complexity, naming, and topology. In John Hassard and John Law (eds.), *Actor Network Theory and After*. Oxford and Malden, MA: Blackwell/Sociological Review.

Martin, Emily 1998. Anthropology and the cultural study of science. *Science, Technology, & Human Values* 23 (1, Special Issue: Anthropological Approaches in Science and Technology Studies): 24–44.

Miller, Daniel 2005. *Materiality*. Durham, NC: Duke University Press.

Mol, Annemarie 1999. Ontological politics. A word and some questions. In John Law (ed.), *Ontological Politics. A Word and Some Questions, Actor Network Theory and after*. Oxford: Blackwell.

——— 2002. *The Body Multiple: Ontology in Medical Practice*. London: Duke University Press.

Oppenheim, R. 2007. Actor-network theory and anthropology after science, technology, and society. *Anthropological Theory* 7 (4): 471–493.

Rabinow, Paul 1996. *Making PCR: A Story of Biotechnology*. Chicago, IL: University of Chicago Press.

Shapin, Steven 1988. Review: Following scientists around. *Social Studies of Science* 18 (3): 533–550.

Stengers, I. 2000. *The Invention of Modern Science*. Minneapolis, MN: University of Minnesota Press.

Strathern, Marilyn 1988. *The Gender of the Gift: Problems with Women and Problems with Society in Melanesia*. London: University of California Press.

——— 1999. *Property, Substance and Effect: Anthropological Essays on Persons and Things*. London: The Athlone Press.

Strum, SS, and B Latour 1987. Redefining the social link: From baboons to humans. *Social Science Information* 26 (4): 783.

Sturdy, Steve 1991. Essay review: The germs of a new Enlightenment. *Studies in the History and Philosophy of Science* 22 (1): 163–173.

Thompson, E.P. 1967. Time, work-discipline, and industrial capitalism. *Past & Present* 38: 56–97.

Traweek, S. 1988. Beamtimes and lifetimes: The world of high energy physicists. Cambridge, MA: Harvard University Press.

Tsing, Anna 2010. Worlding the Matsutake diaspora, Or, Can actor-network-theory experiment with holism? In Ton Otto (ed.), *Experiments in Holism: Theory and Practice in Contemporary Anthropology*. Oxford: Wiley-Blackwell.

Verran, Helen 2001. *Science and an African Logic*. Chicago, IL: University of Chicago Press.

Viveiros de Castro, Eduardo 1998. Cosmological deixis and Amerindian perspectivism. *Journal of the Royal Anthropological Institute* 4 (3): 469–488.

Woolgar, S. 1991. Configuring the user: The case of usability trials. In John Law (ed.), *Configuring the User: The Case of Usability Trials, A sociology of monsters*. London: Routledge.

Zaloom, Caitlin 2003. Ambiguous numbers: Trading technologies and interpretation in financial markets. *American Ethnologist* 30 (May): 258–272.

Chapter 14

The ontological turn

School or style?

Paolo Heywood

Introduction

If there is a word in the contemporary anthropological lexicon that is almost guaranteed to provoke a reaction of some form – whether positive or negative – in listeners and readers, that word is 'ontology'. The movement to which it has become attached has generated reviews of review articles (Pedersen 2012), stimulated a motion in the Group for Debates on Anthropological Theory (Carrithers et al. 2010), been hailed as the basis of a new politics (Holbraad, Pedersen and Viveiros de Castro 2014) and also been challenged as the bastion of an old one (Bessire and Bond 2014; Graeber 2015). Articles with the word in their title often appear near the top of lists of most popular downloaded works in a number of prominent journals, but a great many of those articles are critical in tone, and its use in academic discussions is capable of causing even tolerant eyes to roll. No doubt future readers will find such depth of feeling difficult to imagine, as we might find it hard to believe that many of the other theoretical movements in this book could have once occasioned similar passions (see chapter 1).

Invocations of an 'ontological turn' in anthropology usually refer to a body of work – on disparate ethnographic contexts, and sometimes with differing purposes – that has emerged over the last decade or so in the wake of the publication of an edited volume entitled *Thinking Through Things* (2006). Its most prominent proponents are Eduardo Viveiros de Castro, Martin Holbraad and Morten Pedersen (e.g. Holbraad, Petersen and Viveiros de Castro 2014; Holbraad and Petersen 2017), but a list of recent works influenced by it in some way or another would be extensive (for an example of such a list see Holbraad and Pedersen 2017: 8), and its roots in anthropology go back much further. Equally, it is an on-going phenomenon, and new and significant contributions to it are currently emerging on a regular basis (e.g. Holbraad and Pedersen 2017). It is also a complex phenomenon, and summing it up in a few lines – even in a chapter – is no easy task, but a few of its key features include:

1 A rejection of the concept of 'culture' in the Geertzian sense (see chapter 8) as a symbolic or ideational phenomenon, and as an explanatory concept in anthropology.
2 A related rejection of the distinction between the symbolic and the material and a call for closer attention to the latter.
3 The idea that anthropological analysis should depart from, not explain away, the sometimes strange and unsettling concepts we find in the field.

One of the things I will aim to suggest in this chapter is that there is an important and interesting difference of form between the 'ontological turn' and a number of the 'schools' covered in the other chapters of this book. My aim in doing so is not to make a historical point – and indeed I have no doubt that more capable scholars than I of structuralism (see chapter 2), say, or interpretivism (see chapter 8), might dispute this difference – but to make an analytical one. That point regards the nature and purpose of analysis in anthropology more broadly: in other words, of anthropological theory.

Ontology, as philosopher Willard Quine once put it, roughly speaking, means what there is, as opposed to epistemology, that refers to our knowledge of what there is (1948). But in spite of these deeply metaphysical connotations of its name, the 'ontological turn' does not, in fact, claim to be a theory about the world and what is in it. Instead, its proponents – or at least many of them (Henare et al. 2006: 5; Pedersen 2012; Holbraad and Pedersen 2017) – go to some considerable lengths to argue that what they are doing is outlining a method, or a heuristic (not unlike actor-network theory, see chapter 13). In other words, and somewhat unusually, all that ink and excitement is alleged to concern not claims about the nature of the world, but arguments about how we should usefully approach what we encounter in the field.

So the very word 'ontology' is somewhat misleading. Its use in anthropology is usually not about stating something fundamental about the nature of being or existence, in the sense in which philosophers often use it (e.g. Quine 1948; see also Heywood 2012; Graeber 2015), although it is also a convenient antonym for the kind of 'epistemological' or 'representationalist' anthropology from which the ontological turn seeks to move us, as I discuss below. Instead of being a theory, a picture of the world, it is a tool, like a hammer, to borrow a favourite phenomenological metaphor, wielded – one assumes – with specific aims in mind.

So whilst the majority of this chapter will be occupied with outlining a brief overview of recent discussions of ontology in anthropology and debates over its use and value, it will also aim to draw attention precisely to the question of aims – the aims of anthropological analysis. For that question is one the language of heuristics causes to arise: to say that something is useful lacks meaning without reference to what purpose it is alleged to be useful for (for a much more sustained engagement with heuristics, see Candea 2016).

A number of such purposes spring to mind: good description; conceptual innovation; political change. Yet agreement about their fittingness for anthropology is by no means universal, and in fact this is so even amongst advocates of the use of ontology, some of whom seem, for example, to favour politics, and others intellectual creativity (e.g. Viveiros de Castro 2004; Holbraad 2017).

Needless to say, I will not attempt to propose such a purpose here. I hope merely that one of the questions this chapter gives rise to is what it means to think in terms of heuristics and purposes, rather than theories; in terms of styles or approaches, rather than schools. What I will suggest in the chapter's concluding sections is that the language of heuristics can often have the effect of closing down, rather than opening up, discussion and debate. It can be deployed as a generic response to critical reflections on the validity or otherwise of a conceptual framework: so, for example, if one disagrees with the idea that there is no serious distinction between humans and non-humans, one can be told simply that the objection is wrongheaded, for that idea is not actually a claim about how things are, but a way of thinking that might or might not produce useful results. Whether or not such a distinction actually exists is immaterial.

This form of argument is useful in foreclosing what are alleged to be fruitless discussions of the validity or otherwise of particular positions, but it should also be so in opening up discussions of utility itself. Even if I grant that your stance on the distinction between humans and non-humans is not a representation of the world but a tool for intervening upon it, I may yet disagree with the purpose of your intervention, or its results.

Different beginnings

Difference, or – in its own preferred vocabulary – alterity, is the basic starting point for the ontological turn. This might seem an odd focus for an anthropological movement that emerged in the wake of *Writing Culture* (see chapter 10), and its critique of orientalist exoticism. *Writing Culture* stimulated much suspicion towards anything in anthropological writing that smacked of 'othering' our interlocutors, or seemed to focus on striking or surprising aspects of their thought or culture (some have directed a very similar critique at the ontological turn itself – e.g. Bessire and Bond 2014). For related reasons, it also evinced a fundamental scepticism about anthropological description itself – about the act of 'writing culture'. But whilst the ontological turn is in many ways responding to a set of questions that are similar to those *Writing Culture* addressed, it does so by building on a corpus of literature that raised those questions around the same time, but in an orthogonal manner. Key anthropological sources of inspiration of the turn – Roy Wagner, Eduardo Viveiros de Castro and Marilyn Strathern – were and remain eminently concerned with the politics of anthropological representation. Yet their response to our apparent inability adequately to represent others without imposing upon them our own frameworks of analysis was not to give up on difference but to revolutionise the ways in which we might think about difference itself.

These thinkers had a fundamental impact on the ways in which the ontological turn – and contemporary anthropology more broadly – conceptualises the articulation of difference through the opposition between nature and culture. Although it is not unusual to hear anthropologists of a variety of schools claim that that opposition should be or has been abandoned, thinkers such as Strathern, Wagner and Viveiros de Castro all demonstrated not only that the nature/culture distinction underpins the basis of the anthropological project, claims to the contrary notwithstanding, but also some ways in which it might genuinely be reconceptualised.

Take a comparatively minor example from Strathern's opus, that readers will have already encountered in chapter 12: Strathern's chapter in *Nature, Culture, Gender,* which she edited alongside Carol MacCormack (Strathern 1980). She points out not only that Hageners do not possess concepts of nature and culture that map onto ours, but that the nearest concepts that approach nature and culture are not related to one another through the same dynamic of domination (for a more detailed discussion, see chapter 12).

On the surface that might appear simply another instance of the standard anthropological manoeuvre of relativising something we take for granted – in this case, the nature/culture distinction and its relationship to maleness and femaleness, and this was indeed the role it played in relation to feminist anthropology. But the implications go beyond this. For the very basis of that relativising manoeuvre *is* the nature/culture distinction. Without that distinction, it becomes impossible to say, as we do as a consequence of almost every other such relativising move, that 'in their culture there is no X'. This puts us in something of a double bind: for many anthropologists, the point of the relativising manoeuvre is to parochialise our assumptions about the ways things are, and to show that they are otherwise elsewhere in the world, in other cultures; but here we have parochialised the concept of culture, and thus that manoeuvre itself. If we simply assume they are wrong in an effort to retain the distinction as the basis of our relativism, we seem to betray its purpose; but we cannot treat the distinction as yet another object to be relativised, precisely because it is the basis for all of our other relativisms. Strathern, in other words, is overturning the fundamental basis upon which we articulate difference: nature remains the same, whilst culture varies. Strathern's contributions to debates on these topics continue today, but even this simple example should suffice to make clear how important they have been.

Viveiros de Castro's work on Amazonian conceptions of difference makes a similar point and has had a similarly significant effect on the ontological turn, and indeed he continues to contribute to key anthropological debates surrounding these issues (e.g. 1998; 2003; 2004; 2014). 'Western' cosmologies, according to Viveiros de Castro's model, conceive of the relation between nature and culture as involving natural similarity and cultural diversity. So, we think of ourselves as 'humans' because we share the same (natural) bodies as one another, even though we may think and believe a range of different (cultural) things. Likewise, we all share the same (natural) world, but we have various and sometimes conflicting (cultural) ideas about it. Perspectivist Amerindians, on the other hand, conceive of all persons (including a number of cosmologically significant animals and objects) as possessing the same human culture, or vision of the world: jaguars, in the classic example, see themselves as human, their paws as hands, the blood they drink as beer, and so on. The difference lies in what they perceive to be these things, and the bodies that do the perceiving: if I have a human body I will see the same things that you see when you see beer, houses and rice; if I have the body of a jaguar, on the other hand, I will not. Furthermore, bodies can be exchanged in certain kinds of contexts, much as we talk of exchanging points of view or ideas. In other words, difference is located in 'nature', rather than in 'culture'. Hence his descriptions of Amerindian perspectivism as 'multinaturalism', rather than 'multiculturalism'. This also evokes the same difference of differences as Strathern, and indeed points to the distinction between 'epistemological' anthropology and the ontological turn. The former would interest itself in different ideas about or views upon the world, whilst taking the meaning of 'world' for granted; the latter, on the other hand, raises the question of what 'world' or 'nature' mean for the people involved. Again, the twist comes in the implications this cosmology has for its own anthropological description: as in the case of the example from Strathern, the epistemological – or ontological, as it will become – question here is how to account for this difference itself without recourse to the very notion of culture that it undermines. We cannot say 'in their culture, difference is natural' without already trumping their understanding of difference ('natural') with our own ('cultural').

As should by now be clear, these thinkers – and others like them such as Roy Wagner (1975) – are dealing, in particular ways, with the same 'crisis of representation' to which *Writing Culture* responded by urging anthropology – at least in its occasionally extreme rhetorical varieties – to abandon the project of representation altogether in favour of a focus on 'writing' and the fictions it produces. Rather than take this path, however, they led their anthropological heirs down a different one: when confronted with the unsustainability of the divide between representation and reality, epistemology and ontology, nature and culture, that underpinned much anthropological thinking before them, instead of washing their hands of both, they reconfigured their understandings of them until the divide was no more. Moreover, they often did so by drawing on the very ethnographic material they were seeking to analyse. In other words, their arguments instantiated themselves in a manner that has now become characteristic of ontological anthropology: the contents of their arguments are focused on collapsing the difference between representation and that which is represented; but the form that these arguments take is an example of precisely that collapse, as it makes use of ethnography in order to modify theory, thus producing 'things' that are a mixture of both concepts and objects. So, for example, Viveiros de Castro's description of perspectivism challenges our notions of cultural representation; but, precisely in so doing, the argument itself becomes more than a cultural representation, more than description. Hence the title of the edited volume that in many ways inaugurated anthropology's ontological turn: *Thinking Through Things*, the aim of which was 'to treat meaning and thing as an identity' (2006: 3).

Ontology, epistemology or both?

Hopefully, even on the basis of this brief summary of some of its most important antecedents, it should already be clear that the word 'ontology' is not – at least not in the obvious sense – being employed here in order to point to some deeper level of reality than that with which anthropology had previously occupied itself. Discussions of 'being' or 'multiple worlds' may lead those with even a passing familiarity with philosophy to assume that the ontological turn is making some quite serious and metaphysical claims about the nature of reality. I will return to this point when discussing critical perspectives on the turn, but for the sake of avoiding confusion from the outset, it is worth making clear that, like its progenitors, the ontological turn is aiming to collapse distinctions between ontology and epistemology, nature and culture, not to refocus anthropology on a neglected side of those distinctions. The word 'ontology' does serve to distinguish this sort of approach from what advocates of the turn would call 'epistemological' approaches, that were previously dominant, but not because they wish to dispense with epistemology in favour of ontology. To make the same point in a more banal fashion: none of the most significant advocates of the ontological turn are suggesting that Amazonians or Melanesians somehow live in different universes to ourselves, let alone that they belong to different natural species; neither are they claiming that our condition is one of universal solipsism, in which differences between people and peoples are so insuperable as to prevent any kind of communication or understanding, or that boundaries between things we used to call 'cultures' but now should call 'ontologies' are somewhere out there in the world.

This brings us to the question of heuristics, for what the ontological turn proposes, rather than being a metaphysics, is a method. This, incidentally, is a useful point of contrast between the approaches I have so far been describing and that of another prominent anthropologist often associated with the ontological turn, Philippe Descola. His work (e.g. 2013), like that of Wagner, Strathern and Viveiros de Castro, is targeted at unsettling the nature/culture binary, and indeed he describes a set of ethnographic materials very similar to that of Viveiros de Castro under the rubric of 'animism'. In 'naturalist' cosmologies ('Western' cosmologies, broadly speaking), according to his model, 'exteriority' (bodies, effectively) is continuous, whilst 'interiority' (minds) is discontinuous. In the case of animism, exteriority is discontinuous whilst interiority is continuous. He also adds two further types: 'totemism', in which both are continuous, and 'analogism' in which neither are. The reason I treat this example only too briefly though, is that for Descola, whilst these categorisations avoid the nature/culture binary as an explanatory basis, they nevertheless remain categorisations (or 'types', as Bruno Latour puts it in a 2009 review of a debate between Descola and Viveiros de Castro); whereas for the thinkers we are primarily concerned with in this chapter the purpose of their descriptions (remember they are heuristics, with purposes) is to effect some form of alteration to our conceptual schema. Rather than being about extending our capacity to classify or typologise different cultures, they are concerned with showing how some differences can undermine the whole classificatory edifice of culture itself.

For again, the point is not to substitute one theory for another, but to allow the encounter between ethnography and theory itself to generate new forms of analysis. Deciding *in advance* that representation and reality are distinct precludes that possibility. The things being generated by this method, whether one calls them ontologies or not, are always the products of very specific encounters with difference and alterity (Henare et al. 2006; Holbraad and Pedersen 2009; Pedersen 2012). They occur not simply by default when doing ethnography, or because we have crossed some invisible ontological barrier, but on occasions when the conceptual repertoire we have at our disposal is obviously unable adequately to account for a particular phenomenon.

What exactly is meant by 'unable to account for' is not always completely clear, though it might be helpfully thought of as entailing a mixture of political misrepresentation *à la* Marxist false consciousness ('they don't really understand what they're doing' – see chapter 8) and explanatory failure ('it's impossible to understand the meaning of this phenomenon without recourse to some external, anthropological concept'). An oft-cited example of such failure of accounting is the deployment of the word 'belief' (e.g. Henare et al. 2006: 5; Holbraad 2012: 27). Since Needham first critiqued the concept (1972), anthropologists have pointed out that belief as it is used in anthropology is effectively a synonym for error: 'it is only the unbeliever who believes that the believer believes', as Pouillon puts it (1993: 26). To refer to something as a belief is to oppose it to knowledge, that we the anthropologist possess, and also simply to recast it in our own terms, rather than to explain it. The Nuer may 'believe' that *kwoth* are spirits, and they are entitled to do so, but in describing this as a 'belief' we are implicitly asserting our knowledge that this is not the case, not to mention mis-describing the Nuer, who do not 'believe' that *kwoth* are spirits, they *know* them to be so and would not recognise or understand our description of it as a 'belief'. Note the similarity of this problem to the Hagen or Amazonian cases: how do we describe a phenomenon when it appears to undermine the basis of our description, without making the phenomenon appear illusory or mistaken? The point that advocates of the ontological turn make is that situations like this – in which it appears that we cannot make sense of an indigenous phenomenon without recourse to describing it as false – are evidence of a failure of our conceptual repertoire (e.g. Wastell 2006: 87; Holbraad 2012: 72).

It is such situations that constitute encounters with alterity and with which the ontological turn is primarily concerned. Hence, the opening up of questions of ontology is a contextual artefact of this encounter, not a feature of the world out there.

An excellent example of how this process works is provided by Martin Holbraad (2009; 2012), one of the editors of *Thinking Through Things*, and a major force behind the ontological turn.

The ethnographic problem Holbraad is faced with is as follows: Cuban diviners claim that the verdicts they deliver to clients are indubitable. That is, divinatory statements of the kind 'you are bewitched' cannot be false and are not open to doubt. To the sceptic who points out that people in Cuba and elsewhere often do, in point of fact, doubt the truth of divinatory verdicts, Holbraad elegantly responds that what they doubt is not the truth of the verdict but its divinatory nature: accepting the claim of practitioners that their verdicts are indubitable means accepting it as an analytic statement true by its own virtue, because doing anything else is already a denial of the claim.

Clearly however, as Holbraad goes on to show, a statement such as 'you are bewitched' does not sound like an indubitable truth to us: I *may* be bewitched and the statement may be true, but I may equally not be bewitched (2012: 71). In fact, 'you are bewitched' appears to share all the characteristics of a regular representational statement of fact: its truth depends on certain facts about the world. Regular statements of fact are, of course, by their nature open to doubt: the facts may be one way, or they may be the other. Thus the problem: divinatory truths are indubitable, and at the same time appear to represent a certain state of affairs in the world, thus laying themselves open to doubt.

The failure of analysis, therefore – presuming that this logical absurdity is indeed a failure – must rest with one or other of those facets of divinatory truth. Given the earlier point that not to accept fully the indubitable nature of verdicts logically entails reducing them to beliefs and being locked into what Holbraad calls a '"smarter-than-thou" stance vis-à-vis divinatory practice' (2012: 55), the facet to be doubted must be the possibility of doubt itself.

Any claim to truth that aims to represent a state of affairs in the world must by definition be open to the possibility of doubt: if I say that it is warm in my office and cold outside, or that the 'ontological turn' is a method and not a theory, I am trying – probably – to convey a picture of the world, and that picture may be correct, but it may equally be mistaken, and it is part of the nature of such claims that there is nothing in them to make them *absolutely* and *undoubtedly* true. So, it follows that divinatory truth must be non-representational – it does not aim to convey a picture of the world. As that ethnographic concept – a truth that has no representative relation to states of affairs in the world, and whose opposite is not falsehood – has no equivalent within our analytical repertoire, Holbraad must invent one. That he does in the form of 'inventive definition' or 'infinition'. As it is itself thus invented, 'inventive definition' is both a description of divinatory verdicts and an instance of itself (2009).

Infinitions are non-representational truth statements. Like Roy Wagner's concept of 'invention' (1975), they derive their truth not from their possession of an external relationship to things in the world to which they can be applied either correctly or incorrectly, but from the transformative effects they have on the concepts to which they relate. Thus, as in one of Holbraad's examples, to say – or rather to infine – that 'Wagner is a genius' (2012: 44) is not to connect two pre-existing entities via an external relationship of meaning, but to transform both entities into something new ('Wagner-the-genius'). In the same way, divinatory verdicts are precisely not open to doubt because they do not represent a state of affairs in the world but modify the objects to which they apply: 'You are bewitched' 'transforms me from a person who stands in no particular relation to witchcraft into a person who is being bewitched' (2009: 88).

It should be clear that what is being described here is a method or an approach. Towards the close of the monograph (2012: 255), Holbraad lays out explicitly the schema involved, which I paraphrase here:

Step 1: Describe your ethnographic material with familiar concepts, and do so with the aim of representing it in the most accurate way possible.

Step 2: Check for instances in which those descriptions do not seem to make sense, or appear inaccurate. So, if you have to describe people being 'irrational', or 'believing' something your readers will think to be false, your familiar concepts have probably failed. In the example above, this situation arises because 'you are bewitched' cannot be both a representational statement (as it appears to us) and an indubitable one (as it is claimed to be).

Step 3: Try to figure out what is causing the contradiction or conflict in your description, and which of your assumptions you need to reconfigure. In the case of Cuban divination, the concept in question is that of truth.

Step 4: Redefine the concepts in question – with help from philosophers or other anthropologists – until your description is cogent and no longer generating contradictions.

Step 5: Ensure that your new description accurately represents your ethnographic material. So inventive definition, as a modified concept of truth that does away with the contradictions outlined above, draws on the work of philosophers (like Deleuze 1994) and anthropologists (like Wagner, Strathern and Viveiros de Castro) and renders the concept of indubitable divinatory verdicts into something we can make sense of, rather than a logical absurdity.

To return again to the question of heuristics, note again the quite clear sense that this is a method, not a description of a certain state of affairs, and note also the clear statement of aims: the articulation of 'true representations'. So, and somewhat paradoxically, for Holbraad the purpose of redefining a concept (in this case truth itself) in relation to ethnography is actually to provide truer (in the straightforwardly representationalist sense) pictures of the world around us.

Critical perspectives

As I noted at the beginning of this chapter, the ontological turn has been the target of a significant quantity of critique since its inception. Some of these critics begin with the use of the word itself, and the confusion it is liable to generate amongst those who use it. For example, is it, as a motion at the Group for Debates in Anthropological Theory proposed, simply doing the same work as the word 'culture' (Carrithers et al. 2010)? If so, surely it will suffer from a number of the same drawbacks that any anthropological conception of a 'unit' will: Where does it begin and end? What are its essential characteristics? Some writing within the ontological turn seems to have embraced a version of this position. So Viveiros de Castro, for example, in his more recent writings suggests that there is a political value to what he calls 'tactical quintessentialism' (2011b: 165), a sort of ontological version of the old 'identity politics' notion of 'strategic essentialism', in which aspects of the identity of a (typically subaltern) group or collective are treated as essential, for political purposes. This quintessentialism serves a purpose in the project of deconstructing our Euro-American assumptions, and that of what he elsewhere refers to as the 'theory of peoples' ontological auto-determination' (2011a: 128).

Elsewhere (2014), Pedersen, Holbraad and Viveiros de Castro, have described their approach as constituting a new form of politics. They argue that the production of difference it precipitates through the encounter between ethnography and conceptualisation should be thought of as the very stuff of politics, rather than more traditional objects of political anthropology: 'to think is to differ', as they put it. In this version of the ontological turn, its purpose as a heuristic is something of an admixture of conceptual innovation and political change. Indeed, their argument is essentially that the two are inseparable: politics *is* thinking differently in the manner that the ontological turn allows.

Yet the politics of the ontological turn has also come in for critique. Some (e.g. Bessire and Bond 2014) take issue with precisely the ways in which they see it as ignoring things to which anthropologists of politics have long pointed. They suggest that its quest for alterity leads it to ignore the very real and material inequalities and relations of domination both within the societies upon which it focuses, and between those societies and our own.

A slightly different but nonetheless politically inclined critique is made by David Graeber (2015). Graeber highlights the ways in which ontological-turn-inspired accounts are heavily reliant on specific conceptual premises. So Holbraad's analysis, for example, takes as its starting point the idea that divinatory oracles are always correct. Clearly a lot of Cuban diviners think this, and equally clearly Holbraad is not claiming that nobody thinks otherwise (see above). His argument is that it is simply constitutively essential to Cuban divination that it be unfalsifiable (in the same way that, say, having a legal framework is essential to a liberal democracy – even if laws get broken sometimes). Yet, as Graeber points out, this is effectively to argue that we should take certain kinds of statements as somehow 'authoritative' and then treat them as if they are generative of reality, a position that is hard to distinguish from classical philosophical idealism (2015: 23). As a position, it also demands that we privilege such authoritative statements, instead of starting from the fact that most of our interlocutors are likely to disagree with one another, and often themselves, about such statements. They are usually, in other words, as unsure as we are about the nature of reality.

Other critiques of the ontological turn focus more on its methodological and/or metaphysical implications. James Laidlaw (2012), in a review of one of the most prominent ontological-turn inspired monographs, Morten Axel Pedersen's *Not Quite Shamans* (2011), has taken issue with the slippage in meaning between 'ontology' as fact out there in the world, and 'ontology' as synonym for culture or cosmology, a slippage that Pedersen happily acknowledges (2012). As we have seen, it is a founding premise of the ontological turn that there should be no distinction between things we might once have thought of as real (facts out there in the world) and others we would have thought of as representations (cultures or cosmologies).

In my own work and other work with Laidlaw, I have also queried this premise (2012; 2013; forthcoming), suggesting that it is in conflict with the ontological turn's purported methodological openness, and its claim to be just a heuristic device. Even if every consequent move is made in response to ethnographic contingency, the basic and foundational claim that ontology and epistemology are somehow one and the same thing does not itself originate from any ethnographic material, and is thus nothing if not epistemological. Not only does that make the claim itself somewhat self-refuting, it also suggests that as an approach it will have some problems accounting for people who are not in its own preferred way 'alter' to ourselves. Although she descends from a different – but very much related – methodological school, that of actor-network theory (see chapter 13), we might take the work of Annemarie Mol on a Dutch atherosclerosis clinic as an example of this problem (2002). She takes an 'ontological' approach to disease, but, like her anthropological cousins, by this she means that she understands disease as a processual admixture made from both concepts (epistemology) and objects (ontology). The Dutch doctors she studies though begin from precisely the opposite premise: that disease is a thing out there in the world, and representations of it are not.

Another implication of this point relates to one also made by Michael Scott (2014; see also 2013; Heywood 2012): that for all of its purported focus on difference and alterity, the actual descriptions the ontological turn produces tend to look remarkably similar, despite coming from geographically very distant parts of the world. So how is it that a method premised on producing new concepts directly in relation to particular ethnographic material should be constantly producing *the same kind of* concepts (cf. the similarity between Holbraad on infinition and Wagner on invention) in relation to *different kinds of* ethnography?

In response, Holbraad (2017) has argued that to see such concepts as similar in spite of their different ethnographic sources is to fall precisely into the error of distinguishing between concepts and ethnography from which the ontological turn seeks to rescue us. He suggests that if one follows the injunction to see them as the same kinds of things consistently, one cannot understand 'infinition' or 'invention' without reference to the ethnography from which they came, and thus, to that extent, as different.

Pedersen makes a similar point in his response to both mine and Laidlaw's arguments (2012), arguing that to see the ontological turn as possessing a hidden, abstract theoretical frame is to confuse theory for method: instead of having a vision of the world, he reiterates, the ontological turn has an approach to it, one that is quite capable of producing arguments that might well refute his own, or those of Holbraad, depending upon the ethnographic material in question.

Conclusion

This is an apposite point at which to conclude by returning to the question I raised in the introduction to this chapter: What does it mean to think of an analytical position as a heuristic, and not as a theory? One thing it seems to entail is having a single response to a great many critical arguments: namely that they are simply misguided, because the content of the critique is targeted at some substantive position that can then be abandoned or side-stepped because, after all, it is only a heuristic. I think that response meets its limit in the idea of recursivity itself – that ethnographic concepts can and should alter our conceptual schemas – that surely cannot be abandoned as a metaphysical position without the ontological turn ceasing to mean anything distinctive, but this is not the place to reiterate that argument (see Heywood 2012; forthcoming).

Instead, I simply wish to point out what heuristics should also do, in addition to allowing those employing them to avoid criticisms based on substantive positions. They should raise the question of what exactly it is that they are aiming to achieve, and whether that is something that we wish our conceptualisations to do. They should do this not only because such questions logically follow from the use of heuristics (again – tools are meaningless objects without the purpose they are there to accomplish) but also because purposes are things that we can and should agree and disagree about and debate. Stimulating debate – as this book, and all of the ink spilled over the ontological turn, attest to – is surely one of the things good anthropology, whether school or style, should do.

References

Bessire, L. and D. Bond 2014. Ontological anthropology and the deferral of critique. *American Ethnologist* 41: 440–456.

Candea, M. 2016. De deux modalités de comparaison en antropologie sociale. *L'Homme* 218: 183–281.

Carrithers, M., M. Candea, K. Sykes, M. Holbraad and S. Venkatesan 2010. Ontology is just another word for culture: Motion tabled at the 2008 meeting of the group for debates in anthropological theory. *Critique of Anthropology* 30: 152–200.

Deleuze, G. 1994. *Difference and Repetition*. London: Athlone Press.

Descola, P. 2013. *Beyond Nature and Culture*. Chicago, IL: University of Chicago Press.

Graeber, D. 2015. Radical alterity is just another way of saying 'reality'. *HAU: Journal of Ethnographic Theory* 5: 1–41.

Henare, A., M. Holbraad and S. Wastell (eds.) 2006. *Thinking Through Things: Theorising Artefacts Ethnographically*. London: Routledge.

Heywood, P. 2012. Anthropology and what there is: Reflections on 'ontology'. *Cambridge Anthropology* 30: 143–151.

——— forthcoming. Making difference: Queer activism and anthropological theory. *Current Anthropology*.

Holbraad, M. 2009. Ontography and alterity: Defining anthropological truth. *Social Analysis* 53: 80–93.

——— 2012. *Truth in Motion: The Recursive Anthropology Of Cuban Divination*. Chicago, IL: University of Chicago Press.

——— 2017. The contingency of concepts: Transcendental deduction and ethnographic expression in anthropological thinking. In P. Charbonnier, G. Salmon and P. Skafish (eds.), *Comparative Metaphysics: Ontology After Anthropology*, pp. 133–159. London: Rowman and Littlefield.

Holbraad, M. and M. Pedersen 2009. Planet M: The intense abstraction of Marilyn Strathern. *Anthropological Theory* 9: 371–394.

——— 2017. *The Ontological Turn: An Anthropological Exposition*. Cambridge: Cambridge University Press.

Holbraad, M., M. Pedersen and E. Viveiros de Castro 2014. The politics of ontology: Anthropological positions. *Cultural Anthropology,* 13 January, online: www.culanth.org/fieldsights/462-the-politics-of-ontology-anthropological-positions.

Laidlaw, J. 2012. Ontologically challenged. *Anthropology of this Century* 4, online: http://aotcpress.com/articles/ontologicallychallenged/.

Laidlaw, J. and P. Heywood 2013. One more turn and you're there. *Anthropology of This Century* 7, online: http://aotcpress.com/articles/turn/.

Latour, B. 2009. Perspectivism: 'Type' or 'bomb'? *Anthropology Today* 25: 1–3.

Mol, A. 2002. *The Body Multiple: Ontology in Medical Practice*. Durham, NC: Duke University Press.

Needham, R. 1972. *Belief, Language, and Experience*. Oxford: Blackwell.

Pedersen, M. 2011. *Not Quite Shamans: Spirit Worlds and Political Lives in Northern Mongolia*. Ithaca, NY: Cornell University Press.

——— 2012. Common nonsense: A review of certain recent reviews of the 'ontological turn'. *Anthropology of This Century* 5, online: http://aotcpress.com/articles/common_nonsense/.

Pouillon, J. 1993. *Le cru et le su*. Paris: Le Seuil.

Quine, W.V. 1948. On what there is. *The Review of Metaphysics* 2: 21–38.

Scott, M. 2013. The anthropology of ontology (religious science?). *Journal of the Royal Anthropological Institute* 19: 859–872.

——— 2014. To be a wonder: Anthropology, cosmology, and alterity. In A. Abramson and M. Holbraad (eds.), *Framing Cosmologies: The Anthropology of Worlds*, pp. 31-54. Manchester: Manchester University Press.

Strathern, M. 1980. No nature; No culture. In C. McCormack and M. Strathern (eds.), *Nature, Culture, Gender*, pp. 174-222. Cambridge: Cambridge University Press.

Turner, T. 2009. The crisis of late structuralism. Perspectivism and animism: Rethinking culture, nature, spirit, and bodiliness. *Tipiti* 7: 3–40.

Viveiros de Castro, E. 1998. Cosmological deixis and Amerindian perspectivism. *Journal of the Royal Anthropological Institute* 4: 469–488.

——— 2003. *And*. Manchester: Manchester University Press.
——— 2004. Perspectival anthropology and the method of controlled equivocation. *Tipiti* 2: 3–22.
——— 2011a. Zeno and the art of anthropology: Of lies, beliefs, paradoxes, and other truths. *Common Knowledge* 17: 128–145.
——— 2011b. Zeno's wake. *Common Knowledge* 17: 163–165.
——— 2014. *Cannibal Metaphysics: For a Post-Structural Anthropology*. Minneapolis, MN: University of Minnesota Press.
Wagner, R. 1975. *The Invention of Culture*. Englewood Cliffs, NJ: Prentice-Hall.
Wastell, S. 2006. The 'legal thing' in Swaziland: *Res judicata* and divine kingship. In A. Henare, M. Holbraad and S. Wastell (eds.), *Thinking Through Things: Theorising Artefacts Ethnographically*, pp. 68-92. London: Routledge.

Chapter 15

Persons and partible persons

Marilyn Strathern

The final topic serves as something by way of a conclusion to the book. Controversies over personhood offer a means of reflection on the way anthropologists make and use the concepts without which there would be no dialogue. Like other chapters in this book, this chapter touches on broad areas of anthropological theorizing of which it is necessarily but a digest. However, this chapter creates a focus for the reader through drawing mainly on two recent collections of essays: 'The anthropology of personhood, redux: Views from Christianity', edited by Bialecki and Daswani, and 'Gender and person in Oceania', edited by Morgain and Taylor. Between them, these recapitulate and extend much of the current state of play, especially in relation to changing conditions of social life; they also offer ethnographic materials. Other works are referenced to indicate that many paths and avenues, many lines for pursuit, lead out from and back into even this small part of the field.

Prologue

Concepts that travel are of particular interest to social anthropology. For if it is as travellers that they appear, then they have obviously come from somewhere, by contrast with those formulations that seem already at home. Comparative potential cannot be realized in the same way. To make it of comparative use, 'person', familiar within the scope of everyday language, would have to be defamiliarized; foreign to regular English usage, by contrast, 'partible person' would need to explain its origins, and not least any further paths of migration. Half-way between, anthropological reflections on 'the person' as an abstract category already hint at an entity on the move, creating a field of its own.

The person has long woven in and out of social science discourse, occasionally erupting as an explicit focus of attention, notably through Mauss in his 1938 Huxley Memorial Lecture and its English regeneration 50 years later (Carrithers et al. 1985; Fortes 1987). Latterly, over a period that some relate to the falling away of interest in formal kinship studies, others to a postmodern moment, personhood has come to occupy a broad conceptual space within social anthropology. Diverse theorizings about its role in social process entail diverse locations that set its analytical contours in certain moulds. This is the issue taken up here. In order to create an analytically robust category, accounts of the person have always done battle with different investments in the concept; recent debate over the partible person is but an example.

The field of the person

As an analytical concept, the person inhabits a field of concepts. For Euro-Americans, the field flowers with a whole range: (in English) person, individual, self, agent, subject, human being. These may be variously distributed according to the task in hand; they may also be regarded as cohering, such that if it is 'person' that emerges as the comprehensive point of reference, components of the range appear in terms of it, namely as 'aspects of the person'. These aspects then organize our sense of what a person is through particular attributes, as when an individual person is ascribed inner consciousness or undivided body. Attributes are organizing insofar as they give the construct in question certain contours. One of the themes of this chapter is the way in which concepts or constructs are themselves structured. Thus we may note that (analytically speaking) each component of the range – person, individual and so forth – may equally be understood as a discrete object of enquiry, and thus in another way appears in terms relevant to one of them, namely 'individually'.

It was with reference to the idea of human universals that Mauss wanted to show that, as a category of the human mind, notions of person might yet have a social history (see chapter 1). He does so by a double defamiliarization: on the one hand, citing various ethnographic and historical inflections of the concept; and on the other, introducing (into the field) variants of the term person itself, which endure in anthropological discussion in their French or Latin form. *Personnage*, *persona*, *personne*, followed by the self (*moi*), mark different modes of conceptualization; the human subject of them all being termed 'an individual'. These modes are individuals as characters or role-players/as legal or forensic entities/as endowed with inner soul or conscience/as the cultivators of consciousness, respectively. He intends us to understand such conceptualizations as thoroughly social: how human thought moves on, he says, through societies and their metamorphoses. His organization of the field is itself still in movement.

If, as many anthropologists would today take for granted, the individual as a person is a thoroughly social entity – personhood being 'the emergent form of the self as it develops within a context of social relations', and 'individual selves [being] social in their very constitution' (Ingold 1994: 744–745) – this is in part because they endorse a perspective from social life (that is, as an analytical position, from which in truth all constructs are 'social'). Nonetheless, it is useful to distinguish this sense of the individual as a person from the person as an individual. The latter's individualism entails a historical–cultural conceptualization of person or self as endowed with specific attributes such as autonomy or human dignity. This is 'individualism' as a value (Robbins 2015, after Dumont). Rapport (2010) takes a famous argument from Melanesia, concerning the Gahuka-Gama concept of person (Read 1955), to the effect that these people *lack* a concept of the 'individual person', that is, of a person as an individual, and the attributes accompanying it, only recognizing socially defined positions where distinct personalities arise from combinations of social relations. Rapport's own view is that whether or not they pursue 'individualism' as a value, they are bound, like all human beings, by an irreducible 'individuality' that shows in the discreteness of mind and body, agency and consciousness.

The 'lack' is illuminating. Is a contemporary outcome of Mauss's field of concepts a double discomfort with, on the one hand, analyses that yield everything to 'social relations' and their occluding of individualized agency and, on the other, an awkward admission about re-instating thereby a traditional–modern dualism? It is a discomfort often expressed.

A Melanesianist has recently rephrased it in the words of the French sociologist Théry: 'What then of the agency of the individual and their ability to act for themselves ... in traditional societies where the higher value is that of relationships?' (2009, cited in Lepani 2015: 51). What of self-awareness, of the ability to make one's acts or words one's own, to recognize them as one's own and answer for them? This last, a forensic attribute (self-accountability), also demarcates the importance of protecting the 'unitary person' in modern administrations: 'In our culture the prime need is individual freedom' (Douglas 1995: 85; see Rapport 2012). The impression is given that an oppressiveness of sorts can be read into social relations. It is as though it were in the face of imprisonment by convention or institution that, as an analytic, the person should be freed to flower in that whole field of constructs where self-consciousness and autonomy also grow.

Relational persons

The phrase 'relational person' will serve to summarize approaches that stress the embeddedness of persons in relations. For concomitantly with a renewed interest in personhood, a new kind of conceptual space seems available for the 'relational'; an epithet applying to all manner of links, ties and connections, whether concrete or abstract. Here, however, relations – including but not exclusive to social relations – have a benign cast. Needless to say, the positive value often attached to relations is as analytically unhelpful as a negative value; the question is the work the concept does. It cannot, of course, be considered by itself alone.

A brief digression on the way constructs participate in one another may be helpful. We have already encountered the individual as person and the person as individual: I would see the whole field as an assemblage of merographic relations. The epithet (merographic) need not detain us; it simply points to certain epistemic strategies, ways of organizing knowledge, enabled by English and other European language usage. Thus, connections can be made between parts of social life in a way that sustains the individuality of each, for anything may be distinguishable as a part of something else, so nothing is ever simply part of a whole because another perspective, or interpretation, may redescribe it as part of something else (Strathern 1992: 72–73). Contextualizing, taking multiple viewpoints, switching perspectives (e.g. if a concept is part of a field, the field is part of the concept): these analytical devices familiar to anthropology sustain such relations. In effect, it is relations between constructs that position them so that they seemingly work by themselves. This is evident in the conceptual field of the person. Person, self, agent and so forth: any of them may be taken as the singular starting point for discussing any other. Moreover, any can be rendered distinctive by other perspectives; for example, whether one takes the person as an autobiographical self, a legal individual or an internationally acknowledged human being. Just so, the field itself may be re-contoured through constructs that concern psychoanalysis, zoology, the state, et alia. Anything in these Euro-American formulations, it would seem, is individualizable through connections; anything, too, is thus connectable; indeed, relations run riot. So it sometimes perplexes anthropologists how anyone could have imagined that everything was not already relational.

Apropos personhood, there are at least two dimensions to calling an approach relational. First is the invitation to anthropologists to keep their minds open to the interconnectedness of phenomena, whether such openness is regarded as inherited from structuralism (relations between relations – see chapter 2) or as offering an escape from positivist apprehensions of society or culture. For instance, a relational approach may render relations not as 'other' to a 'self' (the person as self), but as intrinsic to the self as an intersubjective entity (the self as a person). Here the concept of the social may even be a distraction: thus Toren (2012) points rather to an irreducible aspect of human ontogeny, namely the co-constitution of persons over their lifetimes.

A second reason behind the prevalent deployment of relational is ethnographic enlightenment, to bring into theoretical purview lessons learnt from numerous fields of study. Anthropologists have long shared with sociologists diverse understandings of 'social networks' (not to be confused with actor networks – see chapter 13), and early/mid-20th century 'roles' and 'statuses' were nothing if they were not relational concepts describing how persons were embedded in relations with others. But ethnographic elucidation often gives them fresh emphasis. Bonnemère uses a 'relational approach' in her study of ritual processes that transform Ankave people over their life cycle in order to extract the theoretical significance of 'relational statuses' from other approaches to gender identity. These statuses are distinguishable from the person ('individual subject') occupying them. (The crux for her is that such relations 'remain external to the person even as changes in the person's status depend on them' [2014: 740].) A further strand comes in with some anthropologists' almost urgent need to specify a 'relationism' on a par – whether through analytical parity or analytical privilege (Candea 2011) – with Western or Euro-American 'individualism'. This latter placement or location of concepts (relationism/individualism) demands further exposition.

Perhaps part of the urgency of relationism lies in its general implications for recognizing how all manner of phenomena bear on one another, not least at a time of crisis perceived as ecological. Indeed, the appeal to relationism has been salutary in many ways too evident to rehearse here – both in the task of describing heterogeneous realities not encompassed by Euro-American cosmologies and in binding analyses together, as was always underscored by the conceptual work of sociocentric analysis. It is the supposition that people everywhere participate in one another's identities that drives Sahlins's vision of kinship, and of kin persons as 'relationally constructed'; he underscores the category mistake of egocentric (rather than sociocentric) kinship thinking that renders 'the relationships of kinship as the attributes of singular persons' (2013: 27). But the everywhere seems especially evident in particular somewheres. Englund and Yarrow (2013) observe the pivotal place Melanesian ethnography holds in Robbins's exposition of relationism as opposed to individualism. They would be in sympathy with the fact that there is, however, more to it than that. In being 'relationist, Melanesian cultures value the creation of relationships over that of other cultural forms' (Robbins 2004: 292). That value is not of the same order as the stress Robbins would put on 'one of Dumont's most fundamental assumptions: that all human life grows out of social relations' (2015: 173), which is the source of his holism. The latter is the vantage point from which (Western) 'individualism' appears as a specific ideology. Holism is thus at once a value found in some societies *and* a theoretical concept (for a human condition) that encompasses individualism and Melanesian relationism alike. Investigating which value system appears transcendent is shadowed by the same issue in conceptual vocabulary. But if we explicitly require that the concepts we deploy should themselves convey a sense of an encompassing interconnectedness (that seems the role of holism apropos levels of value), there are many ways to do it, and we have stumbled into another field.

This is the field in which the 'social', 'sociocentric', 'mutual', 'relational', 'holistic' (the French would add 'collective') jostle for light. Across anthropological accounts, each is variously aligned with the others and, as with the field of the person, each may be taken as a starting point for analysis. Thus, 'the relational' appears to have the limelight in questions over whether to describe relationism as co-present with individualism or as diagnostic of a theoretically distinguishable cosmology.

This abstract summoning of individualism is the point at which flowery fields turn into battlegrounds. Why do some feel that they have to fight it, that relations will wobble if they are not held together? Perplexing indeed how anyone familiar with the stress Euro-Americans put on making relations could imagine everything is not already relational. Perhaps the problem is precisely that making the supposition explicit cannot be the same as taking it as an implicit assumption. For those looking for it, that implicit place appears occupied by individualism. It appears in the contours of their (Euro-Americans') very tools of analysis. Through the organizing lens of merographic thinking, relations seemingly spring from, and return to, what is taken for granted: concepts – such as the person – that are all too readily individualized. This brings us to a construct that is not quite the same as the relational person.

Partible persons

Whether or not there is anything specific about Melanesian forms of relationism is a question that can only be noted here. It remains true that Melanesia has been a significant contributor to controversies over concepts of personhood. Controversy continues, and the reader should probably take everything that follows as controversial. Two collections drawn on here – not least for the way they show the importance of simultaneously continuing to theorize and to address new ethnographic ventures – figure Melanesian materials. Robbins's paper came from the same journal issue as 'The anthropology of personhood, redux' (Bialecki and Daswani 2015); the other, 'Gender and person' (Morgain and Taylor 2015), was part of a project instigated by Jolly that also introduced Théry's work to an English-speaking audience.

However their interlocutors formulate things, scholars know their analytics set them apart. They are unlikely to find people's notions of 'individuals' or 'relations' doing what they need for their analyses. That does not, however, mean they are locked in to only very local milieux or that ideas cannot travel across social contexts or disciplines. It was literature from Melanesia, alongside Amazonia and Indonesia, that convinced Théry that a 'relational' perspective 'is not restricted to the understanding of distant societies. It can readily be taken up in itself extrapolating from them, and reworked, in terms of research centred on our own culture and Western societies' (2009: 5). Indeed, such openness is a prerequisite to the idea of learning (Laidlaw 2014). In Théry's understanding, 'the individual cannot be separated from the concrete "whole" that is society in which they participate as a person, as an agent of human acts' (2009: 13). She contrasts two European positionings of the self, the 'I' who is a someone with their personal attributes, along with a sense of personal identity, and the 'I' of interlocution, the speaking person taking possession of its own acts, its own words (one context of the remarks previously cited). The attributes of the latter are not absolute but (we could say) composite, for 'mine' only exists in relation to 'yours', 'hers' and so forth: such a person cannot be *solely* an 'I' insofar as other grammatical positions are implied. We can in turn make this notion travel. There is a kind of Melanesian 'I' that is put into words: it is heard from the doers of actions, from those we may call agents. People emphasize the autonomy of acting, on having the action in their mind, on the oneness of the act. An act is, in this view, like an individuated concept. Just as Théry distinguishes the grammatical 'I' from the self as a unitary entity, arguably there are cosmologies where the autonomy of taking action (the reflexive claim that one did it) can be distinguished from the kind of subjective accountability or self-fashioning that turns persons into the authors of their acts. An agent may (in 'Melanesian'-speak) own up to an act whose cause lies with others. Lepani gives a present-day Trobriand example of such autonomous acting: a strong-minded woman who 'in no longer caring for her child but rather cooking for her father . . . [and] younger siblings . . . took possession of her narrative identity, not with words but through acts of labour' (2015: 56). Insofar as the impulse for that action is simultaneously understood as originating in and oriented towards others, then it seems to be within such a relational context that the self becomes a reference point. Conversely, Lepani's demonstration makes it abundantly clear that acknowledging the relations at issue does not obliterate the individuality or autonomy of action. It may be helpful to think of it as a 'grammatical' autonomy, thus carrying a 'grammatical' responsibility.

This brings me to the status of the partible person in the anthropological repertoire. As a one-time author of the awkward phrase, I pose the question whether it is still of comparative use. It was originally introduced in response to materials from Melanesia but, as Englund and Yarrow (2013: 133) note of the 'composite person', not just that – and its form is indicative. Unlike the 'relational person', the 'partible person' is not an amalgam. If it is animated it is by the anticipatory potential of what its parts might enable. It is certainly not a hybrid concept as one might derive from bringing together the relational and personal, an intimation of how a society/individual antithesis might be resolved. It offers a different resolution.

'Partible' occupies its own microcosmic field, alongside 'dividual', 'distributed', 'composite', 'multiple', all epithets used of persons (see, for example, Mosko 2015: 362), and it is no surprise that these participate in, and overlap with, one another. Together, they reflect attempts to avoid assuming that before anything else the person is an individual. They cover studies where the focus of concern is personhood as such, as in the anthropological address to Christianity, and those where the treatment of personhood is diagnostic of other issues, as Carsten (2004) suggests with respect to kinship. That said, the dividual in *The Gender of the Gift* (Strathern 1988), as indicated by the descriptive phrase 'partible person', was an address to 'society'. It was an attempt to find a counterpart to the individual of the individual/society dualism. Under the value of individualism, society was at once regarded as absorbing everything anthropologists might want to say about relations and understood as assuming that relations were 'between' individuals. The concept of individual, and the attributes it summoned, thus affected how one might deploy society as an analytical construct, as in the idea of sociality. If the kind of individual projected by certain mainstream views of society in effect pointed to a particular (Euro-American) cosmology, then for other kinds of relational configurations it might be necessary to put forward a concept with different contours. With respect to Melanesia, I borrowed 'person' from existing anthropological usage (as socially understood) and 'partibility' from the flow of 'detached, partible things' in exchange to which Wagner (1977: 631) referred. Like dividual, from another ethnographic location altogether, the awkwardness of the resulting phrase indicated its non-(English) vernacular origins.

Partible person, then, explicitly divided itself off from the dualism of (person as) individual and society (imagined as relations between individuals). Yet there are limits to how far one can go against the grain of language. Other anthropologists have taken the term person to connote 'an individual' as in individualism, even in conceptual form – the concept itself understood as individuated, indexing 'the person' as a discrete, identifiable entity, and thus as an object of independent enquiry. This is not helped by my having described the (Melanesian) composite person as a singularity, even though it was contrasted with the individualizing actions of an agent. Here I was proposing a language for talking about the perpetual alternation of perspectives 'between being the incomplete agent who is activated in relation to another and the complete person, a product of other's interactions' (Strathern 1988: 287), in which it is the latter that is singular, a composite entity derived from multiple relations. This condition of multiple constitution also renders the person partible, namely as an entity that *anticipates* partition, as when an agent acts to shed one set of relations in favour of another in eliciting an orientation to itself. Nonetheless, the general upshot has been that my language seemingly erased the very phenomenon – society/sociality – I had hoped to redescribe.

Let me turn this into a moment of interchange: maybe it was this seeming erasure that led to Sahlins attributing to my argument about the Melanesian person both an egocentric view (I was the target of the comments above), and a confusion between partibility and participation. There is more theoretically at stake, he says, than 'the make-up of individual persons' (2013: 25). Yet this is exactly the point on which I would query his reading of the person as inevitably the person as individual in the first place! So we are in fact in agreement over the problem. The contours of these concepts become a question of analytical choice. Thus, Sahlins would use 'partibility' of any distribution of personal investment in relations (as in role-playing), as distinct from the participation of persons in one another's lives through incorporation or mutual embodiment, as kin everywhere participate, for which he allows the concept 'dividuality'. Others see the two concepts as synonymous. In the debate on the status of the individual, it is the dividual (rather than the partible) that seems the more widely generalizable category.

Sahlins's argument also affords a platform for divergence. At the heart of where he puts dividuality, kinship, I might point to those English if not Euro-American kin configurations that, endorsing a particular theory of society, reproduce the person as individual, where social relations are imagined as relations with other individuals. For this kind of individual, attachment and detachment hardly work in the way they do when 'persons' and 'agents' are more like alternative perspectives on social life (see Schram 2015). The subtraction and addition of relations from/to this 'English' individual comes through a multiplicity of possible viewpoints (relations with other individuals), such that one travels from one arena of knowledge to another. This is not the mathematics (grammar) of what I had understood as dividuality. On the contrary, putting the English situation this way was originally inspired by what seemed to me the rather different relational circumstances of the 'Melanesian' dividual – hence the further specification of partibility. But is there any point now in making a special case for partibility?

From his point of view, Mosko has argued strongly for the distinctiveness of Melanesian partibility as an attribute of personhood with very specific contours. When he writes, where 'people from a Western viewpoint might appear to be exchanging objects . . . in indigenous Melanesian perspective they are rather transacting over bits of themselves as persons' (2010: 219), the references that follow are not confined to Melanesia (and are both before and after Strathern 1988). But, crucially, what became debated as the partible person appears in the unfolding of exchange relations: as 'one who is divisible or *divide*-able into component parts or relations that are transactable ... through processes of elicitive gift exchange' (Mosko 2015: 362, original emphasis). The diagnostic of partibility for Mosko lies in the detachment and attachment of parts. 'Through acting, partible persons are *de*composed, anticipating and evincing the recognition of their externalized capacities through the responses of corresponding [others]' (2010: 218, original emphasis). While, in certain areas of Melanesia, ceremonial gift exchange achieves this in political contexts, and life-cycle exchanges in kinship ones, he does not restrict detachment and attachment to the flow of partible exchange items, the origin of my own usage. Moreover, Mosko has lifted the restrictions even further, to include sacrifice in religious contexts. His expanded model of partible personhood takes him into intriguing waters when it comes to ritual action and Christian theology. The distinctiveness of partibility, or dividuality in this sense, is shown up by its continuing relation to (comparison with) other analytical configurations. Thus, he argues against what appear to him 'relationist' misreadings of Melanesian sociality: apropos the exchange of items:

> Regardless of how intensely people valorize relationships, if the items transacted are not regarded as parts of the transactors as persons, relationist perspectives tacitly recapitulate the subject/object distinction on which possessive individualism in the West is premised.
>
> (2010: 219)

The usage of 'dividual' came from outside Melanesia, it being Marriott's (1976; see Marriott and Inden 1977) term apropos Hindu India. Reflecting on her South Indian fieldwork, Busby (1997) was struck by the consensus among Melanesianists at the time that led to them depicting bodily composition in terms of 'parts', as when maternal blood and paternal semen are described as male and female parts in the reproduction of the person. By contrast, in South India, persons engage with others in flows of substance but the latter 'always refer to the persons from whom they originated: they are a manifestation of persons rather than of the relationships which they create' (p. 273). Busby counterposed two constructions of the person, the one 'internally divided and partible (Melanesia), [the other] ... internally whole, but with a fluid and permeable body boundary (India)' (p. 269). It is the quality of substantive permeability that Werbner (2011) stresses in his account of Apostolic charismatics in Botswana as dividuals. Busby's observation that the one term, dividual, can obscure fundamental (regional) differences remains salutary.

It is reanimated in current questions about the co-presence of 'individualism and relationality' (Morgain and Taylor 2015), as Werbner (2011) has also articulated by bringing together the concepts of dividual and 'relational self' and proposing a model of dividuality and individuality as 'mutually constitutive'. Bialecki and Daswani explicitly argue for taking dividual/individual less as heuristics or as different modes of organizing the subject than as actualizations of real-life problematics worked through in various locales, with diverse dividual and individual crystallizations of the person. Individualism and dividualism emerge 'as dynamics that mutually implicate each other' (2015: 272). For them, the question is how problems are structured, where, for instance, self-fashioning becomes a project, and thus where and when a concept appears significant as an actable-on-capacity (after Humphrey 2008). Yet one cannot tell by simple inspection: the conditions under which any concept becomes *analytically* actable-on will depend on its place in one's field of concepts.

Travelling concepts

Such interchanges are, it seems to me, highly productive of what we might want comparison to do. They underscore the role of anthropologists' larger intentions in their own formulation of problems (as in problematics), which is one reason why so much of this chapter has been taken up with the construction of constructs. The contours of one's concepts matter. Hence the chapter has treated the 'relational person' as an exemplar of the ambition to evoke sociality and intersubjectivity in the making of persons. Yet the dualistic construction of that concept allows it – may even encourage it – to encompass the dualism of society/individual as well; even with the language of persons, it can reduce relationism to a notion of an entity *and* its relations. (The reduction is in analytical potential; it is of course an English colloquialism to speak of individuals situated in networks of relations.) The 'partible person' is a much less tractable concept, but in avoiding that analytical trap appeared (to diverse critics) to forget about society altogether. It is no surprise perhaps that the partible person has not travelled as far as the dividual (Sahlins 2013: 25 gives an itinerary). Indeed, like 'composite' or 'distributed', the 'dividual' travels between the micro-fields of both the relational and the partible person.

For the writers cited here, choice of analytic category invariably points to fundamental formulations of thought and social life. Where Euro-Americans might see divisions as having to be overcome, the notion of partible person addressed the active role that division or partition plays in Melanesian understandings of relations (e.g. Schram 2015: 323). Stasch (2009: 10) talks thus of the 'internal alterity of Korowai social relations'. Now, quite apart from the Euro-American counter-tradition invested in theorizing alterity that Stasch addresses are other mainstream conventions. Thus, in describing its (the partible person's) contours as a concept, I gave it an evident 'relational' cast, as an element in a wider analytical configuration. The excursus on merographic connections showed a further dimension to the way in which Euro-Americans routinely imagine relations between concepts: namely how they participate in or are part of one another. So I can also give the concept a 'partible' cast in the English vernacular: by dividing a construct – any construct – from its relational nexus one is able to take it as a detached entity with the potential for travelling. Indeed, this might be an apt metaphor for apprehending aspects of Melanesian personhood, even though in English this is not a salient attribute in the way persons are generally celebrated or indeed constructed.

As to that difference, what we may have learnt from a reminder of an originally restrictive ('Melanesian') reading of partibility is how particular kinds of partitioning work. It makes a difference whether divisions are taken for granted or must be acted upon, and they may be divisions at any scale of social life. In the words of Morgain and Taylor (2015: 7), Théry's challenge is to 'individualistic understandings of human nature . . . by attacking the dualist vision that underpins such understandings . . . in which the human person is seen to be composed of two distinct entities' (here, self and body). If, indeed, there is a dualism in secular post-Enlightenment thinking rooted in identitarian approaches to the individuated nature of phenomena, then the numerous occasions on which entities are opposed only to be merged, hybridized or otherwise rendered co-present or – as in the modernist society–individual duo – co-produced, reveals its particular creativity. The revelations may prompt concept-makers to move their constructs around again. Like shifting the contours of the grammatical 'I' to include the 'grammatical' act, the performance of such analytical moves can sometimes tell us more than a direct address to Euro-American knowledge practices would have let us know.

What those practices do promote is a horizon of seemingly limitless movement, on which the creative task is to put new limits. Needless to say, anthropologists are always picking up concepts and running with them – what would be the point of concepts that did not travel? When they become springboards for new ventures, in this dynamic, old conceptual fields are inevitably left behind.

Acknowledgements

Sedimented in this chapter are ideas that have travelled from diverse sources, although the references are highly selective. Special thanks for the criticisms and illuminations offered in the works of Michael Carrithers, Margaret Jolly, Lisette Josephides, Bruce Kapferer and Alan Rumsey. Serge Tcherkézoff kindly sent me a copy of Irène Théry's inaugural lecture. At the time of writing, I held a Leverhulme Emeritus Fellowship for which I express much appreciation.

References

Bialecki, Jon and Girish Daswani (eds.) 2015. The anthropology of personhood, redux: Views from Christianity, *HAU: Journal of Ethnographic Theory*, 5 (1): 271–404.

——— 2015. Introduction: What is an individual? The view from Christianity. *HAU: Journal of Ethnographic Theory* 5 (1): 271–294.

Jolly, Margaret 2015. *Braed praes* in Vanuatu: Both gifts and commodities? *Oceania* 85 (1): 63–78.

Lepani, Katherine 2015. 'I am still a young girl if I want': Relational personhood and individual autonomy in the Trobriand Islands. *Oceania* 85 (1): 51–62.

Morgain, Rachel and John Taylor 2015. Transforming relations of gender, person, and agency. *Oceania* 85 (1): 1–9.

Mosko, Mark 2015. Unbecoming individuals: The partible character of the Christian person. *HAU: Journal of Ethnographic Theory* 5 (1): 361–393.

Robbins, Joel 2015. Dumont's hierarchical dynamism: Christianity and individualism revisited. *HAU: Journal of Ethnographic Theory* 5 (1): 173–195.

Schram, Ryan 2015. A society divided: Death, personhood, and Christianity in Auhelawa, Papua New Guinea. *HAU: Journal of Ethnographic Theory* 5 (1): 317–337.

Taylor, John and Rachel Morgain (eds.) 2015. Gender and person in Oceania, *Oceania*, vol. 85.

Other references

Bonnemère, Pascale 2014. A relational approach to a Papua New Guinea male ritual cycle. *JRAI* (N.S.) 20 (4): 728–745.

Busby, Cecilia 1997. Permeable and partible person: A comparative analysis of gender and body on South India and Melanesia. *JRAI* (N.S.) 3 (2): 261–278.

Candea, Matei 2011. Endo/Exo. *Common Knowledge* 17 (1): 146–154.

Carrithers, Michael, Steven Collins and Steven Lukes 1985. *The Category of the Person: Anthropology, Philosophy, History. Cambridge:* Cambridge University Press. (Includes translation by W.D. Halls of Mauss's *Une catégorie de l'esprit humain: La notion de personne, celle de 'moi'. JRAI* 68, 1938.)

Carsten, Janet 2004. *After Kinship*. Cambridge: Cambridge University Press.

Douglas, Mary 1995. The cloud god and the shadow self, *Social Anthropology* 3 (2): 83–94.

Englund, Harri and Thomas Yarrow 2013. The place of theory: Rights, networks, and ethnographic comparison. *Social Analysis* 57 (3): 132–149.

Fortes, Meyer 1987. *Religion Morality and the Person: Essays on Tallensi Religion*, J. Goody (ed.). Cambridge: Cambridge University Press.

Humphrey, Caroline 2008. Reasembling individual subjects: Events and decisions in troubled times. *Anthropological Theory* 8 (4): 357–380.

Ingold, Tim 1994. Introduction to 'social life': Becoming a person. In T. Ingold (ed.), *Companion Encyclopedia of Anthropology: Humanity, Culture and Social Life*, pp. 744–747. London: Routledge.

Laidlaw, James 2014. *The Subject of Virtue: An Anthropology of Ethics and Freedom*. Cambridge: Cambridge University Press.

Marriott, McKim 1976. Hindu transactions: Diversity without dualism. In B. Kapferer (ed.), *Transaction and Meaning*, pp. 109–142. Philadelphia, PA: ISHI Publications.

Marriott, McKim and Ronald Inden 1977. Towards an ethnosociology of South Asian caste systems. In K. David (ed.), *The New Wind: Changing Identities in South Asia*, pp. 227–238. The Hague: Mouton.

Mosko, Mark 2010. Partible penitents: Dividual personhood and Christian practice in Melanesia and the West. *JRAI* (NS) 16 (2): 215–240. (Plus responses from Robbins, Knauft, Barker, Errington and Gewertz; see also the article by Street in the same volume.)

Rapport, Nigel 2010. Individualism. In A. Barnard and J. Spencer (eds.), *The Routledge Encyclopedia of Social and Cultural Anthropology*, 2nd edition, pp. 378 –382. London: Routledge.

——— 2012. *Anyone: The Cosmopolitan Subject of Anthropology*, New York, NY: Berghahn Books.

Read, Kenneth 1955. Morality and the concept of the person among the Gahuka-Gama, *Oceania* 25 (4): 233–282.

Robbins, Joel 2004. *Becoming Sinners: Christianity and Moral Torment in a Papua New Guinea Society*. Berkeley, CA: University of California Press.

Sahlins, Marshall 2013. *What Kinship Is – And is Not*. Chicago, IL: University of Chicago Press.

Stasch, Rupert 2009. *Society of Others: Kinship and Mourning in a West Papuan Place*. Berkeley, CA: University of California Press,

Strathern, Marilyn 1988. *The Gender of The Gift: Problems With Women and Problems With Society in Melanesia*. Berkeley, CA: University of California Press.

——— 1992. *After Nature: English Kinship in the Late Twentieth Century*. Cambridge: Cambridge University Press.

Théry, Irène 2009. Gender: A question of personal identity or a mode of social relations? Inaugural lecture, Centre M. Bloch, EHESS, Berlin, translated by S Anderson-Morton and S Tcherkézoff. (Accessed 21 August 2016: www.pacific-dialogues.fr/pdf/4-IT_ConfBerlin_2009_def.pdf.) (Drawn from Théry, Irène 2007, *La distinction de sexe: Une nouvelle approche de l'égalité*. Paris: Odile Jacob.)

Toren, Christina 2012. Imagining the world that warrants our imagination: The revelation of ontogeny. *Cambridge Anthropology* 30 (1): 64–79.

Wagner, Roy 1977. Analogic kinship: A Daribi example. *American Ethnologist* 4 (4): 623–642.

Werbner, Richard 2011. The charismatic dividual and the sacred self. *Journal of Religion in Africa* 41 (2): 201–226.

Index

Printed in Great Britain
by Amazon

86742202R00154